Cryptography,
Information Theory,
and Error-Correction

Cryptography, Information Theory, and Error-Correction

A Handbook for the 21st Century

Aiden A. Bruen
Mario A. Forcinito

A JOHN WILEY & SONS, INC., PUBLICATION

Published by John Wiley & Sons, Inc., Hoboken, New Jersey.
Published simultaneously in Canada.

For general information on our other products and services please contact our Customer Care Department within the U.S. at 877-762-2974, outside the U.S. at 317-572-3993 or fax 317-572-4002.

Wiley also publishes its books in a variety of electronic formats. Some content that appears in print, however, may not be available in electronic format.

Library of Congress Cataloging-in-Publication Data:

Bruen, Aiden A., 1941–
    Cryptography, information theory, and error-correction : a handbook for the 21st century
/ Aiden A. Bruen, Mario A. Forcinito
       p. cm.
    Includes bibliographical references and index.
    ISBN 0-471-65317-9 (cloth)
       1. Computer security.   2. Telecommunications systems—Security measures.   3.
Cryptography.   I. Forcinito, Mario, 1962–   II. Title.

QA76.9.A25B79 2005
005.8—dc22   2004058044

10 9 8 7 6 5 4 3 2 1

Contents

II   Information Theory   159

Preface

It is our privilege and pleasure to welcome all our readers to the dynamic world of cryptography, information theory and error correction. Both authors have considerable industrial experience in the field. Also, on the academic side Dr. Bruen has been a long-time editor of leading research journals such as "Designs, Codes and Cryptography". Prior to his appointment in Calgary he worked in mathematical biology at Los Alamos. The book is an outgrowth both of presentations to industry groups and of a lecture course at the University of Calgary. The course was for undergraduate and graduate students in Computer Science, Engineering and Mathematics.

In addition to the academic topics in that course, we also include material relating to our industrial consulting work and experience in writing patents on the topics in the title of the book. In particular we describe revolutionary new algorithms in chapter 24 for hash functions and symmetric cryptography including quantum cryptography. These have been patented and have already made their way into industry.

This book can be read at many different levels. For example, it can be used as a reference or a text for courses in any of the three subjects or for a combined course. To this end we have included over three hundred worked examples and problems, with answers or solutions as needed. But we were determined to make the work highly accessible to the general reader as well. We hope that the exposition fulfills this goal. Large sections of this book have been written in such a way that little is required in the way of mathematical background. In places this was difficult to do but we believe that the effort has been worthwhile.

The three topics become more and more entwined as science and technology develop. In our opinion, the time when the three topics can be treated in isolation is rapidly drawing to a close. For example, if you search the internet for cryptographic information it is more and more likely that you will run up against terms such as entropy, CRC checksums, random number generators and the like. [Digressing: the main undergraduate course in computer science which is concerned with data structures — and we have all taught it — covers Huffman codes and compression at length but the word entropy is never mentioned. This is a shame].

Thus it seemed quite appropriate to us to try to write a complete but highly accessible account of the three subjects stressing, above all, their interconnections and their unity. These interconnections can be hidden if one relies only on separate accounts of the three subjects. In addition, as part of information theory, we discuss some potential applications in cell biology. In the last chapter we present some new, exciting algorithms which combine all three of the subjects.

This is not the first time that a book combining the three subjects has been attempted. Several very good recent books, specializing in cryptography, have a few chapters on the other subjects. But our goal was to give a full in-depth account. We should mention that other books have handled nicely two of the three topics. A splendid book, published in 1988 by Dominic Welsh gives an account of all three of the subjects. However, a lot has happened since 1988. Also, our focus, emphasis and level of detail is different.

Let us briefly explain the 3 subjects.

Cryptography. This is an ancient subject concerned with the secret transmission of messages between two parties denoted by A and B. This could be done if A, B shared a secret language, say, not known to outsiders. More generally they can communicate in secret by sharing a common secret "key". Then A uses the key to scramble the message to B, who unscrambles the message with a copy of the same key that is owned by A. We may think of military commanders sending secret messages to each other or home movie providers sending movies to authorized customers. Apart from secrecy there are also crucial questions in cryptography involving authentication and identification.

Information Theory. This subject, also known as Shannon Theory after Claude Shannon, the late American mathematician and engineer, gives precise mathematical meaning to the term "information". This leads to answers to such questions as the following:

How much compression of data can be carried out without losing any information?

What is the maximum amount of information that can be transmitted over a noisy channel?

This fundamental question is answered precisely in Shannon's famous channel capacity theorem which was discovered around 1948.

Error Correction. We introduce redundancy ["good redundancy"] for the transmission of messages, as opposed to the "bad redundancy" which was banished using compression. In this way we try to ensure that the receiver decodes accurately within the bounds of the Shannon capacity theorem mentioned in the previous section. The wonderful pictures of far-away planets, that have recently been made available, are just one example of what error-correcting codes can do. With a modern modem we can both compress as well as encode and decode to any required degree of accuracy.

Interconnections. These are spelled out in detail in the text but let us give a few short informal connections. How secure is your cryptographic password? It depends on how hard it is to guess it, i.e. it depends on its entropy as measured in Shannon bits. We then need information theory to properly discuss this.

In cryptography, A is sending information in secret to B, but what exactly is information and how is it measured? Again we need information theory.

Suppose that A is sending a secret key K over a channel to B in order to encrypt, at some future date, a secret message M with K and transmit it to B. Now, a basic property of K is this. If the transmission of K is off by even one big then B will end up with a message that is completely different from the intended message M . The bottom line is that a transmission error could be catastrophic. The best way to guard against this is for A to use robust error-correction when sending the cryptographic key K to B.

The great Claude Shannon made the following fundamental point. In error-correction, the receiver B is trying to correctly decode what the transmitter A has sent to B over a "noisy channel". Compare this to the cryptographic situation where A is sending secret messages to B. They must contend with the eavesdropper — the evil Eve — who is listening in. We can think of Eve as receiving a "noisy" version of M and trying to decipher, or decode M. We are back to coding theory. [Parenthetically, we mention that Shannon designed an interesting theory of the stock market by regarding the market as a very noisy channel!].

We must point out that this point of view of Shannon is extremely useful and not just as a formal device. We drive the point home with several problems in Chapter 16 where the analogy becomes quite striking. Moreover, in Chapter 24, A and B and Eve may have the same information to start out with, yet A and B have to come up with a way of beating Eve and publicly generating a secret key using a technique known as "Privacy Amplification".

Here is yet another basic interconnection. Random numbers and pseudo-random numbers are the work-horses of cryptography, especially symmetric cryptography. One of the best ways of generating them is with shift-registers. In fact, as is pointed out in Schneier [Sch96], "stream ciphers based on shift registers have been the workhorse of military cryptography since the beginning of electronics". But shift registers are central in information theory as they are great proving-grounds [or grave-yards] for questions on entropy. To understand entropy you have to confront shift registers. But — and here is the astonishing part — these shift registers, over any field, correspond exactly to cyclic linear codes which are at the heart of error-correction. For the expert, Reed–Solomon codes, and not just their error-correction, are merely special kinds of shift registers in disguise!.

We move on now to a more conventional-type preface and address some standard questions.

Intended Readership. This is a book for everyone and can be used at many different levels. We are writing for many different kinds of readers.

1. All-rounders or renaissance types who have taken some mathematics or computer science or engineering [or none of the above] and who want to find out about these topics and have some fun.

2. Undergraduates or graduate students in mathematics, computer science or engineering.

3. Instructors of algebra and linear algebra who would like some real life practical applications in their courses, such as shift registers.

4. Biologists who may be interested in our discussions of such topics as biological compression and the channel capacity corresponding to the genetic code.

5. IT workers, venture capitalists and others who want an overview of the basics.

6. Academics looking for a good source of important (and doable) research problems.

7. Philosophers and historians of science who want to move on from quantum theory and relativity to a new, practical area which also, incidentally, has strong connections to quantum mechanics.

Rewards for Readers.   If you make a good effort at understanding this book and working out some of the problems you will be well rewarded. This book covers everything you need. In particular, you will elevate your skills and mathematical maturity to a new level. You will also have an excellent background — better than that of most practitioners — in these areas. You will be ready to think about a career in cryptography or codes or even information theory. The market, especially in such areas as data compression is hot. You will be very well-placed for advanced work in cryptography, error-correction or information theory.

Our Goals.   We want to help develop your skills and inspire you to new heights. Let this book be your inspiration. Master it and then get out and write those patents!

Possible Courses Using this Book.   There is more than enough material for a stand-alone course at the undergraduate or graduate levels, in any of the three areas. The extensive list of problems and worked examples will be a big help. For those few chapters that don't have problems there are opportunities for many fun group-projects geared towards reporting on patents, publications etc. We would recommend some "poaching" among the three parts of the book in such a course. A one year combined course would also work well.

A Course for Non-Specialists.   Most of Part I, apart from the Chapter on elliptic curves, requires very little mathematical background but covers a lot of ground in cryptography.In information theory, we highly recommend Chapter 10 which gives a panorama of information theory and interesting related topics such as the "MBA problem" on weighings. The chapter on topics related to the genetic code does not require much background, and should be of considerable interest. We also recommend Chapter 18 introducing coding theory. Chapter 20 tells the amazing story of how the famous perfect Golay code G_{11} was first published in a Finnish soccer magazine in connection with the football pools in that country. Chapter 24 describes what appears to be a breakthrough in symmetric cryptography, error correction, and hash functions. We highly recommend it!

Level, Mathematical Style, Proofs, Exercises. We have made a considerable effort to ensure that the chapters are as accessible as possible. In terms of style, our motto, which is the opposite of many mathematicians and engineers, is this: "Never use a symbol if you can get away with a word."

What about proofs? It really depends. If the proof enhances the ideas we try to present it. Also, some results, such as the Shannon source-coding result are so astonishing that we have to give the details. However, in the case of the noisy channel theorem we have a different approach. From teaching, we found it considerably more effective to give five or six different approaches rather than to just give the standard official proof.

This book was not written just for theoreticians. Much of our time was spent in designing good problems and solution. We urge our readers to take advantage of them.

Mathematical Prerequisites. Honestly? We try to cover everything "on the fly" along with one special chapter on specialized topics but here is a short summary of what we need.

Calculus: a small amount having to with the concavity of a graph [second derivative] and function maxima [first derivative, end points].

Linear algebra: Multiplying matrices, subspaces, invertibility and determinants.

Elementary probability and statistics: Mean and variance, Bernoulli trials, the normal curve, law of large numbers.

Algebra: A small amount of material on groups, finite fields, modular arithmetic.

Here and there we go over the top. For example, a bit of Fourier analysis for the Shannon sampling theorem is needed. But generally speaking, the above list covers most of the material and we do discuss the needed background as we go along.

What's New. Most of the Chapter have a "New, Noteworthy" heading where we try to summarize such matters. However, here is a brief summary of "what's new" in the book. The topics are listed in no particular order.

- An in-depth integrated discussion of cryptography, information theory and error-correction emphasizing their interconnections, including new, clear, accessible proofs of major results, along with new results.

- A discussion of RSA that clears up several issues and shows how, for example, a given encryption index may have several decryption indices: Also,an indication of a possible new attack on RSA.

- A study of potential applications of information theory in cellular biology.

- An overview of important practical considerations in modern cryptography and communication theory.

- A whole new treatment of "perfect secrecy", including a refutation of the standard assertion concerning the equivalence of perfect secrecy and the one-time pad, together with a proof of the equivalence of perfect secrecy and latin squares.

- A highly accessible summary of information theory and its applications for non-specialists for non-specialists.

- A detailed look at hash functions from the point of view of linear codes.

- A detailed discussion of shift registers in cryptography, information theory and error correction including several new results and their application to the Berlekamp–Massey theory of Reed–Solomon and BCH decoding.

- A clarification of several points of confusion in the literature relating to security.

- A presentation of five different approaches to Shannon's noisy channel theorem.

- A detailed discussion of the sampling theorem and Shannon's fundamental band-limited capacity formula to the effect that $C = B \log(1 + \frac{S}{N})$, using precise statistical and geometrical techniques.

- A look at some of the history of cryptography and coding theory including a brief biography of Claude Shannon and an account of the original discovery of the Golay code in a Finnish soccer-pools magazine.

- A description of invariant theory and combinatorics applied to coding theory with particular reference to "the computer algebra theorem of the twentieth century" i.e. the nonexistence of a plane of order 10 and related work of one of the authors.

- Connections between MDS codes, secret-sharing schemes, Bruck Nets and Euler's "famous problem of the 36 officers".

- A brief description of research work due to the author and two co-authors solving, in the main, the fifty year old problem of finding the longest MDS code.

- A streamlined approach to Reed–Solomon codes via MDS codes.

- A highly accessible account of the decoding of Reed–Solomon codes.

- A major breakthrough in symmetric (and quantum) cryptography using some new research due to the authors and David Wehlau: the work has been patented and is being used in industry.

Missing Topics. We seem to have covered all the essential topics. We meant to discuss convolutional codes but ran out of space. But they can be covered from the shift register point of view and feed back shift registers have been covered in considerable detail. We also wanted to put in some computer code. We plan on putting some on the website if there is a demand for it.

Professional Acknowledgements. First of all we would like to thank Ted Bisztriczky, Marguerite Fenyvesi and the Department of Mathematics at the University of Calgary, for their support. Richard K. Guy has offered wise counsel and mathematical erudition. We are very grateful to Joanne Longworth, Department of Mathematics for her typing and editorial work including the diagrams. In the same vein we want to thank Dr. Guo of the CPSC department. Richard Cannings was a major catalyst for our involvement with industrial cryptography. Thank you Richard. We also thank, in a similar way, Gerald Stariula for getting us involved and sharing important ideas with us. Dr. David Wehlau has already been acknowledged in Chapter 24. The late Gian Carlo Rota of MIT had a major positive influence and encouraged us to write the book. Professor Bruce Rothschild has been a major source of help and inspiration. The first author would like to thank his brother, Bernard Bruen of New York for his help. He also thanks his sister-in-law, Margit Veldi, an IT worker in Montreal for valuable consultations. Rachael Bevan Bruen kindly supplied important biological insights. Prof Keith Nicholson deserves our gratitude as does the Wiley editorial staff including Susanne Steitz, Steve Quigley and Danielle Lacourciere.

We come to the students-Apichart Intrapanich, Eric Lenza, Reza Pasand, Dave Richardson, Zoron Rodiquez, Paul Tarjan and Feng Zhang. The book could not have been written without their help. In particular, Eric and Paul made new mathematical discoveries which have been included in the book. They also constructed first drafts of chapters. In addition, along with Dave and Zoron they cheerfully criticized, re-drafted, corrected and proof-read. Our thanks to all.

Other friends and colleagues to whom we are grateful include David Torney, Bob Davis, Jose Rueda, Chris Fisher, Tim Alderson, and Richard Biggs who write chapter 5.

Personal Acknowledgements. For some personal acknowledgements, the first author would like to thank his loving wife Katri Veldi, their three children [Trevor, Robin and Merike] and Trevor's wife Rachael for their love and support. He would also like to mention his siblings, Phil, Antoinette and Bernard, and their spouses. The book is also partially dedicated to the memory of his late parents, Edward A Bruen and Bríd Bean de Brún.

The second author would like to thank his beloved wife Claudia Martinez for all her support and the many hours of patient typing of early versions of Part I and their son Dante for contributing a daily dose of joy. The book is also partially dedicated to his parents Alberto Forcinito and Olga Swystun de Forcinito.

Book Website, Corrections.   We will maintain a website for the book at

$$www.SURengineering.com.$$

We have done our best to correct the errors but, inevitably, some will remain. We invite our reads to submit errors to mario@SURengineering.com We will post them, with attribution, on the website.

About the Authors.   Aiden A. Bruen was born in Galway, Ireland. He read mathematics for his Undergraduate and Master's degree in Dublin. He received his doctorate at the University of Toronto, supervised by F. A. Sherk. At Toronto he also worked with H. S. M. Coxeter, E. Ellers and A. Lehman. He is a Professor of Mathematics and Statistics at the University of Calgary.

Mario A. Forcinito was born in Buenos Aires, Argentina where he took his Bachelors degree in Engineering. He obtained his doctorate in Engineering at the University of Calgary under the supervision of M. Epstein. He is an industrial engineering consultant in Calgary.

Update.   New results are constantly being obtained. As this book was going to press, it was reported in the Toronto Globe and Mail that "encryption circles are buzzing with news that mathematical functions embedded in common security applications have previously unknown weaknesses".

In particular, the report cites security vulnerabilities, discovered by E. Biham and R. Chen of the Israel Institute of Technology, in the SHA-1 hash function algorithm. SHA-1 is used in popular programs such as PGP and SSL and is considered the "gold-standard" of such algorithms. Certified by NIST (the National Institute of Standards and Technology) it is the only algorithm approved as a Digital Signature Standard by the US government.

In Chapter 24 we discuss the construction of hash functions from error-correcting codes.

Dedications

Dedicated to my beloved wife Katri and to the memory of my late parents, Edward A
Bruen and Bríd Bean de Brún (AAB)

Also dedicated to my beloved wife Claudia and to my parents, Alberto Forcinito and Olga
Swystun de Forcinito (MAF)

Part I

Mainly Cryptography

Chapter 1

Historical Introduction and the Life and Work of Claude E. Shannon

Goals, Discussion We present here an overview of historical aspects of classical cryptography. Our objective is to give the reader a panoramic view of how the fundamental ideas and important developments fit together. This overview does not pretend to be exhaustive but gives a rough time line of development of the milestones leading to modern cryptographic techniques. The reader interested in a complete historical review is advised to consult the definitive treatise by Kahn [Kah67].

Following this we discuss the life and work of Claude Shannon, the founding father of modern cryptography, information theory and error correction.

1.1 Historical Background

Cryptology is made up of two Greek words: *kryptos*, meaning "hidden," and *ology*, meaning "science." It is defined in [Bri97] as the science concerned with communications in secure and usually secret form. It encompasses both cryptography (from the Greek *graphia* meaning "writing") and *cryptanalysis*, or the art of extracting the meaning of a cryptogram.

Cryptography has a history that is almost as long as the history of the written word. Some four millennia ago (see [Kah67] p. 71) an Egyptian scribe recorded in stone the first known hieroglyphic symbol substitution in the tomb of Khnumhotep II, a nobleman of the time. Although the intention in this case was to exalt the virtues of the person, rather than to send a secret message, the scribe used for the first time one of the fundamental

3

elements used by cryptographers throughout the ages, namely, **substitution**. He used unusual hieroglyphic symbols, known perhaps only to the elite, in place of the more common ones.

In substitution, the sender replaces each letter of a word in a message by a new letter (or sequence of letters or symbols) before sending the message. The recipient, knowing the formula used for the substitution—**the secret key**—is able to reconstruct the message from the scrambled text that is received. It is assumed that only the recipient and the sender know the secret key.

The other main cryptographic technique used is **transposition** (or permutation), in which the letters of the message are simply rearranged according to some prescribed formula that would be the secret key in this case.

The Greeks were the inventors of the first **transposition cipher**. The Spartans [Kah67], in the fifth century B.C. were the first recorded users of cryptography for correspondence. They used a secret device called a *scytale* consisting of a tapered baton around which was spirally wrapped a strip of either parchment or leather on which the message was written. When unwrapped, the letters were scrambled, and only when the strip was wrapped around an identically sized rod could the message be read.

Today, even with the advent of high-speed computers, the principles of substitution and transposition form the fundamental building blocks of ciphers used in **symmetric cryptography**.

To put it in a historical perspective, **asymmetric** or **public key** cryptography was not invented until the 1970s. Exactly when it was invented, or who should take most of the credit, is an issue still in dispute. Both the NSA[1] and the CESG[2] have claimed priority in the invention of public key cryptography.

Cryptography has had several reincarnations in almost all cultures. Because of the necessity of keeping certain messages secret (i.e. totally unknown to potential enemies) governments, armies, ecclesiastics, and economic powers of all kinds have been associated throughout history with the development of cryptography. This trend continues today.

The Roman general Julius Caesar was the first attested user of substitution ciphers for military purposes ([Kah67] p. 83). Caesar himself recounted this incident in his *Gallic Wars*. Caesar found out that Cicero's station was besieged and realized that without help he would not be able to hold out for long. Caesar had a volunteer ride ahead with an encrypted message fastened to a spear, which he hurled into the entrenchment. Basically, Cicero was told to keep up his courage and that Caesar and his legions were on their way.

In the cipher form used by Caesar, the first letter of the alphabet "A" was replaced by the fourth letter "D", the second letter "B" by the fifth, "E", and so on. In other words, each original letter was replaced by a letter three steps further along in the alphabet. To

---

[1] United States National Security Agency
[2] Britain's Communications Electronics Security Group

this day, any cipher alphabet that consists of a standard sequence like Caesar's is called a Caesar alphabet even if the shift is different from three.

Not much mention is made of the coding abilities of Augustus Caesar, the first Emperor of Rome and nephew of Julius Caesar. His cipher involved a shift of only one letter, so that for the plain text (that is, the original text), A was enciphered as B.

Mention of cryptography abounds in early literature: Homer's *Iliad* refers to secret writing. The *Kama-sutra*, the famous textbook of erotics from the Indian subcontinent, lists secret writing as one of the 64 arts or yogas that women should know and practice ([Kah67] p. 75). One of the earliest descriptions of the substitution technique of encryption is given therein. One form involves the replacement of vowels by consonants and vice versa.

In Hebrew literature there are also examples of letter substitution. The most prevalent is the **atbash** technique. Here the first and last, second and second last, (and so on) letters of the Hebrew alphabet are interchanged. An example can be found in the Hebrew Bible. Kahn ([Kah67] p. 77) cites Jeremiah 25:26 and Jeremiah 51:41, where the form SHESHACH appears in place of *Babel* (Babylon).

In Jeremiah 51:41, the phrase with SHESHACH is immediately followed by one using "Babylon." To quote:

How is SHESHACH taken!
And the praise of the whole earth seized!
How is Babylon become an astonishment
Among the nations!

Through Aramaic paraphrases of the Bible, it is clear that SHESHACH is the same as Babel. With the atbash technique, the second letter of the Hebrew alphabet "b" or *beth* becomes the repeated sh or shin, the next to last letter in the alphabet. Similarly the "l" of *lamed* becomes the hard ch, or kaph of SHESHACH. As Babylon appears below in the passage, the use of atbash here was not to actually hide the word but perhaps just a way for the scribe to leave a trace of himself in the work he was copying.

The first people to clearly understand the principles of cryptography and to elucidate the beginnings of cryptanalysis were the Arabs [Kah67]. While Europe was in the Dark Ages, Arab arts and science flourished and scholars studied methods of cryptanalysis, the art of unscrambling secret messages without knowledge of the secret key. A complete description of this work, however, was not published until the appearance of the multivolume *Subh al-a'sha* around 1412.

European cryptology was being developed around this time in the Papal States and the Italian city-states [Kah67].

The first European manual on cryptography (c1379) was a compilation of ciphers by Gabriele de Lavinde of Parma, who served Pope Clement VII. The Office of *Cipher Secretary* to the Pope was created in 1555. The first incumbent was Triphon Bencio de Assisi. But considerably before this, in 1474, Cicco Simonetta wrote a manuscript that was entirely devoted to cryptanalysis.

Cryptanalysis was to have tragic consequences for Mary, Queen of Scots. It was the decipherment of a secret message to Anthony Babington supposedly planning an insurrection against Elizabeth I [Lea96] that resulted in her tragic end. Having obtained this evidence, Sir Francis Walshingham, the head of Queen Elizabeth's secret service, sent his agent back to Fotheringay Castle to intercept and copy more of Mary's secret messages, with the result that Mary and all her coconspirators in the plot were finally arrested. As a result of the trial all were executed, but only Mary was beheaded. Walshingham later claimed that his agents had found the keys to as many as fifty different ciphers in Mary's apartments. (There has long been a conjecture that Mary was actually innocent and that the evidence was planted to remove this inconvenient rival to the English throne.)

The architect Leon Battista Alberti, born in Florence in 1404, is known as "the Father of Western Cryptology." In 1470 he published *Trattati in Cifra*, in which he described the first cipher disk. His technique led to a generalization of the Caesar cipher, using several shifted alphabets instead of just one alphabet. This gave rise to the so-called Vigenère cipher discussed in Chapter 2. (This is actually a misattribution as de Vigenère worked on auto-key systems.)

In 1563 the Neapolitan Giovanni Battista Porta published his *De Furtivis Literarum Notis* on cryptography, in which he formalized the division of ciphers into transposition and substitution.

Moving up several centuries we find that cryptography was widely used in the American Civil War. The Federal army [Bri97] made extensive use of transposition ciphers in which a key word indicated the order in which columns of the array were to be read and in which the elements were either plain text words or code word replacements for plain text. Because they could not decipher them, the Confederacy, sometimes in desperation, published Union ciphers in newspapers, appealing for readers to help with the cryptanalysis.

To make matters worse for the Confederate army, the Vigenère cipher that they themselves used was easily read by the Union army.

Kahn reports ([Kah67] p. 221) that a Vigenère tableau was found in the room of John Wilkes Booth after President Lincoln was shot. Because there was actually no testimony regarding any use of the cipher, could this have been a convenient method of linking Booth and the seven Southern sympathizers with the Confederate cause?

Lyon Playfair, Baron of St. Andrews, recommended a cipher invented in 1854 by his friend Charles Wheatstone to the British government and military. The cipher was based in a digraphic[3] substitution table and was known as the *Playfair Cipher*. The main difference here when compared with a simple substitution cipher is that characters are substituted two

---

[3] *di* meaning two, *graph* meaning character or symbol

at a time. Substitution characters depend on the positions of the two plain text characters on a secret 5×5 square table (the key) whose entries are the characters of the alphabet less the letter "J".

In 1894, Captain Alfred Dreyfus of the French military was accused of treason and sent to Devil's Island, because his handwriting resembled that in an encrypted document that offered military information to Germany. To prove his innocence, the note had to be cryptanalyzed. To be certain that the decipherers' work was correct, an army liaison officer with the Foreign Ministry managed to elicit another similarly encrypted note in which the contents were known to him. The plain text then showed that Dreyfus had not written the encrypted document, but it took several more years before he was to "receive justice, re-instatement and the Legion of Honour" ([Kah67] p. 262). Early in the twentieth century, Maugborne and Vernam put forth the basis for the cipher known as the one-time pad. Although—as was proven later by Shannon—this cipher is effectively unbreakable, its use is somewhat restricted because, in practice, a random key that is as long as the message must be generated and transmitted securely from **A** to **B**. Soviet spies used this cipher, and it is said that the phone line between Washington and Moscow was protected with a one-time pad during the Cold War era.

Edward Hugh Hebern [Bri97] of the United States invented the first electric contact rotor machine. In 1915 he experimented with mechanized encryption by linking two electric typewriters together, using 26 wires to randomly pair the letters. In turn, this led to the idea of rotors that could mechanize not only substitution but alphabet shifts as well. The function of the rotor was to change the pairing of letters by physically changing the distribution of electric contacts between the two typewriters. By 1918 he had built an actual rotor-based encryption machine.

At about the same time (1918–19) three other inventors, the German Arthur Scherbius, the Dutchman Hugo Koch, and the Swede Arvid Damm, were filing patents of rotor-based encryption machines. The Scherbius idea, which included multiple rotors, materialized in the first commercial models having four rotors, ENIGMA A and ENIGMA B, in 1923. Ironically, Hebern only filed for patent protection in 1921, received one in 1924, and lost a patent interference case against International Business Machines in 1941.

Later modifications to the Scherbius machine including a reflector rotor and three interchangeable rotors were implemented by the Axis forces during World War II.

Rotor-based machines give the possibility to implement polyalphabetic substitution ciphers[4] with very long keys or *cycles* in a practical way. With the advantage of mechanization, the ability of widespread deployment of cryptographic stations and widespread use became a reality. This translated into a bigger volume of messages (potentially all messages) being encrypted. However, the increase in traffic gave more cipher text for cryptanalysts to ana-

---

[4]A polyalphabetic cipher uses several substitution alphabets instead of one.

lyze, and the probability of operators making a deadly mistake in the management of keys was multiplied.

The timely breaking of the ENIGMA cipher by the Allies was due in part to inherent weaknesses in the encryption machine, mismanagement of keys by the operators, and lots of mechanized, analytical work. The cipher was first broken, with only captured cipher text and a list of daily keys obtained through a spy, by the Polish mathematician Marian Rejewski. One of the important players in the mechanization of ensuing breaks was the English mathematician Alan Turing, who also contributed to the establishment of the basis for what is today called Computation Theory.

As a side note, after World War II many of the ENIGMA machines captured by the Allies were sold to companies and governments in several countries.

Another very interesting cryptographic technique of a different kind was used by the U.S. military in the Pacific campaign in World War II. Secret military messages were encrypted by translating them from English to the Navajo language. For decryption at the other end, of course, the Navajo was translated back into English. Some words describing military equipment did not exist in the original Navajo language, but substitutes were found. For example, "tanks and planes" were described with the Navajo words for "turtles and birds." To avoid the possibility of the enemy getting a handle on the code, the whole system was committed—by means of an intensive training program—to the memory of the translators or "Code Talkers." This code was never broken.

Immediately after World War II, Shannon was publishing his seminal works on information theory. Almost simultaneously, thanks to the efforts of Ulam, von Neumann, Eckert and Mauchly another key technological development was starting to make major progress, namely the introduction of the newly invented digital computer as a mathematical tool [Coo87].

Because of the importance of his contributions to the issues in this book, we present here a brief biography of Claude Shannon, before finishing the chapter with a review of modern developments.

1.2   Brief Biography of Claude E. Shannon

Claude Shannon has been described as the "father of the information age." His discoveries are everywhere. Waldrop [Wal01] gives an excellent example.

> Pick up a favorite C.D. Now drop it on the floor. Smear it with your fingerprints. Then slide it into the slot on the player and listen as the music comes out just as crystal clear as the day you first opened the plastic case. Before moving on with the rest of your day, give a moment of thought to the man whose revolutionary ideas made this miracle possible: Claude Elwood Shannon.

Figure 1.1: Claude E. Shannon. Courtesy of Lucent Technologies Inc./Bell Labs

Computers give us the power to process information, but Shannon gave us the capacity to understand and analyze information. Shannon demonstrated the unity of text, pictures, film, radio waves, and other types of electronic communication, and showed how to use these media to revolutionize technology and our way of thinking.

1.3 Career

He was born in Petoskey, Michigan in 1916. His father was a businessman who later became a judge and his mother was a high school teacher.

As a youngster he was interested in, and became adept at, handling all kinds of contraptions such as model airplanes and boats as well as learning the workings of the telegraph system.

At the age of 20 he graduated with degrees in mathematics and electrical engineering from the University of Michigan.

In the summer of 1936, Claude joined the MIT Electrical Engineering Department as a research assistant to work on an analog computer (as opposed to our modern digital computers) under the supervision of Vannevar Bush. Bush's analog computer, called a differential analyzer, was the most advanced calculator of the era and was used mainly for solving differential equations. A relay circuit associated with the analyzer used hundreds of relays and was a source of serious study by Shannon, then and later.

During the summer of 1937, Shannon obtained an internship at Bell Laboratories, and returned to MIT to work on a Master's thesis.

In September of 1938 he moved to the Mathematics Department there and wrote a thesis in genetics with the title *An Algebra for Theoretical Genetics*. He graduated in 1940 with his PhD in mathematics and the S.M. degree in electrical engineering.

Dr. Shannon spent the academic year of 1940–41 at the Princeton Institute, where he worked with Herman Weyl, the famous group theorist and geometer.

Subsequently, he spent a productive fifteen years at the Bell Laboratories in New Jersey, returning to MIT in 1956, first as a visiting professor and then, in 1958, as Donner Professor of Science. This was a joint appointment between mathematics and electrical engineering. Here he did not teach ordinary courses but gave frequent seminars. According to Horgan [Hor90], he once gave an entire seminar course, with new results at each lecture!

He retired from MIT in 1978 but continued to work on many different problems including portfolio theory for the stock market, computer chess, juggling, and artificial intelligence. He died in 2001 at the age of 84 a few years after the onset of Alzheimers disease.

1.4   Personal—Professional

Dr. Shannon's Master's thesis (Transactions American Institute of Electrical Engineers 57, 1938, 713–723) won him the Alfred Noble Prize along with fame and renown. The thesis has often been described as the greatest Master's thesis of all time; many feel that this may in fact understate the case.

At MIT he was idolized by both students and faculty. Golomb [Gol02] reports that Shannon was "somewhat inner-directed and shy, but very easy to talk to after the initial connection had been made."

In his spare time, Shannon built several items including THROBAC (Thrifty Roman numerical Backward-looking Computer), which was actually a calculator that performed all the arithmetic operations in the Roman numerical system. He also built Theseus, a mechanical mouse, in 1950. Controlled by a relay circuit, the mouse moved around a maze until it found the exit. Then, having been through the maze, the mouse, placed anywhere it had been before, would go directly to the exit. Placed in an unvisited locale, the mouse would search for a known position and then proceed to the exit, adding the new knowledge to its memory.

Shannon was the first to develop computerized chess machines and kept several in his house. He built a "mind-reading" machine that played the old game of penny-watching.

As juggling was one of his obsessions, he built several juggling machines and worked hard on mathematical models of juggling. He was famous for riding his unicycle along the corridors at Bell Laboratories, juggling all the while. On the more practical side, Shannon was also interested in portfolio management and the stock market, which he connected to information theory, treating the market as a noisy channel!

Over the course of his career, Dr. Shannon received umpteen awards, honors, prizes, honorary degrees, and invitations. In the end it all became too much, and he "dropped out." To quote Waldrop [Wal01] "he turned down almost all the endless invitations to lecture or to give newspaper interviews. He didn't want to be a celebrity. He likewise quit responding to much of his mail. Correspondence from major figures in science and government ended up forgotten and unanswered in a file folder he labeled "Letters I've procrastinated too long on." Dr. Shannon did attend one other Shannon lecture in Brighton, England in 1985 (delivered by Golomb), where the shy genius created quite a stir. As Robert McEleice recalls (see [Hor90]) : "It was as if Newton had showed up at a physics conference."

1.5 Scientific Legacy

Circuits   Shannon's Master's thesis (see above and [Sha48]) was the first work to make him famous. He became intrigued by the switching circuits controlling the differential analyzer while working for Vannevar Bush. He was the first to notice that the work of a mathematics professor named George Boole in Cork, Ireland, done a century earlier, yielded the necessary mathematical framework for analyzing such circuits.

"On" and "Off" could be represented by "1" and "0". The Boolean logical operations of AND, OR correspond exactly to a circuit with two switches in series or in parallel, respectively. He demonstrated that any logical statement, no matter how complex, could be implemented physically as a network of such switches. He also showed how the crucial Boolean decision operation could be implemented in a digital system, marking the main qualitative difference between a calculator and the powerful digital computers to follow.

Cryptography   Shannon published just one paper in cryptography, namely, [Sha49]. Its contents had appeared in a war-time classified Bell Laboratories document, which was later declassified. The beginning sentence is very revealing. It reads as follows:

> The problems of cryptography and secrecy systems furnish an interesting application of communication theory.

Indeed, this is precisely the point of view that inspired the authors of this book! We believe it is unrealistic to separate the study of cryptography from the study of communication theory embodying error correction and information theory.

To illustrate this, Shannon points out that just as in error correction, where the receiver tries to decode the message over a noisy channel, so also in cryptography a receiver (this time, Eve, the eavesdropper) tries to decode the message over a noisy channel, the noise being the scrambling by the key which obfuscates the plain text into the cipher text!

In this paper, Shannon discusses at length his two famous principles of confusion and diffusion, described in detail in Chapter 4. Although most authors in cryptography cite the paper, it is not clear that they have actually read it. As an illustration, many authors claim that in this paper Shannon states that the Vernam cipher is "essentially the only cipher offering perfect security." But this is incorrect, and Shannon never says such a thing. We discuss perfect security in detail in Part II of the book where it is shown that, under appropriate conditions, perfect security corresponds precisely to a Latin square. Shannon's paper makes it quite clear that he was aware of this phenomenon, although he did not explicitly state it.

In the paper, Shannon clearly differentiates between computational and unconditional security. Whether or not he "missed" public key cryptography is far from clear. However, in [Mas02] Massey points out that Hellman, of Diffie–Hellman fame, has credited the following words from Shannon's paper as the inspiration for their discovery:

> The problem of good cipher design is essentially one of finding difficult problems .... We may construct our cipher in such a way that breaking it is equivalent to ... the solution of some problem known to be laborious.

Of course, the jury is still out, as Massey [Mas02] points out, on whether one-way functions, the foundations of public key cryptography, really exist. The reader is referred to Chapter 3 and 4 on this point.

Shannon Theory: Information Compression and Communication.

Shannon's revolutionary paper [Sha49] on information theory electrified the scientific world and has dominated the area of communication theory for over fifty years. No other works of the twentieth century have had greater impact on science and engineering.

First of all, Shannon unified what had been a diverse set of communications—voice, data, telegraphy, and television. He quantified and explained exactly what information means. The unit of information is the Shannon bit. As Golomb [Gol02] so elegantly puts it, this is the "amount of information gained (or entropy removed) upon learning the answer to a question whose two possible answers were equally likely, a priori."

In the above, we can think of entropy as "uncertainty" analogous to entropy in physics (which is the key idea in the second law of thermodynamics).

An example would be the tossing of a fair coin and learning which turned up—heads or tails. If the coin were biased, so that the probability of a head was p (and the probability of a tail was $1 - p$) with $p \neq 1/2$, the information gained, on learning the outcome of the toss, would be less than one. The exact amount of information gained would be

$p \log(1/p) + q \log(1/q)$, where $q = 1 - p$ and where we take logs to the base 2. $\quad$ (1.1)

Note that when $p = \frac{1}{2}$ and $q = \frac{1}{2}$ this works out to be 1. However, if for example, $p = 2/3$ we gain only approximately 0.918 Shannon bits of information on learning the outcome of the coin toss.

It can be mathematically proven that the only information function that gives sensible results is the appropriate generalization to a probability distribution of Formula (1.1) above. Formula 1.1 ties into the fundamental notion of entropy (or uncertainty). There are many examples of redundancy in the English language, i.e. the use of more letters, words, or phrases than are necessary to convey the information content being transmitted. As Shannon points out, the existence of redundancy in the language is what makes crosswords possible.

This redundancy can be reduced in various ways. An example is by writing acronyms such as "U.S." for "United States". When information is to be electronically transmitted we remove redundancy by data compression. Shannon's formula for data compression is intimately related to entropy, which is in turn related to the average number of yes-no questions needed to pin down a fact. Shannon showed that it is possible to obtain a bound for the maximum compression that is the best possible. The actual technique for compressing to that ultimate degree is embodied in the construction of the so-called Huffman codes, well known to all computer science undergraduates. Later, other compression techniques followed, leading to modern technologies used in, for example, mp3s (music compression). This part of Shannon's work is also connected to the later work of Kolmogorov on algorithmic complexity and the minimum-length binary program needed for a Turing machine to print out a given sequence.

But this was only the beginning. Shannon then went on to prove his fundamental result on communication, based on entropy and the mathematical ideas delineated above. He showed that any given communications channel has a maximum capacity for reliably transmitting information which he calculated. One can approach this maximum by certain coding techniques—random coding and now turbo coding—but one can never quite reach it. To put it succinctly: *Capacity is the bound to error-free coding.* Thus, for the last 50 years, the study of error correction has boiled down to attempts to devise techniques of encoding in order to come close to the Shannon capacity. We will have much to say about this bound in Parts II and III of this book.

Shannon's work, theoretical and practical, still dominates the field and the landscape. To quote Cover in [Cov02]:

> This ability to create new fields and develop their form and depth surely places Shannon in the top handful of creative minds of the century.

Few can disagree with this assessment. Indeed, in Part III of this book we describe new protocols in cryptography and error correction based squarely on C. E. Shannon's work in information theory.

1.6 Modern Developments

From the vantage point of post-Shannon understanding and availability of serious computing power, several cipher systems were developed to take advantage of both. Worthy of mention are those created at the IBM Watson Research Lab in the 1960s under the leadership of Horst Feistel that are collectively known as Feistel's ciphers. A modified version of the cipher originally known as Lucifer became the United States Data Encryption Standard (DES) in 1977[5]. DES found world wide acceptance and has been in use for the last 20 years. Because of its short key-length (56 bits), it is nowadays possible in certain cases to conduct a brute-force attack on the entire key-space. However, it is still being used as a component of Triple DES, a 3-key cipher.

In 1976, the concept of public key cryptography was created and made public by Diffie, Hellman, and Merkle. In the early days of public key systems, a public key algorithm called the "Knapsack algorithm" was invented by Ralph Merkle. In 1977 Rivest, Shamir, and Adleman created (with the help of D. E. Knuth and others) the algorithm known as RSA that was first published in 1978. In recent years, both the British Communications Electronics Security Group (CESG) and the United States National Security Agency (NSA) have released classified documents showing that they invented public key cryptography including analogs to the Diffie–Hellman key-exchange and RSA in the early 1970s.

Regardless of priority claims, the invention of public key cryptography brought to the fore the possibility of on-line key generation and secret exchanges between two entities not known to each other, using a public channel. This new ability stressed the need for more secure authentication protocols, as anybody can generate valid keys and signatures.

Following an earlier idea of Weisner, Charles H. Bennett conducted experimental work exploring the possibility of using a stream of single photons transmitted over fiber-optic cable as a way to create a key between **A** and **B**. This leads to Quantum Cryptography. In the early 1990s, Bennett, Brassard, et al. developed a key reconciliation algorithm intended to be used in the context of Quantum Cryptography. The algorithm exchanges information over a public channel in such a way that **A** and **B** have high probabilities of reaching a common key, while an eavesdropper will not be able to retrieve enough information to reconstruct the key.

In 1991, Phil Zimmermann released the first version of PGP (Pretty Good Privacy) as freeware. PGP offered high security to the general citizen, and as such it has become a

---

[5]It was published as Federeal Information Processing Standard (FIPS) PUB-46.

worldwide standard for normal E-mail needs at a very low cost.

In 1999, Rijndael, a block cipher developed by Joan Daemen and Vincent Rijmen was selected as the Advanced Encryption Standard (AES)[6], which is a new standard for symmetric cryptography.

The last years of the twentieth century witnessed an explosive growth of global need for secure communications as more governments, private companies, and individuals relied on E-mail, the Internet, and wireless links for the communication of sensitive information. With this unprecedented and widespread need for data security, cryptography has become an essential tool for economic development.

---

[6]Published as FIPS standard 197.

Chapter 2

Classical Ciphers and Their Cryptanalysis

Goals, Discussion In this chapter, we survey some historical ciphers that have been used since antiquity. (They are all symmetric ciphers. Public key ciphers were not invented until the 1970s and are discussed in Chapter 3.) Although the ciphers presented here are obsolete, they still provide good examples of cryptographic procedures. For example the Vigènere cipher, being a block cipher, is a forerunner of modern block ciphers such as DES. From these classical ciphers we can also learn about various attacks in cryptography. This subject is pursued more fully in Chapter 7.

New, Noteworthy We discuss the Vigenère cipher and show how it can be broken by finding the length of the key-word and then the key-word itself. We explain clearly the simple principles involved without getting bogged down in lengthy formulae. We also give a detailed but accessible description of the famous Enigma system used in World War II both from the mechanical and the mathematical point of view.

2.1 Introduction

Since the early stages of human civilization, there has been a need to protect sensitive information from falling into the wrong hands. To achieve such secrecy, mankind has relied on a branch of mathematics known as **cryptography**, which is the study of designing methods to securely transmit information over non-secure channels. To achieve this goal, one must first **encipher**, or scramble, the intended message to prevent an eavesdropper from obtaining any useful information should the message be intercepted. The message (referred to as **plain text**) is scrambled into **cipher text** with a predetermined key known

17

to both sender and receiver. The encrypted message is constructed in such a way so as to be resilient against attack, while allowing the intended recipient to **decipher**, or unscramble, the message with ease. The methods we will be investigating in this section to accomplish this task may be outdated and in some cases obsolete, but they can still provide us with valuable insight into some techniques that are still in use today.

2.2   The Caesar Cipher

Figure 2.1: Caesar Cipher Wheel

While Julius Caesar was building his empire, he needed a method to transmit vital messages without risk of the enemy obtaining any crucial information. To achieve this goal, he employed one of the first known ciphering methods. The idea was to substitute each letter of the plain text with the letter appearing three spaces to the right in the alphabet i.e. a is enciphered to D, b is enciphered to E, and z is enciphered to C (the alphabet wraps around). Thus, "six" is enciphered to "VLA".

In practice, this can be easily achieved with a simple device consisting of two disks, such as is shown in Figure 2.1. Both disks have the alphabet engraved on their perimeter, and they can rotate with respect to each other. If we assign the inner disk to represent the plain text alphabet and the outer disk to represent the cipher text alphabet, enciphering is accomplished simply by rotating the outer disk by three letters counterclockwise and reading off the cipher text corresponding to the plain text. To decipher the message, one need only reverse the procedure. The "key" of the cipher is just the number of letters that the outer disk is shifted by, and is denoted by k. Both sender and recipient are in possession of this common secret key.

For a numerical explanation, suppose we label $a, b, \ldots, z$ by the numbers $0, 1, \ldots, 25$. Using Caesar's key of three, the plain text message "six" is enciphered as follows:

$$\text{six} \longrightarrow (18, 8, 23) \xrightarrow{k=3} (18 + 3, 8 + 3, 23 + 3) = (21, 11, 0) \longrightarrow \text{VLA}$$

Note that in the above example, $23 + 3 = 26$. We replace 26 by $Rem[26, 26]$, which is the remainder when we divide by 26 (see Chapter 19 for details). In this case, the remainder is 0, corresponding to the letter A. Similarly, $24 + 11$ becomes $Rem[35, 26] = 9$. The number 9 corresponds to the letter J.

To decipher the message, reverse the operation (shift left by k spaces):

$$\text{VLA} \longrightarrow (21, 11, 0) \xrightarrow{k=3} (21 - 3, 11 - 3, 0 - 3) = (18, 8, 23) \longrightarrow \text{six}$$

In this case, we have $0 - 3 = 23$. If x is a negative number then it is replaced by $26 + x$. So, for example, -3 gets replaced by $26 + (-3) = 23$. The reasoning is that $Rem[-3, 26] = 23$ since $-3 = 26(-1) + 23$. Alternatively, (see Chapter 19) we have $-3 \equiv 23 \pmod{26}$.

The Caesar cipher is a simple example of a type of cipher known as a **monoalphabetic** cipher. Monoalphabetic ciphers belong to a class of ciphers known as **substitution ciphers**, in which each of the plain text characters in the message is replaced by another letter. Mathematically speaking, the enciphering process of a monoalphabetic cipher can be represented by the mapping of a plain text character to a cipher text character

$$x \longrightarrow Rem[x + k, 26]$$

Similarly, deciphering is represented by the mapping

$$x \longrightarrow Rem[x - k, 26]$$

where k is the **cipher key**, with $1 \le k \le 25$. In the case of the Caesar cipher, $k = 3$. If $x - k$ is negative, then, as explained above, we use $Rem[x - k, 26] = 26 + (x - k)$.

To break such a cipher, one can decrypt the message by trying all 26 keys (this is referred to as an exhaustive search). For long messages, the likelihood of a cipher text decrypting to two intelligible messages is small.

We mention here briefly affine ciphers. They are similar to Caesar ciphers in that they are simple substitution ciphers, but they differ in that enciphering involves not only addition, but multiplication as well.

2.3   The Scytale Cipher

The Scytale cipher was introduced around 500 B.C. by the Spartans, who used this rather simple but effective method to send crucial planning data between generals and bureaucrats. Both the sender and receiver were in possession of cylindrical tubes of the same diameter. To encode the message, the sender would wrap a thin strip of paper around the tube, with the paper spiraling its way down the length of the tube. The message was then written on the strip, with letters being written one beneath the other until the end of the tube was reached. (The message was then continued by starting a new column of letters, and this process was repeated until the message was finished.) To encode, the sender would simply unwrap the paper, leaving a thin strip of unintelligible letters. To decode, the receiver only had to wrap the paper around a tube of the same size and read the message off in columns.

It is often much simpler to duplicate the Scytale process with pencil and paper. Using a preselected number of rows (this number is the cipher key), write the message in columns. Then, 'unwrap' the message by writing a string of characters consisting of the concatenated rows. For example, the message "THE ENEMY WILL ATTACK AT DAWN" is encrypted as follows:

$$
\begin{array}{cccc}
T & M & A & A \\
H & Y & T & T \\
E & W & T & D \\
E & I & A & A \\
N & L & C & W \\
E & L & K & N \\
\end{array}
$$

$$\downarrow$$

$$T\,M\,A\,A\,H\,Y\,T\,T\,E\,W\,T\,D\,E\,I\,A\,A\,N\,L\,C\,W\,E\,L\,K\,N$$

Knowledge of the cipher key reduces the decryption process to a trivial matter. Using the fact that the key for this example is six, count up the total number of characters and divide by six. Doing so yields the period of the sequence, which is four. Thus, by taking the first, fifth, ninth,... characters, one can reconstruct the columns. Upon completion, the message can be read off, column by column.

The Scytale concept can be modified to create very complex ciphers. By arranging the plain text in varying matrix patterns and 'unwrapping' in different ways, messages can be scrambled very effectively. The Scytale cipher, along with its variants, belongs to a class of ciphers called **transposition ciphers**, in which all plain text characters are present in the cipher text but they appear in a substantially different order.

2.4   The Vigenère Cipher

The next classical cipher of great interest is known as the Vigenère cipher. Although it is a relatively simple concept, it has influenced many ciphers used throughout history, including some that are still in use today. The idea is to modify the notion of Caesar ciphers, which were covered in Section 2.2. Instead of a single cipher key, we make use of an entire keyword, which is a key in the form of a string of letters chosen by the user. For example, suppose we decide to use the word "encode" as our keyword. The enciphering process is carried out as follows. The keyword is repeated as many times as necessary to span the length of the message. Each letter corresponds to a number between 0 and 25 (so $a = 0, b = 1, \ldots, z = 25$). Then, to encipher, we add the corresponding numbers of the keyword and the message, subtracting 26 if the sum is bigger than 25. Then we switch back to letters to get the cipher text. For example, if the message letter is y ($= 24$), and the keyword letter is E ($= 4$), then the cipher text letter corresponds to $24 + 4 = 28$. However, 28 is larger than 26, so we subtract 26 to get 2, which is C. Thus, y enciphers to C.

Keyword:	E	N	C	O	D	E	E	N	C	O	D	E	E	N	C
Message:	t	h	e	s	k	y	i	s	f	a	l	l	i	n	g
Cipher text:	X	U	G	G	N	C	M	F	H	O	O	P	M	A	I

Note that each letter of the plain text gets shifted by varying amounts, unlike the Caesar cipher. Interpreting this as many monoalphabetic ciphers acting on individual characters, it is easy to see why the Vigenère cipher is referred to as a **polyalphabetic cipher**.

To decipher, simply subtract the value of the respective key letters from each cipher text letter. For example, to decrypt the cipher text "XUGGNCMFHOOPMAI" from above, we use the keyword "ENCODE" as follows:

$$X - E \quad = 23 - 4 \quad = 19 \qquad\qquad = t$$
$$U - N \quad = 20 - 13 \quad = 7 \qquad\qquad = h$$
$$G - C \quad = 6 - 2 \quad = 4 \qquad\qquad = e$$
$$G - O \quad = 6 - 14 \quad = -8 = 18 \quad = s$$

Repeat the process until all plain text characters have been determined. This enciphering process is easy if one has knowledge of the key. However, it can be difficult to break the cipher without such information. This will be investigated in Section 2.7.

The Vigenère cipher is known as a type of **block cipher**, where the block length is equivalent to the key length. In a block cipher procedure, the plain text or message is encrypted block by block rather than character by character.

There are a few important remarks to be made regarding the use of the Vigenère cipher. First, because the same plain text character enciphers to different characters depending on the position, the cryptanalysis of such a cipher is much more complex than for a simple monoalphabetic substitution cipher. We also point out that the Vigenère cipher was invented to hide the frequencies of letters in the English language. The Caesar cipher, for example, does not do this. Also, if the key phrase "VIGENERECIPHERX" had been used instead of "ENCODE", in our previous example, the encrypted message would have had **perfect secrecy**. Perfect secrecy is achieved if the cipher text yields no information about the plain text, and this occurs, roughly speaking, when the keyword is as long as the message itself. Such a secure system can be obtained by using **one-time pads**, which we investigate later in the book.

2.5   The Enigma Machine and Its Mathematics

Figure 2.2: A German Enigma machine

During World War II, German troops were able to march unopposed through much of Eastern Europe. At the heart of this war machine was an encryption scheme that allowed commanders to transfer crucial planning data with near total secrecy. Before the invasion of Poland, three Polish cryptologists named Marian Rejewski, Henry Zygalski, and Jerzy Różycki were able to crack the majority of the Enigma code used by the German army. Fearing capture during the German invasion, they escaped to France, bringing with them vital secrets about the Enigma machine.

Mechanically speaking, the Enigma machine consists of three removable rotors, a keyboard, a reflector plate, and a plugboard (2.2)). Each rotor has 26 electrical contacts on each face (with each one representing a value between 0 and 25), and wires connecting contacts on opposite faces in a variety of ways. The rotors rotate clockwise and are geared in such a way that the settings of the first rotor change with each plain text character that is enciphered and cycles through the values 0 to 25. Upon the transition between 25 back to 0, the second rotor rotates 1/26th of a turn. Following the transition between 25 back to 0 on the second rotor, the third rotor rotates 1/26th of a turn. This ensures that if the same character is sent twice in a row, it will likely be enciphered as two different cipher text letters. To increase the number of possible permutations, the rotors can be interchanged with one another. The reflector plate is a device with 26 contacts on the face adjacent to the last rotor, wired in such a way that the contacts are connected in pairs. Once a signal is sent to the reflector, it is sent through the corresponding wire and back into the third rotor. The plugboard consists of a series of sockets, and the board changes the identity of the input character based on the following conventions: if the given socket contains a plug, the character's identity is changed; if the socket is empty, the character remains unchanged. This device simply provides another set of permutations, meant to increase the complexity of the enciphering scheme. A basic block diagram of the system is depicted in Figure 2.3.

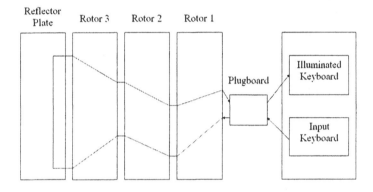

Figure 2.3: Block Diagram of the Enigma Machine

To use the machine, an operator inputs the desired plain text into a keyboard, one character at a time. An electrical signal is passed from the keyboard through the rotors, which are connected in series, until the charge reaches the reflector plate. Then, the signal is passed back from the plate through the rotors and back into the keyboard, where a separate panel consisting of light bulbs is illuminated. Each light bulb corresponds to a cipher text letter, which is recorded by the operator. As the signal passes through each rotor, the plain text character is continually substituted, depending on the daily settings of the rotor and the specific wiring between contacts, which govern the permutations of substitutions that are possible. When the enciphering process is complete, the operator sends the cipher text via radio to the intended receiver, who also possesses an Enigma machine. The receiver can then decode the message, given that the initial settings and the permutation sets of the machines are coordinated, by simply typing in the cipher text into the machine. The plain text message then appears on the illuminated keyboard.

It is worth noting some of the deficiencies in the machine design, as they made it possible for Allied cryptanalysts to eventually break the cipher. However, these deficiencies are beyond the scope of this discussion. The reader is referred to Trappe and Washington [TW01] and Hardy and Walker [HW03] for more details.

We will now investigate the machine from a mathematical standpoint. Each rotor is represented by a set of permutations containing all letter values between 0 and 25. The transition of each set runs left to right, with each bracket representing a wrap-around or cycle. The first, second, and third rotors have unique permutation sets denoted α_1, α_2, and α_3 respectively (each representing the possible transitions between letters). To aid in our analysis, we introduce the variables r_1, r_2, and r_3 to represent the initial rotor settings (taken to be the character currently located at the top of the rotor). For the purposes of this analysis, we will be ignoring the role of the plugboard. Finally, the reflector plate is modeled as a set of permutations between pairs of characters, denoted by β. The goal is to track a signal as it leaves the input keyboard, travels though the rotors and reflector, and back to the illuminated display. To determine the appropriate cipher text for each given plain text letter, we will calculate the shift of each rotor, the resulting reflector permutation, and reflected signal shifts until we end up with a final cipher text character.

To show how the enciphering process works, consider the modified system shown in Figure 2.4.

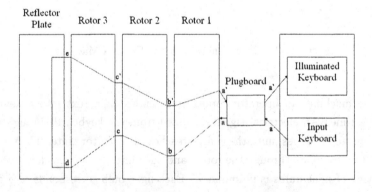

Figure 2.4: Simplified Enigma Model

The idea is to keep track of each intermediate substitution, in order to determine the final cipher text character. To illustrate the encoding process, consider the following example:

Example 2.1 *Suppose the permutation sets of each rotor and reflector are defined as:*

$$\begin{aligned}
\alpha_1 &= (0\ 15\ 6\ 10\ 14\ 8\ 19\ 17\ 22\ 18\ 11)\ (1\ 2\ 9\ 13\ 21\ 25)\ (3\ 4\ 23\ 5\ 24\ 7\ 12\ 16\ 20) \\
\alpha_2 &= (0\ 7\ 9\ 4\ 6\ 18\ 23\ 25\ 8)\ (1\ 17\ 19)\ (2\ 20\ 10)\ (3\ 12)\ (5\ 11\ 13\ 21)\ (14\ 22\ 15\ 16\ 24) \\
\alpha_3 &= (0\ 2\ 4\ 7\ 16\ 17\ 19\ 5)\ (1\ 6\ 3\ 8\ 21\ 24\ 11\ 13\ 9\ 10\ 25\ 12\ 14\ 15)\ (18\ 23\ 20\ 22) \\
\beta &= (0\ 4)\ (1\ 7)\ (2\ 9)\ (3\ 16)\ (5\ 20)\ (6\ 8)\ (10\ 19)\ (11\ 17)\ (12\ 25)\ (13\ 18)\ (14\ 24)\ (15\ 22)\ (21\ 23)
\end{aligned}$$

So, with α_1, 0 gets moved to 15, 15 gets moved to 6, 11 moves to 0, etc.

Each permutation set possesses an inverse, which "undoes" the action of said permutation, as follows.

$$\begin{aligned}
\alpha_1^{-1} &= (11\ 18\ 22\ 17\ 19\ 8\ 14\ 10\ 6\ 15\ 0)\ (25\ 21\ 13\ 9\ 2\ 1)\ (20\ 16\ 12\ 7\ 24\ 5\ 23\ 4\ 3) \\
\alpha_2^{-1} &= (8\ 25\ 23\ 18\ 6\ 4\ 9\ 7\ 0)\ (19\ 17\ 1)\ (10\ 20\ 2)\ (12\ 3)\ (21\ 13\ 11\ 5)\ (24\ 16\ 15\ 22\ 14) \\
\alpha_3^{-1} &= (5\ 19\ 17\ 16\ 7\ 4\ 2\ 0)\ (15\ 14\ 12\ 25\ 10\ 9\ 13\ 11\ 24\ 21\ 8\ 3\ 6\ 1)\ (22\ 20\ 23\ 18)
\end{aligned}$$

The initial settings are defined with $r_1 = 22$ (i.e. the letter at the top of rotor 1 is V), $r_2 = 7$, $r_3 = 12$.

For the signal traveling toward the reflector plate, the substitutions through the rotors are represented mathematically as:

$$\begin{aligned}
b &= Rem[a + r_1, 26]^{\alpha_1} \\
c &= Rem[b + r_2, 26]^{\alpha_2} \\
d &= Rem[c + r_3, 26]^{\alpha_3}
\end{aligned}$$

where raising a term to the exponent α_1 means locating the term in the permutation set and replacing it with the number to the right of the term. If there is a bracket adjacent to the term, wrap around to the beginning of the subset. For example, with our settings as above, $3^{\alpha_1} = 4$ and $8^{\alpha_2} = 0$.

Since the reflector has contacts which are only connected in pairs, we get:

$$e = (d)^\beta$$

Once e has been output from the reflector, we follow the signal back to the keyboard:

$$\begin{aligned}
c' &= Rem[e^{\alpha_3^{-1}} - r_3, 26] \\
b' &= Rem[c'^{\alpha_2^{-1}} - r_2, 26] \\
a' &= Rem[b'^{\alpha_1^{-1}} - r_1, 26]
\end{aligned}$$

After the successful completion of the cipher text substitution, we need to update the rotor settings to take into account the rotation(s) that may have taken place:

r_1 is redefined as $Rem[r_1 + 1, 26]$. If $r_1 = 25$ and we add 1, the new r_1 becomes 0 and r_2 is advanced by one.

Similarly, if $r_2 = 25$ and we add 1, the new r_2 becomes 0 and r_3 is advanced by one. Let's see what happens when we encode the letter 'k', which has numerical value 10.

$$
\begin{aligned}
a &= 10 \\
b &= Rem[a + r_1, 26]^{\alpha_1} &&= Rem[10 + 22, 26]^{\alpha_1} &&= 6^{\alpha_1} &&= 10 \\
c &= Rem[b + r_2, 26]^{\alpha_2} &&= Rem[10 + 7, 26]^{\alpha_2} &&= 17^{\alpha_2} &&= 22 \\
d &= Rem[c + r_3, 26]^{\alpha_3} &&= Rem[22 + 12, 26]^{\alpha_3} &&= 8^{\alpha_3} &&= 21
\end{aligned}
$$

Reaching the reflector, we get $e = (d)^{\beta} = 21^{\beta} = 23$

Now following the signal back through the rotors, we obtain:

$$
\begin{aligned}
c' &= Rem[e^{\alpha_3^{-1}} - r_3, 26] &&= Rem[23^{\alpha_3^{-1}} - 12, 26] &&= Rem[18 - 12, 26] &&= 6 \\
b' &= Rem[c'^{\alpha_2^{-1}} - r_2, 26] &&= Rem[6^{\alpha_2^{-1}} - 7, 26] &&= Rem[4 - 7, 26] &&= 23 \\
a' &= Rem[b'^{\alpha_1^{-1}} - r_1, 26] &&= Rem[23^{\alpha_1^{-1}} - 22, 26] &&= Rem[4 - 7, 26] &&= 23
\end{aligned}
$$

Therefore, the first cipher text character corresponds to 23, and is thus 'X'.

Now, we must update the rotor settings: $r_1 = 23, r_2 = 7, r_3 = 12$.

If the settings were such that r_1 was 25, the updating process would proceed as follows: $r_1 = 0, r_2 = 8, r_3 = 12$.

As mentioned above, an interesting aspect about the Enigma enciphering scheme is the fact that deciphering a message follows the exact same process.

2.6   Frequency Analysis

The idea behind the use of frequency analysis in cryptanalysis is that all human languages have underlying statistical patterns and redundancies that can be exploited to help break a variety of ciphers. For the English language, it is well documented that the distribution of the most frequent characters is remarkably similar throughout texts of diverse style and length, as indicated in Table 2.1.

Frequency analysis can be used for cryptanalysis. However, one needs a lot of craft in its use, along with any information that can be gathered about the contents of the message and the sender.

2.7   Breaking the Vigenère Cipher, Babbage–Kasiski

Now that the Vigenère cipher has been defined, we will show how to use the frequencies of letters in English, to break this cipher. We have 2 tasks.

Letter	Frequency(%)	Letter	Frequency(%)
e	12.7	m	2.4
t	9.1	w	2.3
a	8.2	f	2.2
o	7.5	g	2.0
i	7.0	y	2.0
n	6.7	p	1.9
s	6.3	b	1.5
h	6.1	v	1.0
r	6.0	k	0.8
d	4.3	j	0.2
l	4.0	x	0.1
c	2.8	q	0.1
u	2.8	z	0.1

Table 2.1: Frequencies of Letters in the English Language [TW01]

- Find the keyword length.

- Find the keyword itself.

We have 2 methods to find the length of the keyword. The first method, the **Babbage-Kasiski method**, attempts to find repeated sucessive triplets (or 4-tuples, or 5-tuples, etc) of letters in the cipher text. The second method treats the English language like a stationary or even ergodic source (see Chapter 11).

We will use 2 fundamental principles in carrying out our tasks.

- 'E' is the most frequent letter of the english language.

- Informally, written English tends to "repeat itself". This means that the frequencies of a passage starting in position 1 are similar to the frequencies of the passage starting in position k when we slide the text along itself.

Once we obtain n, the key-length, we can find the keyword itself. We do this by using the first fundamental property above. Namely, we exploit the statistics of the letters in English, or pairs of letters (i.e. digrams), or trigrams, etc.

The second principle has two important interpretations. For the Babbage-Kasiski method, this means that if we find a repeated letter (or sequence of letters) in the cipher text there is a good chance that it comes from a given letter (or sequence of letters) in the plain text that has been enciphered by a given letter (or letters) in the key. Thus there is a reasonable expectation that the distances between such repeated sequences of letters equals the key-length or a multiple of the key-length.

The second principle has an important implication in terms of our second method, called the method of "coincidences", as well. The basic idea is explained in an example below.

The Babbage-Kasiski method   To demonstrate this method for finding n, suppose we received the following cipher text.

EHMVL	YDWLP	WIWXW	PMMYD	PTKNF	RHWXS
LTWLP	OSKNF	WDGNF	DEWLP	SOXWP	HIWLL
EHMYD	LNGPT	EEUWE	QLLSX	TUP	

Our first task is to search for repetitions in the above text. For a small cipher text, the brute-force method is not too difficult. We focus on trigrams, and highlight some as follows.

EHMVL	YDWLP	WIWXW	PMMYD	PTKNF	RHWXS
LTWLP	OSKNF	WDGNF	DEWLP	SOXWP	HIWLL
EHMYD	LNGPT	EEUWE	QLLSX	TUP	

After having found these, we compute the distances between the trigrams.

$$
\begin{array}{lcl}
\text{EHM} & - & 60 \\
\text{WLP} & - & 25,\ 15,\ 40 \\
\text{XWP} & - & 39 \\
\text{MYD} & - & 45
\end{array}
$$

We note that with the exception of 39, 5 divides all of the distances. In fact, if we proceed with frequency analysis, it turns out that we can decipher this message with a key-length of 5. The codeword is "ladel", and the plain text is "Thor and the little mouse thought that they should douse the house with a thousand liters of lithium." (Who said that secret messages had to have a clear meaning!) Frequency analysis is used in our next example below.

It is purely by chance that we had the repeated trigram WXP - this repeated trigram was not the result of the same three letters being enciphered by the same part of the keyword. This highlights the fact that the above method is probabilistic.

The method of coincidences   We will now use the second principle of "coincidences", to find the length of the keyword. The sequence of plain text letters in positions 1 to n, $n+1$ to $2n$, $2n+1$ to $3n$ etc. should be approximately the same, especially if n is large. It follows that a similar result holds true for the corresponding sequences of cipher text letters. Thus, if we slide the entire cipher text along itself and count the number of coincidences (i.e. agreements) at each displacement, then on average the maximum number of coincidences will occur after an integer multiple of the keyword length n (i.e. the max occurs for some λn, $\lambda = 1, 2, 3, \ldots$). This technique can be illustrated with the following example. We will first determine the period n and use it to then determine the keyword from the following cipher text passage.

VVHZK	UHRGF	HGKDK	ITKDW	EFHEV	SGMPH
KIUWA	XGSQX	JQHRV	IUCCB	GACGF	SPGLH
GFHHD	MHZGF	BSPSW	SDSXR	DFHEM	OEPGI
QXKZW	LGHZI	PHLIV	VFIPH	XVA	

To find n, write the cipher text on a long strip of paper and copy it onto a second strip of paper beneath the first strip. For a displacement of 1, move the top strip to the right by one letter and count the number of character coincidences. Repeat this process for several displacements until the maximum possible displacement is obtained. The shift shown below corresponds to a displacement of 3:

$$V \quad V \quad H \quad Z \quad K \quad U \quad H \quad R \quad G \quad F \quad H \quad G \quad K \quad \ldots \quad \ldots$$
$$V \quad V \quad H \quad Z \quad K \quad U \quad H \quad R \quad G \quad F \quad H \quad G \quad K \quad D \quad K \quad I \quad \ldots \quad \ldots$$

By repeating this process for a number of displacements, we obtain Table 2.2.

Displacement	Number of Coincidences
1	4
2	4
3	9
4	12
5	5
6	2
7	7
8	7

Table 2.2: Number of Character Coincidences Corresponding to Displacement

From our results, the maximum number of occurrences appears for a displacement of 4. Since we know that the maximum displacement probably occurs for a scalar multiple of the period, the period is likely either 2 or 4.

Remark In applying the second principle, we are using a probabilistic argument. That is, in the above example, we cannot be certain that the period is either 2 or 4; however, we can say with high probability that it is likely to be either 2 or 4. If we were unable to decipher the text with a keyword length of 2 or 4, we would then try with the next highest number of coincidences, which occurs for displacement 3.

Finding the Keyword Now that know how to find n, we examine how to find the keyword itself. Suppose, for example, that our key-length is 7. Consider the plain text character in the $1^{st}, 8^{th}, 15^{th}, \ldots$ positions (i.e. characters at a distance of 7 spaces). If a particular letter occurs in positions 1 and 8, the cipher text letters in positions 1 and 8 will be the same (because we use the same key letter to encipher both characters). How can we deduce which cipher text characters correspond to which plain text characters? In the English language, the most frequently used letter is 'e'. If we restrict ourselves to the letters of the message in positions $1, 8, 15, \ldots$, this will still be the case. Therefore, the most frequent cipher text letter in positions $1, 8, 15, \ldots$ will probably have come from the enciphering of the letter 'e'. Thus, by computing the number of occurrences of each cipher text letter at intervals of 7 letters, we can determine the most frequently occurring cipher text letter and assign it to the plain text letter 'e'. Hence, we will have determined the first letter of the keyword. Similar remarks apply to positions $\{2, 9, 16, \ldots\}$, $\{3, 10, 17, \ldots\}$. In general, if we know the period n, we can capture the key with frequency analysis.

We will first try the case where the period is 4 and we will determine the character frequencies for the $\{1^{st}, 5^{th}, 9^{th}, \ldots\}$ letters, $\{2^{nd}, 6^{th}, 10^{th}, \ldots\}$ letters, and so on. Taking the $1^{st}, 5^{th}, 9^{th}, \ldots$ letters, we get:

$$\text{VKGKT EVPUG JVCCP GDGPD DMGKG PVPA}$$

from which we obtain the following table of frequencies:

A	B	C	D	E	F	G	H	I	J	K	L	M
1	0	2	3	1	0	6	0	0	1	2	0	1

N	O	P	Q	R	S	T	U	V	W	X	Y	Z
0	0	5	0	0	0	1	1	4	0	0	0	0

Since G is the most frequently occurring letter, we make the assumption that 'e' enciphers to G. Thus the first key letter might be 'C'. Similarly, for the second set of letters (i.e. the $2^{nd}, 6^{th}, 10^{th}, \ldots$ letters) we obtain the following table:

A	B	C	D	E	F	G	H	I	J	K	L	M
0	1	0	1	0	5	2	4	2	0	1	0	1

N	O	P	Q	R	S	T	U	V	W	X	Y	Z
0	1	0	1	0	4	0	1	2	1	0	0	1

Here there are three letters that could likely decipher to 'e'. To determine which letter is the right choice, consider each character separately. If 'e' enciphers to 'F', we have a key letter of 'B'. If 'e' enciphers to 'H', we have a key letter of 'D'. Finally, if 'e' enciphers to 'S', we have a key letter of 'O'. Now we must examine each choice case by case. A key letter of 'B' means that the most frequently occurring letters besides 'e' are 'g' and 'r'. Similarly, a key letter of 'D' means that the most frequently occurring letters in the plain text (in positions $2, 6, 10, \ldots$) besides 'e' are 'c' and 'p'. Finally, a key letter of 'O' means that the most frequently occurring letters besides 'e' are 't' and 'r'. From the results shown in Table 2.1, it seems that 'O' is then the more likely key-letter. Therefore, we conclude that the second key letter is 'O'.

For the $3^{rd}, 7^{th}, 11^{th}, \ldots$ letters, we obtain the following frequency table:

A	B	C	D	E	F	G	H	I	J	K	L	M
1	1	0	1	1	2	2	8	0	0	2	2	0

N	O	P	Q	R	S	T	U	V	W	X	Y	Z
0	0	0	2	0	0	0	1	0	2	2	0	1

For this set of letters, the most frequently occurring letter is H. Therefore, we make the assumption that 'e' enciphers to 'H'. This corresponds to a key letter of 'D'.

Finally, for the $4^{th}, 8^{th}, 12^{th}, \ldots$ letters, we compute the frequencies to be:

A	B	C	D	E	F	G	H	I	J	K	L	M
1	0	1	0	2	0	1	2	5	0	0	1	1

N	O	P	Q	R	S	T	U	V	W	X	Y	Z
0	0	1	0	3	3	0	0	1	1	3	0	2

From this table, we deduce that because 'e' likely enciphers to 'I', our fourth and final key letter is 'E'.

Putting all of this together, we have determined that the period n is four and the corresponding keyword is the word 'code'. This gives the message "The Vigenère cipher was created in the sixteenth century and was considered by everyone to be unbreakable until the twentieth century."

The method used above, though simple to use, is very effective in determining the keyword of a given cipher text passage. The reader should be aware that there may be times when it may take some more work to pin the keyword down, because of multiple period choices and ambiguities that may occur in the frequencies of cipher text letters.

Remark In examining these methods for breaking the Vigenère cipher, we have stated many times that these methods are only probabilistic; that is, they are only likely to work, not guaranteed. It is possible that we could go through the above process, only to decipher a given cipher text incorrectly. The question of how many messages encipher to a given cipher text is discussed in Chapter 15, and it turns out that we can roughly expect there to be only one intelligble message fitting with a given cipher text when the cipher text has more than 28 letters.

2.8 Modern Enciphering Systems

With the advent and ubiquity of computer-based encryption systems, cryptanalysis has shifted the emphasis from attacks based purely on the ciphering scheme to attacks on other aspects of cryptosystems, such as authentication, key exchanges, and digital signatures. We will detail some of these procedures later in the book, after the basis for modern ciphers, authentication, and digital signatures are developed.

2.9   Problems

1. Encipher the plain text message 'encode' using the following cipher schemes:

   (a) Caesar cipher with $k = 3$.
   (b) Caesar cipher with $k = -5$.

2. Using the Scytale cipher with $k = 2$, encipher the plain text message 'encode'. Compare your results with those from Problem 1.

3. Decode the following cipher text, given that it has been produced by a Caesar cipher: **JRYYQBAR**.

4. Is any additional security gained by enciphering a message by using two monoalphabetic ciphers with different keys in succession?

5. Encipher the first sentence of Homer's "Odyssey" by using a Scytale cipher with key $= 7$.

   TELL ME O MUSE, of that ingenious hero who travelled far and wide after he had sacked the famous town of Troy.

6. Explain how frequency distribution can be used to determine whether a given entry of cipher text was produced with a substitution cipher or a transposition cipher.

7. Use the Vigenère cipher with keyword 'ODYSSEY' to encipher the first sentence of Homer's "Odyssey". Compare the cipher text with the results obtained from Problem 5.

8. Decipher the message below, using the fact that the keyword 'CIPHER' was used to encode it.
   VPXZG ZRPYM JGIHF XFDZT HOZHW CLOEQ EHALV MMNDS IF

9. Suppose Caesar sends a message to one of his generals, and the message contains only one letter. What can you say about the message's security and why?

10. Does the Enigma machine perform substitution enciphering or transposition enciphering? Explain.

11. Ignoring the plugboard, how many possible initial settings are there for the three-rotor Enigma machine?

12. Using the rotor and reflector permutation sets, along with the initial rotor settings, from the example in Section 2.5, encipher the following message: **'move out!'**.

13. Find the period of the given cipher text, given that it was enciphered by the Vigenère cipher (see Section 2.7).

    LVCKO GXKRR ITSKC XIPOG GZLCB GYXFC AYGDI RBMAU CFYAC FIPGM
    RRXFO JPSIB WELDI QDJPO USORA IHGCX PSFSD MMXEL NEJSX RVIJE
    GISXA KRZOH MXI

14. Using the results of Problem 13, determine the keyword used to encipher the passage above.

15. Use the Enigma Machine with initial settings given in Section 2.5 to decipher the following message: **YDDMYU**.

16. Take a page of your favorite book and estimate the number of characters on the page. Count the number of times the letters e, t, a, and o appear on that page and calculate their relative frequencies. Compare your results with Table 2.1.

2.10   Solutions

1. (a) encode $\rightarrow (4, 13, 2, 14, 3, 4) \overset{k=3}{\rightarrow} (7, 16, 5, 17, 6, 7) \rightarrow$ HQFRGH.

    (b) encode $\rightarrow (4, 13, 2, 14, 3, 4) \overset{k=-5}{\rightarrow} (25, 8, 23, 9, 24, 25) \rightarrow$ ZIXJYZ.

2. Since $k = 2$, the number of rows is two. Thus the message is encoded as:

    E C D $\rightarrow$ ECDNOE
    N O E

3. We use a brute-force attack, with results shown in the following table. The value k correponds to a Caesar shift of magnitude k.

k	J	R	Y	Y	Q	B	A	R
1	I	Q	X	X	P	A	Z	Q
2	H	P	W	W	O	Z	Y	P
3	G	O	V	V	N	Y	X	O
4	F	N	U	U	M	X	W	N
5	E	M	T	T	L	W	V	M
6	D	L	S	S	K	V	U	L
7	C	K	R	R	J	U	T	K
8	B	J	Q	Q	I	T	S	J
9	A	I	P	P	H	S	R	I
10	Z	H	O	O	G	R	Q	H
11	Y	G	N	N	F	Q	P	G
12	X	F	M	M	E	P	O	F
13	W	E	L	L	D	O	N	E
14	V	D	K	K	C	N	M	D
15	U	C	J	J	B	M	L	C
16	T	B	I	I	A	L	K	B
17	S	A	H	H	Z	K	J	A
18	R	Z	G	G	Y	J	I	Z
19	Q	Y	F	F	X	I	H	Y
20	P	X	E	E	W	H	G	X
21	O	W	D	D	V	G	F	W
22	N	V	C	C	U	F	E	V
23	M	U	B	B	T	E	D	U
24	L	T	A	A	S	D	C	T
25	K	S	Z	Z	R	C	B	S
26	J	R	Y	Y	Q	B	A	R

After investigating the entries in the table, the only intelligible message exists for $k = 13$. The plain text message is 'well done'.

4. No. For example, if a message is enciphered with a key of 4, and the resulting cipher text is enciphered again using a key of 8, the final cipher text will be the same as if the message was enciphered with a key of 12.

5. Since the key is 7, we know that we need 7 rows. For the number of columns, count up the total number of characters and check whether it is divisible by 7. Since it is not, we must add 4 Zs to the end. Therefore, we have 91 characters, and $\frac{91}{7} = 13$. Thus we need 13 columns. Writing the message out in columns, we get the following matrix:

T	M	H	N	R	A	F	I	R	A	E	T	R
E	U	A	I	O	V	A	D	H	C	F	O	O
L	S	T	O	W	E	R	E	E	K	A	W	Y
L	E	I	U	H	L	A	A	H	E	M	N	Z
M	O	N	S	O	L	N	F	A	D	O	O	Z
E	F	G	H	T	E	D	T	D	T	U	F	Z
O	T	E	E	R	D	W	E	S	H	S	T	Z

After unwrapping, we obtain the encrypted message:

TMHNR AFIRA ETREU AIOVA DHCFO OLSTO WEREE KAWYL EIUHL
AAHEM NZMON SOLNF ADOOZ EFGHT EDTDT UFZOT EERDW ESHST Z

6. Transposition ciphering will produce cipher text with the same frequency distribution as the English language. Substitution ciphering, with a polyalphabetic cipher for example, will yield frequencies that can be much different, since the plain text letters are actually changed instead of being reordered. Therefore, if the distribution is "flattened," we can assume that a substitution cipher was used. If it does not, we can assume that a transposition cipher was used.

7. The computations for the first few letters are shown.

Keyword:	O	D	Y	S	S	E	Y	O	D	Y
Plain text:	T	e	l	l	m	e	o	m	u	s
Cipher text:	H	H	J	D	E	I	M	A	X	Q

Continuing the process, the corresponding cipher text is:

HHJDE IMAXQ WGJRV DRAFK CBLMM KLCFR UZGXP OYCDW HDOUY
FVAGR HYXLI PVHFS VVYQN CVLLC TDKGM WRCZL GXXPC B

8. We obtain the plain text by subtracting the cipher text from the keyword:

Keyword:	C	I	P	H	E	R	C	I	P	H
Cipher text:	V	P	X	Z	G	Z	R	T	Y	M
Plain text:	t	h	i	s	c	i	p	h	e	r

Working the rest of it out, we obtain the following message.

'this cipher is easy to break if one knows the keyword'

9. The message has perfect security, because the message could be any of the 26 letters of the alphabet. That is, knowledge of the cipher text does not give any information regarding the message. Alternatively, one can think intuitively that "the key is as long as the message."

10. Substitution enciphering, because each letter of the plain text message is obtained by substituting different cipher text characters. Transposition enciphering is the reordering of the same letters, whereas substitution enciphering doesn't necessarily use the same characters.

11. 3 rotors with 26 possible initial settings each $= 26^3$ initial settings $= 17576$. Since the three rotors can be interchanged, there are six ways to order them. Therefore, we have a total of $6 \times 26^3 = 105,456$ different initial settings

12. The corresponding cipher text is **JCJDUKJ**.

    The initial settings are:

    $\alpha_1 = $ (0 15 6 10 14 8 19 17 22 18 11) (1 2 9 13 21 25) (3 4 23 5 24 7 12 16 20)

    $\alpha_2 = $ (0 7 9 4 6 18 23 25 8) (1 17 19) (2 20 10) (3 12) (5 11 13 21) (14 22 15 16 24)

    $\alpha_3 = $ (0 2 4 7 16 17 19 5) (1 6 3 8 21 24 11 13 9 10 25 12 14 15) (18 23 20 22)

    $\beta = $ (0 4)(1 7)(2 9)(3 16)(5 20)(6 8)(10 19)(11 17)(12 25)(13 18)(14 24)(15 22)(21 23)

    $\alpha_1^{-1} = $ (11 18 22 17 19 8 14 10 6 15 0) (25 21 13 9 2 1) (20 16 12 7 24 5 23 4 3)

    $\alpha_2^{-1} = $ (8 25 23 18 6 4 9 7 0) (19 17 1) (10 20 2) (12 3) (21 13 11 5) (24 16 15 22 14)

    $\alpha_3^{-1} = $ (5 19 17 16 7 4 2 0) (15 14 12 25 10 9 13 11 24 21 8 3 6 1) (22 20 23 18)

    Also, the initial settings were defined as:

    $r_1 = 22$, $r_2 = 7$, $r_3 = 12$

    Using the given information, the cipher text was obtained as follows:

    1st letter: a = 12 (m)

    $b = Rem[a + r_1, 26]^{\alpha_1} = Rem[12 + 22, 26]^{\alpha_1} = 8^{\alpha_1} = 19$

    $c = Rem[b + r_2, 26]^{\alpha_2} = Rem[19 + 7, 26]^{\alpha_2} = 0^{\alpha_2} = 7$

    $d = Rem[c + r_3, 26]^{\alpha_3} = Rem[7 + 12, 26]^{\alpha_3} = 19^{\alpha_3} = 1$

    Reaching the reflector, we get $e = (d)^{\beta} = 1^{\beta} = 7$.

    Now following the signal back through the rotors, we obtain:

    $c' = Rem[e^{\alpha_3^{-1}} - r_3, 26] = Rem[7^{\alpha_3^{-1}} - 12, 26] = Rem[4 - 12, 26] = 18$

    $b' = Rem[c'^{\alpha_2^{-1}} - r_2, 26] = Rem[18^{\alpha_2^{-1}} - 7, 26] = Rem[6 - 7, 26] = 25$

    $a' = Rem[b'^{\alpha_1^{-1}} - r_1, 26] = Rem[25^{\alpha_1^{-1}} - 22, 26] = Rem[10 - 7, 26] = 3$

    Therefore, the first cipher text character is D.

    Now, we must update the rotor settings: $r_1 = 23, r_2 = 7, r_3 = 12$.

    For the remaining characters, proceed through the same process. Remember, when r_1 changes from 25 back to 0, update r_2 to 8 (this occurs after the fourth cipher text character is computed).

13. After writing the cipher text on two strips of paper, we obtain the following table:

Displacement	# of Coincidences
1	4
2	0
3	3
4	3
5	3
6	6
7	1
8	4

Here we note that the maximum number of coincidences occurs for a displacement of 6. Therefore, the period is probably either 3 or 6, because the displacement producing the largest number of coincidences is a scalar multiple of the period.

14. To complete the problem, we will try a period of 3 first. If it doesn't succeed, we will try the second choice of 6. With our results, we will find the most common letter and assume it deciphers to 'e'. If there are ties for the most frequent character, we will investigate each case individually to determine the most probable choice.

Starting with the first letter of the keyword, we create a table of cipher text frequencies: The 1st, 4th, 7th, ... letters of cipher text are:

LKXRS   XOZBX   ZDBUY   FGROS   WDDOO   ICSDX   NSVES   KOX.

From this, we compute:

A	B	C	D	E	F	G	H	I	J	K	L	M
0	2	1	4	1	1	1	0	1	0	2	1	0
N	O	P	Q	R	S	T	U	V	W	X	Y	Z
1	5	0	0	2	5	0	1	1	1	5	0	2

Here we note that the most frequent cipher text letters are O, S, and X. Now we have to consider each letter to determine which is most likely the key letter. If O deciphers to 'e'(yielding a key letter of 'K') then S → i, and X → n. Looking at Table 2.1, the number of occurrences of 'i' and 'n' in the cipher text are reasonable. Alternatively, if S deciphers to 'e' (yielding a key letter of 'O') then O → 'a' and X → 'j'. However, five occurrences out of 38 letters is far too high for j (the frequency of j is 0.002%). Finally, if X deciphers to 'e', (yielding a key letter of 'T') then O → 'v' and S → 'y'. Again, five occurrences for each 'v' and 'y' in 38 letters are far too high to be correct. Therefore, 'K' is the most probable key letter.

We use the same reasoning for the next two key letters. For the 2nd, 5th, 8th, ... letters, we compute:

A	B	C	D	E	F	G	H	I	J	K	L	M
1	0	1	0	3	2	3	2	8	2	2	1	3

N	O	P	Q	R	S	T	U	V	W	X	Y	Z
0	1	0	0	2	0	0	1	1	0	4	1	0

Here we have an overwhelming choice for 'e', namely 'I'. Thus if I deciphers to 'e', we have a key letter of 'E'.

Similarly, for the 3rd, 6th, 9th, ... letters, we compute.

A	B	C	D	E	F	G	H	I	J	K	L	M
3	1	5	0	0	2	4	0	1	2	0	2	2

N	O	P	Q	R	S	T	U	V	W	X	Y	Z
0	0	5	1	4	2	1	0	0	0	0	1	1

This gives us two likely choices for 'e', although two others are very close. If 'C' deciphers to 'e', we have a key letter of 'Y'. If this is the case, them G → 'i', R → 't', and P→ 'r', all of which have a reasonable number of occurrences. If, on the other hand, P deciphers to 'e', then the key letter is 'L'. This would mean that C → 'r', G → 'v', and R → 'g'. However, the number of occurrences of both 'v' and 'g' are too high to be realistic. Therefore, we arrive at a key letter of 'Y', producing a keyword of 'key'.

To make sure that 3 is the period, one can use the newly acquired keyword to decipher the message. For long messages, it is nearly impossible that two separate messages would appear out of the same piece of cipher text. Thus if the first key works, we are done. If not, then the period is likely 6 instead. This problem illustrates the ambiguities one can run into when attempting to break the Vigenère cipher and serves as a reminder to use the methods outlined here with diligence and care.

15. Repeat the exact same process as in Problem 16, inputting the letters YDDMYU with the same initial settings as before. The resulting output is the message 'attack'.

16. If the book contained typical english text, then the frequencies should be very simlar to the table.

Chapter 3

RSA, Key Searches, SSL, and Encrypting Email

Goals, Discussion This chapter is pivotal, but no mathematical background is required. We avoid making essential use of number theory in the text although it can be used to shorten the calculations. By the end of this chapter the reader will be well-versed in the main public key algorithm, namely RSA, as well as its applications to e-commerce with SSL and to encrypting email[PGP and GPG]. The reader will also be clear on the two kinds of cryptography (public key and symmetric) and will know about searching key-spaces. Some cryptographic attacks,both mathematical and real-world are discussed here and in Chapter 7.

One important difference between the two kinds of encryption is this: If an eavesdropper Eve immediately tries to guess the secret message she can verify whether or not the guess is correct whenever a public key system such as RSA is being used. However, this is not the case with symmetric encryption.

We discuss in detail one public key algorithm, namely the RSA algorithm, named after its inventors (or co-inventors) Rivest, Shamir, and Adelman. We show how anyone can learn to carry out RSA on a calculator.

Let us briefly explain the idea. Alice wants to send a secret message to Bob. Bob has chosen a number N and another number e (for encryption). Very often e is three or seventeen. The pair $[N, e]$ represents *Bob's public key* and is listed in a "public key directory." Alice represents the secret message as a number M lying between 1 and $N - 1$. To encrypt or scramble the message M, Alice multiplies M by itself e times, gets the remainder after dividing by N, and transmits the result to Bob. The result is called the cipher text C. An eavesdropper, noting C, realizes that the message itself must be equal to the eth root of C, or $C + N$, or $C + 2N$, or $C + rN$ for some unknown r. Eve (the eavesdropper) cannot find M as there are too many values of r to try. *It is a remarkable fact that if Eve can guess a single value of r, say $r = r_0$, such that the eth root of $C + rN$ is a whole number lying between 1 and $N - 1$, then this whole number must indeed be M.* Moreover, it can be shown that if for any positive integer r the eth root of $C + rN$ is a whole number v, then the remainder of v upon division by N must be M (see Chapter 19).

The recipient Bob, however, can calculate M immediately from a formula involving his private key consisting of a "decryption index" $\dot{d}$ along with two prime numbers p, q. The reason is that N is the product of p and q. Bob knows p and q. Anybody else, even knowing N, cannot determine what the factors p, q are in a reasonable amount of time. We can think of the public key as being located in the transmitter and a private (secret) key as being located in the receiver.

Eve can try guessing the message without knowing d by guessing r_0. Alternatively Eve can try guessing p and q from which she can calculate d. In other words Eve can try to guess the private key and then determine the message.

We detail some potential weaknesses with public key algorithms such as RSA. However, this algorithm is still the main public key algorithm. Its security, when carefully implemented, seems to still be strong after 25 years of constant use.

One of the most useful facts in cryptography is that in a **brute-force attack** on a key-space (one where we try all possible keys), the correct key is likely to be found after trying about half the total number of keys. Proofs of this statement seem to be rare. In this chapter we provide a short, simple proof of this fact.

Of course, only in special situations, such as when dealing with public key algorithms, can we check, with certainty, that we have the right key.

New, Noteworthy The encryption exponent e must be chosen to have no factors in common with $p - 1$ and no factors in common with $q - 1$. One reason for assuming this is so that d can then be calculated. However, a more basic reason is that this condition must be satisfied in order that two different messages get two different encryptions. This is discussed in the problems. Another interesting fact is that for a given $[N, e]$, **the decryption index need not be unique!** We provide several examples. This is important because some attacks on RSA are possible if d is small; we refer the reder to Chapter 7. So, if d is not unique, this makes it more difficult to guard against this attack.

What we mean by "not unique" is that there may be more than one value of d such that the remainder of C^d, on division by N, is M. The reason is that instead of working with $(p - 1)(q - 1)$, we can work with any number t that is divisible by $p - 1$ and $q - 1$, as explained in the algorithm description. It is always possible to find $t < (p-1)(q-1)$ so that the calculations are simplified, and we get a shortcut even if the resulting d is the same.

We give new mathematical insights on symmetric encryption and on such questions as the security of algorithms such as DES. The answer is that it all depends on the context and

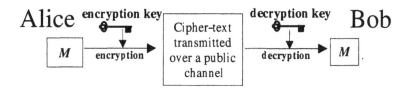

Figure 3.1: General Encryption

the assumptions being made. Even among experts there appears to be a misunderstanding of some of these topics. We present a discussion of a factoring algorithm in Chapter 10.

3.1 Background

Cryptography is an important part of information security. In the last several years, the subject has become ever more important, both academically and commercially.

On the academic side, most colleges and universities offer undergraduate courses in security and privacy. Recent announcements by the IEEE Computer Society described computer security and bioinformatics as the two most popular areas of computing.

3.2 The Basic Idea of Cryptography

Cryptography is an old subject dating back at least as far as 1500 B.C. A technique developed by Porta associated also with Vigenère in the Middle Ages is still at the cutting edge of part of modern cryptography. Additionally, cryptography is closely connected to information theory and error-correction, with many fundamental ideas going back to Claude Shannon. Further details about Shannon and the history of cryptography are provided in Chapter 1.

At the most elementary level, cryptography is the art of keeping messages secret. Imagine that **A**, **B** are two entities who wish to communicate in secret. Assume that **A** wants to send a secret message to **B**. For example, **A** might be a home movie provider and **B** a customer, or **A** might be a military commander who wants to transmit a message involving military secrets to **B**.

The procedure is as follows (Fig. 3.1). First, **A** scrambles the message using a **cryptographic key**. The process of scrambling the message is called **encryption**: alternatively, **A enciphers** the message.

The encryption or enciphering scrambles the message M, that is the plain text, into unintelligible output called the cipher text. Next, the sender **A** transmits in the open (pub-

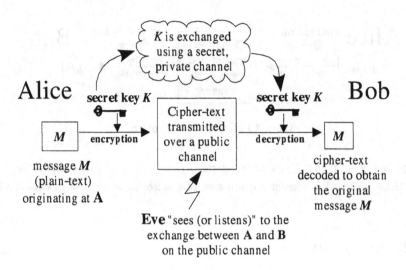

Figure 3.2: Symmetric Encryption

licly) the cipher text C to the receiver **B**. When **B** receives the cipher text, **B** descrambles or deciphers the cipher text with a key that may or may not be the same as the original key used by **A**. **B** can now recover the original message M that was transmitted by **A**.

In summary, the sender **A** encrypts or enciphers the message M into unintelligible cipher text C with an encrypting or enciphering key. The enciphering is done by a specific procedure involving a sequence of steps or rules called the **enciphering algorithm** (or **encryption algorithm**).

Using the **decryption** or **deciphering key**, and using the **deciphering algorithm** (**decryption algorithm**) the receiver **B** then **decrypts** or **deciphers** C and thus recovers the original message M that was transmitted by the sender **A**. Moreover, at least in theory, an intruder Eve cannot access the message M since Eve will not have the decryption key that is needed for decrypting (deciphering, inverting) C to recover M.

Evidently, everything depends on **B** being the sole possessor of the decryption key, apart possibly from **A**. (If the decryption and encryption keys are the same — as they are in symmetric encryption — then **A** also has the decryption key.)

Generally speaking, a key is a mathematical object such as a number (or several numbers) or a string of zeros and ones i.e., a **binary string** such as the binary string (1 1 0 1) of length 4.

The enciphering and deciphering operations are usually mathematical procedures. For example, let's suppose that the enciphering key is the number 7 and that the enciphering operation is "add 7". Suppose the secret message that **A** wants to transmit to **B** is the number 6. (For example **A** might be directing her stockbroker **B** to buy six thousand shares of a given security on the stock market).

Then, **A** calculates the cipher text 13 (= 6 plus 7) and transmits this to **B**. Now **B** knows that the enciphering transformation is "add 7". To undo, or invert this, **B** subtracts 7 from 13 (as this is the deciphering operation) and ends up recovering the original message transmitted by **A**, namely, 6.

It should be mentioned that the cryptographic keys above need not be mathematical objects: in fact, historically, they often were not. A famous example, mentioned in Chapter 1, occurred in World War II when, in effect, the key was an entire language! This was the Navajo language used by the Navajo tribe in Arizona and adapted for encryption purposes by the U.S. armed forces around 1942. Enciphering consisted of translating messages from English into the Navajo language, while deciphering simply meant translating Navajo back to English at the other end. At that time, this symmetric encryption was extremely effective.

Using encryption for storing messages and files is another important function of encryption in today's society. As an example, we mention the encryption of a file - or even the hard disk - in a computer so that, if it is set aside (or stolen) an individual other than the owner cannot access the contents. We can fit this into our previous general situation with the owner of the computer playing the role of both **A** and **B**.

Later on in the book we will discuss the speeds of encryption in more detail. We mention here, parenthetically, that it can take up to two hours to encrypt a hard disk on a modern laptop using the very fast AES algorithm for encryption.

We have been silent on how **A** and **B** get their enciphering and deciphering keys. This is discussed in a later chapter, but it will depend on the kind of encryption being used. The two fundamentally different possibilities for cryptography are as follows.

1. **Symmetric Cryptography**

2. **Asymmetric Cryptography** i.e., **Public Key Cryptography.**

Recall that as before **A**, **B** are the communicating entities and **A** wants to send a secret message M to **B** (Fig. 3.2). In symmetric encryption there are three features:

1. The enciphering key K used by the transmitter **A** is equal to the deciphering key used by the receiver **B**, and this key is known only to **A** and **B**.

2. The enciphering algorithm, converting the plain text to cipher text, is such that the cipher text C can be calculated immediately given M and K.

3. The deciphering algorithm, converting C back to M, can be calculated immediately given C and K.

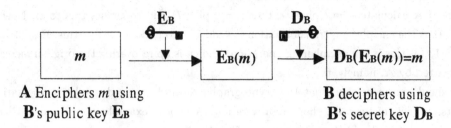

Figure 3.3: Asymmetric or Public Encryption

The security depends on the fact that the secret key K is known only to **A** and **B**. Public-key cryptography works differently (Fig. 3.3). The procedure is as follows:

1. The enciphering key e_B used by **A** (or anybody else) to send a message M to **B** is publicly known, and is called the public key of **B**. However, the deciphering key d_B used by **B** to decrypt the cipher text is known only to **B** and is the **private key** of **B**. So the two keys are quite different.

2. The enciphering procedure, converting the plain text to cipher text C, can be immediately calculated given M and e_B.

3. The deciphering procedure, converting C back to M, can be calculated immediately by **B** using d_B. However it is not possible for somebody else who is not in possession of d_B to convert C back to M in a reasonable amount of time.

   The security of public key cryptography rests on the assumption that it is not feasible to convert C back to M without knowledge of the private key d_B.

   Thus in **public key cryptography** each user **B** in a network has a public key e_B and a private key d_B, which are supplied by a **public key authority (PKA)**.

Symmetric cryptography (encryption) is also called **secret key cryptography** (encryption). The security depends, as stated above, on the assumption that only the communicating parties **A** and **B** know the (common) key. Note that **A**, **B** could also denote groups of entities on a network and that **B** can also send a secret message to **A** using their common secret key.

Historically, cryptography meant symmetric cryptography. Nowadays, the main symmetric algorithms in use are **DES, Triple DES, AES** and the **One-Time Pad**. They are all discussed in detail later on in the book. Our previous example of encryption with the "add by 7" rule is an example of how symmetric encryption might work.

3.3 Public Key Cryptography and RSA on a Calculator

We now turn to some examples of asymmetric or public key cryptography. First, let us explain **RSA**, the main public key algorithm. As before, **A** wants to send a secret message M to **B**. For convenience, let's think of M as being the number 6, say, as in our previous example. We make the encryption more complicated. So instead of saying "add 7" we say "multiply 6 by itself 7 times" i.e., calculate $(6)(6)(6)(6)(6)(6)(6) = 6^7 = 279,936$. As an extra complication, let's take some number N and declare the encryption algorithm to be "multiply 6 by itself 7 times and take the remainder of this number when divided by N to be the cipher text C". As a small working example, let $N = 55$. So our cipher text is the remainder of 6^7 upon division by 55. This remainder is easily calculated, using any calculator, as follows.

We want to find the (unique) remainder z that is left over when we divide 279,936 by 55. So we have

$$279,936 = 55y + z \qquad (3.1)$$

where z is one of $0, 1, 2, 3, 4, \ldots 53, 54$. We are not really interested in the value of y; we just need z. Dividing across by 55 in equation 3.1, we get

$$\frac{279,936}{55} = y + \frac{z}{55} \qquad (3.2)$$

Pushing the divide button on the calculator, we get

$$\frac{279,936}{55} = 5089.745455 \qquad (3.3)$$

This indicates that y is 5089, $\frac{z}{55} = 0.745455$ so $z = 55(0.745455) = 41.000025$. *This is not what we were hoping for: z is supposed to be a whole number, namely, the remainder when 279,936 is divided by 55!* However, the calculator has made rounding errors and we suspect that z is 41 (and y is 5089). This is easily checked. We can verify that Equation 3.1 checks out with $y = 5089$, $z = 41$ since $279,936 = (55)(5089) + 41$.

Principle 1 *To calculate the remainder of 279,936 when divided by 55, perform the division on a calculator and multiply the decimal part by 55. Verify your answer by checking that Equation 3.1 is satisfied. This also works to get the remainder whenever we divide a positive integer (= positive whole number) u by another positive integer v.*

Question How do we know that z is unique? Maybe there are two possible values?

Go back to Equation 3.1 and suppose we have two solutions with y_1, y_2 being positive and z_1 and z_2 both lying between 0 and 54. So we have

$$279,936 = 55y_1 + z_1 \tag{3.4}$$

$$279,936 = 55y_2 + z_2 \tag{3.5}$$

Then $55y_1 + z_1 = 55y_2 + z_2$. Now, if $y_1 = y_2$ it follows that $z_1 = z_2$. So assume that $y_1 \neq y_2$. Call the larger one y_1, so $y_1 > y_2$

We now have $55(y_1 - y_2) = z_2 - z_1$. Since y_1 is at least 1 bigger than y_2, we get that the left side is *at least* 55. Since z_1 and z_2 are between 0 and 54 we see that the right side is at *most* 54. Since $55 > 54$ we conclude that the assumption $y_1 \neq y_2$ leads to a contradiction. Thus $y_1 = y_2$ (and so also $z_1 = z_2$): end of story. *As a consequence, to check Equation 3.1 in the future all we need to do in the case above is to ensure that $279,936 - z$ is divisible by 55.*

Getting back to our main narrative, **A** transmits the cipher text **C** = 41 to **B** having calculated this from the message $M = 6$. How does **B** recover M from 41? **B** knows that $N = 55 = (5)(11)$. Since we are using a public cryptosystem, the enciphering algorithm is public knowledge (in this particular example, the enciphering algorithm is "multiply the message by itself 7 times and take the remainder on division by N": this gives the cipher text 41. **B** calculates the deciphering index d as follows.

There is a unique positive integer d between 1 and 39 such that $7d$ gives a remainder of 1 when divided by 40. In this case, it turns out that $d = 23$ (more on this later) since $(7)(23) = 161$ and 161 leaves a remainder of 1 on division by 40. Here, 40 comes from the fact that $(11 - 1)(5 - 1) = 40$ and 5, 11 are the factors of $N = 55$.

To recover the message, **B** multiplies the cipher text, namely 41, by itself 23 times, gets the remainder on division by 55, and this should give the original message, namely 6. So we are claiming that $41^{23} - 6$ is divisible by 55.

We will use the following principle to get the remainder of a product of two numbers.

Principle 2 *Calculate the product of the two individual remainders and then get its remainder, if necessary.*

If we calculate 41^{23} — or in general any M^e — on a calculator or a computer, we run into overflow problems. To avoid them, we use this principle, combined with the **repeated squaring method**. Here is how this method works in the present case. We first express 23 as a sum of powers of 2. Thus $23 = 7 + 16 = 1 + 2 + 4 + 16$. So, if x is any number we have $x^{23} = x^{1+2+4+16} = (x)(x^2)(x^4)(x^{16})$. Each number in this product is the square of the previous number except for x^{16}, which is the square of the square of the previous number.

Let us calculate 41^{23} and get the remainder upon division by 55.

In detail, $x^1 = 41$, nothing more to do. Then, $x^2 = 41^2 = 1681$ gives a remainder of 31 when divided by 55. Proceeding, instead of calculating $x^4 = 1681^2$ by squaring x^2, we need only calculate 31^2 and get the remainder on division by 55 which is 26. Now, to get the next term in the product (namely x^{16}) instead of squaring x^4 — to get x^8 — and then squaring again to get x^{16} — we need only square 26, get the remainder, square the remainder again, and finally get the remainder on division by 55 which is 36. So, the four remainders for x, x^2, x^4, x^{16} are 41, 31, 26, 36. In principle, now we have to multiply 41 by 31 by 26 by 36 and get the remainder on division by 55. Again, we can take shortcuts by using Principle 2. We can multiply 41 by 31 and get the remainder. We calculate $(26)(36)$ and get the remainder (on division by 55). Multiplying the two remainders together, and getting the remainder, on division by 55, gives us the answer. The two remainders are 6 and 1. Then $(6)(1) = 6$ and **B** ends up recovering the message, which is 6. Note that in the example above, N is the product pq of two distinct primes p and q with $p = 11$ and $q = 5$. The **enciphering index** e is 7, M is 6, the cipher text C is 41, and the **deciphering index** d is 23.

Several remarks are in order.

1. The fact that **B** deciphers C to recover the message M by simply using the deciphering algorithm explained above is proved in Chapter 19.

2. Having found a decryption index the official (hard) way, let's find an easier method. All we need is the unique integer, let's call it d', between 1 and 20, such that $7d'$ gives a remainder of 1 when divided by 20. Why 20? Well, instead of using $(p-1)(q-1)$, we can use any positive integer divisible by both $p-1$ and $q-1$. With $p = 11$ and $q = 5$, we choose the number 20, and get $d' = 3$. Then it is easy to check that the remainder of 41^3 upon division by 55 is 6. It is much easier to use the decryption index 3 instead of the decryption index 23!

3. *The security of RSA rests on the mathematically unproved assumption that, even given C, e, N, an individual (other than **B**) cannot recover M in a reasonable amount of time if p and q are large.*

In technical language the problem of recovering M is said to be computationally infeasible (= **infeasible**) or **intractable**. The enciphering function transforming M is conjectured to be a **one-way function**, i.e., it is easy to calculate C given M but it is impossible to undo this calculation.

Given e and the two factors p, q of N it is easy to calculate d and thus to obtain M from C (see Chapter 19). Thus, if one can solve the problem of factoring N quickly, one can calculate d quickly, and thus M, given C. On the other hand, if we can find d, then we can get M (but also p and q).

It is now time to give a formal description of the RSA algorithm, as follows.

3.4   The General RSA Algorithm

A (=Alice) wants to send a secret message M to **B** (=Bob). Bob has already chosen two large unequal prime[1] numbers p and q. Bob multiplies p and q together to get $N = pq$. Bob also chooses some integer e bigger than 1. The integer e (e for enciphering) must have no factors in common with $p - 1$ and no factors in common with $q - 1$. In other words, the greatest common divisor of e and $p - 1$ is 1 (and similarly for e and $q - 1$). In symbols we write $\gcd(e, p - 1) = 1$ and $\gcd(e, q - 1) = 1$. Thus the only number dividing both e and $p - 1$ is 1 and the only number dividing both e and $q - 1$ is 1. We say also that e is **relatively prime** to $p - 1$ and to $q - 1$. It follows that $\gcd(e, (p - 1)(q - 1)) = 1$.

Because of the conditions imposed on e, namely, that e is relatively prime to $(p-1)(q-1)$, there exists a unique integer d (d for deciphering) that is greater than 1 but less than $(p - 1)(q - 1)$ and is such that the remainder of de when divided by $(p - 1)(q - 1)$ is 1. It is easy for Bob to calculate d, using a method related to the Euclidean Algorithm (see Chapter 19), because Bob knows p, q which are the factors of N.

Bob puts the numbers e and $N(= pq)$ in a public directory under his name. He keeps secret the primes p and q: d is called Bob's private key and the pair $[N, e]$ is Bob's public key.

Now Alice has a secret message M to transmit to Bob. Alice converts M to a number between 1 and $N - 1$ represented in binary (which we also denote by M). If M is too large, Alice breaks M into blocks, each of which is less than N. Let's assume, for simplicity, that M is less than N. Then, Alice enciphers M by calculating the cipher text $C = Rem[M^e, N]$. Note that **Rem[u,v] means the remainder when u is divided by v**, so in other words Alice multiplies M by itself e times and gets the remainder upon division by N. This can be done quickly using the "repeated squaring" method and the principle described earlier. In principle e can be any positive integer relatively prime to $p - 1$ and $q - 1$. However, suppose $Rem[e, (p - 1)(q - 1)] = e_1$. Then it can be shown that $Rem[M^e, N] = Rem[M^{e_1}, N]$, *and so we may as well assume that* $e < (p - 1)(q - 1)$.

When Bob receives $C = M^e$, he in turn multiplies C by itself d times and gets the remainder upon division by N. As explained in our earlier example, the calculation can be simplified. This remainder is in fact equal to M, the original message.

Let us formalize the procedure.

---

[1] Recall that a number is *prime* if it has no factors save itself and 1. So 11 is prime but $6 = 2 \times 3$ is not since 2 and 3 divide 6, i.e., they are *factors* of 6.

The RSA Algorithm.

1. Bob chooses in secret two large primes p, q with $p \neq q$ and sets $N = pq$.

2. Bob chooses e bigger than 1 with e relatively prime to $p - 1$ and to $q - 1$ and with $e < (p - 1)(q - 1)$.

3. Bob calculates the decryption index d where $d < (p - 1)(q - 1)$ is such that the remainder of de on division by $(p - 1)(q - 1)$ is 1. More generally, Bob calculates a decryption index d where $d < t \leq (p - 1)(q - 1)$ is such that the remainder of de on division by t is 1. Here, t is any number divisible by both $p - 1$ and $q - 1$.

4. Bob announces his public key $[N, e]$ and keeps his private key d secret.

5. Alice wishes to send a secret message M and represents M as a number between 0 and $N - 1$. Alice then encrypts the message M as the remainder C of M^e upon division by N and transmits C to Bob.

6. Bob decrypts C by calculating the remainder of C^d upon division by N: this gives the original secret message M.

Remark 1 If p, q are odd then $p - 1, q - 1$ are even, and so each is divisible by 2. So, by choosing t to be the *least* common multiple of $p - 1$ and $q - 1$, we get that $t \leq (p - 1)(q - 1)/2$.

Remark 2 It is beneficial for Bob to also store p and q rather than just d. The reason is that some decryption algorithms work faster if he makes use of p and q rather than just d. Thus the private key is sometimes defined as the triple $[p, q, d]$.

Another example of the RSA algorithm In the example below, we also briefly indicate how more sophisticated number theory can shorten the calculations.

Suppose Bob chooses the primes $p = 863$ and $q = 937$. So $N = pq = 808631$, and $(p - 1)(q - 1) = 806832$. A valid choice for e is 7, as $\gcd(806832, 7) = 1$. Using the Euclidean Algorithm, Bob can also calculate d (see Chapter 19). Bob announces his public key $[N, e] = [808631, 7]$ and finds $d = 461047$. Bob keeps $[p, q, d]$ secret. When Alice wants to send $M = 205632$ to Bob, she computes $C = Rem[205632^7, 808631]$ using the repeated squaring method to find that $C = 256779$. Alice then transmits C in public, and when

Bob receives it, he can either compute $M = Rem[C^d, N] = Rem[256779^{461047}, 808631] = 205632$ directly using the repeated squaring technique, or take a more efficient approach, as follows: Bob calculates $M_1 = Rem[256779^{461047}, p]$ and $M_2 = Rem[256779^{461047}, q]$, then uses the Chinese Remainder Theorem (of Chapter 19) to combine them to find M. Since $Rem[255779, p] = Rem[255779, 863] = 468$, Bob knows that $M_1 = Rem[256779^{461047}, 863] = Rem[468^{461047}, 863]$, and by a theorem due to Fermat[2] this is equal to $Rem[468^{739}, 863] = 238$. Similarly, $M_2 = Rem[256779^{461047}, 937] = Rem[41^{535}, 937] = 429$. Bob then combines M_1 and M_2 to find $M = 205632$.

Remark 3   Instead of using 461047 as the deciphering index, Bob can calculate that the least common multiple of $p-1$ and $q-1$ is $t = 403416$. Then he can find that the remainder of $d'e$ when divided by t is 1, where $d' = 57631$, and use this for a deciphering index instead.

It is conceivable that the **RSA problem** of obtaining M from C is easier than the **factoring problem**. For some methods of obtaining M from C that work in special cases we refer to the problems. *The factoring problem is to obtain p, q given $N = pq$.* Once p, q, e are known it is easy to find the message M from C by calculating d: this is what Bob does. Mathematically nobody has been able to prove that the factoring problem cannot be solved in a reasonable amount of time. Similarly, it has not been shown that M cannot be obtained from C in a reasonable amount of time by some method or another. We point out also that given d we can find p, q, even when d is chosen so that $Rem[ed, t] = 1$, where $(p-1)$ divides t and $(q-1)$ divides t. (see Buchmann [Buc00]). *Thus the problem of finding d is equivalent to the factoring problem.*

Let us return again to our example of symmetric key encryption where the enciphering algorithm was "add 7". To avoid overflow and storage, we fix a large positive integer N. Let the message be some number M between 0 and $N - 1$, i.e., $0 \leq M \leq N - 1$. Our enciphering algorithm now reads: "increase M by 7 and get the cipher text C by calculating the remainder upon division by N". For example if N is 55 and $M = 50$ then $C = 2$. So **A** transmits the cipher text 2. Now **B** must undo (or decrypt or decipher) 2 to get the original message. Before, our decryption algorithm read "subtract 7 from C", i.e., "add the inverse of 7 to C". We do this now. First we must get *the additive inverse of 7 modulo N* i.e., *the inverse of 7 modulo 55* (see Chapter 19). In other words we must find d such that $d + 7$ leaves a remainder 0 when divided by 55. In this case d is 48. Then, to decipher C we increase $C(= 2)$ by 48 and obtain the remainder upon division by 55. In this case we obtain the number 50. This is the original message.

The kind of procedure just described seems remarkably similar to the RSA algorithm. A summary is as follows.

---

[2]Fermat's Theorem says that $Rem[a^{p-1}, p] = 1$ when p is prime and a is not a multiple of p.

RSA Algorithm (Outline)

- **B** selects in private two large primes p, q with $p \neq q$ and sets $N = pq$.

- **A** chooses a message $M, 0 \leq M \leq N - 1$.

- **B** chooses any positive enciphering index e with $\gcd(e, (p - 1)(q - 1)) = 1$ and $1 \leq e < (p - 1)(q - 1)$.

- **A** forms the cipher text C where C is the remainder when M^e is divided by N. Let d be the multiplicative inverse of e modulo $(p - 1)(q - 1)$, so that ed leaves a remainder of 1 upon division by $(p - 1)(q - 1)$. Then, to decipher, **B** raises C to the power d and gets the remainder of C^d upon division by N.

Symmetric Algorithm (Outline)

- **A** and **B** publicly choose any positive integer N.

- **A** chooses a message $M, 0 \leq M \leq N - 1$.

- **A** and **B** secretly agree on an enciphering index e between 0 and $N - 1$.

- **A** forms the cipher text C where C is the remainder when $M + e$ is divided by N. Let d be the additive inverse of e modulo N. Then, to decipher, **B** adds d to C and gets the remainder of $d + C$ upon division by N: this yields M.

3.5 Public Key Versus Symmetric Key

The algorithms seem remarkably similar. However, by contrasting them we will glean the fundamental insights into the difference between public key and symmetric cryptography.

We note the following.

1. The integer e is kept secret in the symmetric case and is known only to **A** and **B**. In the RSA case, e is public knowledge.

2. The choice of the integer N in the symmetric case is unrestricted. In the RSA case, N must be the product of two large unequal primes p, q and furthermore these primes are secret and are known only to **B**. The integer d constitutes the private key of **B** in the RSA case.

3. In this particular case, the symmetric algorithm provides perfect secrecy. *This means that knowledge of the cipher text provides no clue as to the message because any message will fit with the given cipher text. The cipher text does not narrow down the possibilities for the message in any way.* Thus the only way for an intruder to get

the message is by guessing; the eavesdropper Eve can try to guess the message but will have no way of verifying the guess is correct. For example, when $N = 55$ and we take any cipher text C, say $C = 2$ (as above), then any message can be made to fit this cipher text. For example, the message 14 can be made to fit this cipher text by choosing $e = 43$.

Contrast this with the RSA situation. Only one message will fit with a given cipher text. If an intruder tries to guess the message she can verify immediately whether or not the guess is correct. In particular, she can do this by enciphering the guess. Moreover, *given sufficient time and computing resources an adversary is certain to get the message.* (However, the adversary may need a very long time!)

Thus the RSA algorithm affords only **computational security** as do all other public key algorithms. This means that the security comes from the unproven mathematical assumption that no deciphering algorithm can be computed in a reasonable amount of time (technically, in polynomial time) to obtain the message.

However, not all symmetric algorithms enjoy perfect secrecy. For this to happen Shannon proved that roughly, "the key must be as long as the message". Symmetric algorithms like DES, where the key is much shorter than the message, are sometimes "insecure". The relative security offered by such a scheme depends on the length and the nature of the message, for example, a text message encoded in binary. Generally speaking, the knowledge of the cipher text will at least narrow down the possibilities for the message. We discuss this later in this chapter.

One of the main problems with RSA and public key systems is that somebody may come up with a fast method of deciphering or factoring. Also, an empirical rule which has held true for several years, called **Moore's Law**, asserts that *computing power doubles every eighteen months for the same price.*

Thus, one needs longer and longer keys for public key systems to withstand computational attack. Nowadays the industry uses mainly symmetric cryptography. The secret key needed for this can be supplied by a central server (as is the case with the **Kerberos system**) or by a public key methodology such as RSA where the message M is the key to be used for the symmetric key session, i.e., the **session key**. The RSA algorithm is also widely used as the base for secure e-mail applications such as PGP and Internet security protocols such as SSL.

As mentioned, Moore's Law asserts that computing power doubles every 18 months. Because of Moore's Law one needs to work with bigger and bigger primes p, q to try to ensure that factoring N cannot be done quickly. Back in 1998 one needed to have p, q at least 154 digits (512 bits) each. But nowadays this may no longer be enough. In particular Brown [Bro00] points out that the CREST system, introduced by the Bank of England in 1997, which uses keys 512 bits long is now vulnerable to attack by cryptanalysts. Of course, with longer keys, transmission errors become more and more prevalent. Several additional remarks are in order.

1. P. Shor has proved that factoring numbers is computationally feasible with a quantum computer. Thus, if quantum computers ever become a reality, RSA will no longer be viable: it will be completely insecure.

2. The problem of transmission errors has to be addressed given that primes p, q should be at least 154 digits (512 bits) each, ensuring a modulus N of 308 digits (1024 bits) for long-term security.

3. Despite the recent mathematical results of Agrawal et. al. [AKS0?], which shows that one can quickly test for primality[3] of a given number, one still has to generate large primes p, q of roughly the same size which are suitable for RSA use. In particular, they must be resistant to factorization algorithms, one of which is discussed later in the book.

4. For frequent RSA communications, one also needs a fast algorithm for ensuring that e has no factors in common with $p - 1$ and $q - 1$.

5. The RSA algorithm is (relatively) slow, roughly a thousand times slower than the DES symmetric key algorithm.

6. The security of RSA is in jeopardy without extensive "preprocessing" of plain text message units before encryption (see Mollin [Mol02]).

7. Other attacks on RSA include timing attacks and "man in the middle" attacks. These are discussed in Chapter 7.

Apart from guaranteeing the secrecy of the message from **A** to **B** (or **B** to **A**), other fundamental issues in cryptography are as follows.

1. *Authentication.* Roughly, how can **B** be sure that the message came from **A**?

2. *Message Integrity.* How can **B** be sure that the message has not been altered?

3. *Digital Signatures.* How can a user "sign" a message?

4. *Non-repudiation.* How can **B** prove (in court, for example) that **A** sent the message?

Technically, authentication has two aspects. One relates to the verification of the origin of received data. The other refers to verifying the identity of a user. Traditional methods

---

[3]That means to see whether the number in question is a prime number.

include passwords, P.I.N. numbers, etc., but the use of biometric data readers is coming to the fore. We discuss this in Chapter 8.

Message integrity can be achieved using hash functions as described in Chapter 4. Digital Signatures can be carried out using either symmetric or public key encryption. Also described Chapter 4.

3.6   Attacks, Security of DES, Key-spaces

There are many **attacks** on crypto-systems i.e., attempts by an intruder to break the system by recovering the key and the message or the message directly. By far, the attack most difficult to defend against is an impersonation or **man-in-the-middle attack**. A variety of attacks are discussed in Chapter 7.

There is some confusion in the literature as to the security of algorithms for symmetric key cryptography such as DES and AES. Let us take DES as an example. The message is divided into blocks 64 bits long (a bit is a binary digit, i.e., a 0 or a 1). The total number of keys is 2^{56}. The message M is divided up into blocks $(M_1, M_2, M_3, \ldots)$ each with 64 bits and each block is encrypted independently with a common key K to give the cipher text $(C_1, C_2, C_3, \ldots)$ (this mode of operation is commonly referred to as the Electronic Code Book mode, or ECB. Modes of operation are discussed in detail in Chapter 5). If the message M is in fact a random number, (e.g. the statement of account from a bank or a key to be used in a future communicating session or "session key"), then knowledge of C_1 pins down the key K to being one of 2^{56} possibilities and so pins down M to being one of 2^{56} possibilities. However knowledge of $C_2, C_3 \ldots$ *then gives no further information about the message M*. Thus a **brute-force attack** where we try all possibilities *will never reveal the message* since the uncertainty (or entropy -see Chapter 10) is 56 bits.

On the other hand, as discussed by Trappe and Washington [TW01], (see also Problems 1.4, 1.5) suppose that M is known to be a text message, so that the message from **A** to **B** consists of a sequence of letters or numerals written in computer code[4], namely ASCII. Then C_1 narrows down the key for the eavesdropper, Eve, to about 40 bits (= 2^{40} possibilities). Assuming independence between blocks, the cipher texts C_1 and C_2 narrow down the key to around 24 bits or 2^{24} possibilities. So the original message, no matter how long, may be pinned down to around 2^{24} possibilities. Eventually one can get the entire message M in this way by utilizing the cipher texts $C_1, C_2, C_3, \ldots$; some details of this procedure are discussed in Problems 1.4, 1.5 below. Of course if there is only one block, that is, if M has just 64 bits *then the message can never be recovered by Eve except by guesswork*.

Contrast this with any public key algorithm such as RSA (or elliptic curve cryptography). Here, *given sufficient time and computing power, the eavesdropper is certain to recover the*

---

[4]Note that this is very different from expressing a number in binary, e.g., see the Appendix for a conversion table between each character and the corresponding ASCII code 8-bit binary string.

message. In fact, with RSA it is generally quicker to try to solve the factoring problem than to try all possible values of d (brute-force).

A basic issue in cryptography is this: *If we are trying to guess one of n possible passwords in order to log on (or n possible keys say) then how many guesses will we have to make on average until we are successful? The answer is easy to find.*

Theorem 3.1 . *On the average, when trying to guess one of n possible keys, we only need $(n+1)/2$ guesses.*

First, we explain the concept of average value, which is also discussed in Chapter 9. Suppose that, in a class of 6 students, 3 get 70%, 2 get 80% and 1 gets 92%. What is the average grade? One can write this average as $((3)(70) + (2)(80) + (1)(92))/6 = 77$. So the average grade is 77%. We could also calculate the average as $\frac{3}{6}(70) + \frac{2}{6}(80) + \frac{1}{6}(92)$ where $\frac{3}{6}, \frac{2}{6}, \frac{1}{6}$ are the probabilities of getting $70, 80, and 92$ respectively. Now we proceed to the proof.

Proof. The probability of guessing correctly the first time is $1/n$. To get it right in exactly 2 guesses we must get it wrong on the first guess and then, having discarded the unsuccessful guess, we must guess correctly on the next attempt. So the probability of being successful in exactly 2 guesses is $(1-1/n)frac1n-1 = 1/n$. Similarly, the probability of being successful in exactly $3, 4, \ldots, n-1, n$ guesses is also $1/n$. To get the average number of guesses we multiply the number of guesses by the probability and add up. This gives the average number of trials until success to be $(1)(1/n) + (2)(1/n) + (3)(1/n) + \ldots + (n-1)(1/n) + n(\frac{1}{n}) = (1+2+3+\ldots+(n-1)+n)(1/n) = (n+1)/2$. So on average, you get the correct password after $(n+1)/2$ attempts.

Returning to RSA, one of its main advantages is the convenience and low cost, especially for e-commerce. The advantage of symmetric systems is the improved security and the speed. A difficulty with all systems is the **key distribution problem** i.e., the problem of getting the common secret key to **A** and **B**. This is eloquently expressed by Professor Lomonaco when he writes about the Catch-22 of cryptography [Lom98].

Catch-22 of Cryptography : To communicate in secret we must first communicate in secret.

For symmetric encryption, the key has to be given secretly to **A** and **B**. For public key cryptography, the private key has to be given secretly by the public key authority to both **A** and **B**.

We have spoken already of the assumption that the encryption algorithms for public key cryptography are assumed to be mathematical one-way functions. This means that enciphering has the property that its values are easily computed by a computer (i.e., are computed in polynomial time) yet the deciphering algorithm cannot be computed in a reasonable amount of time (even on a computer). In other words, the problem of deciphering the cipher text is intractable.

Of course, we emphasize again that the existence of such mathematical one-way functions is still in doubt since nobody has discovered a mathematical function that is provably one-way.

But one-way functions abound in the physical world, many of them related to time. For example, as Beutelspacher [Beu94] points out, most people are not getting any younger i.e., the aging function is one-way!

Another analogy is the telephone directory of any big city. Here each name gets enciphered to the corresponding telephone number. The deciphering algorithm starts with a number and tries to find the corresponding name, a much more daunting task.

One can also make physical analogies concerning the two kinds of cryptography.

For public key cryptography, consider the following scenario. **A** wants to send a secret message to **B**. A number of mailboxes are available: **A** knows that the mailbox for **B** is number 3 (3 is the public key for **B**). All **A** has to do is to drop the message into mailbox number 3 for **B**. Then **B** (but nobody else) can recover the message since **B** has the key to mailbox number 3.

Another system for public key algorithms is as follows: **A** wants to send a secret message to **B**. He goes to the hardware store and buys a box and a combination lock marked **B** (this corresponds to the public key of **B**). Then **A** puts the message in the box, locks the box with the combination lock and mails it to **B**. When the box arrives, **B** opens it and gets the message because he knows the combination for the lock. Nobody else can open the box to get the message.

The following is an analogy for symmetric encryption: If **A** wants to send a secret message to **B**, we can imagine **A** putting the message in a strong box, locking the box with a key and mailing it to **B**. When **B** receives the box, he opens it with his key and gets the message. Only **A** and **B** have a key for the box, so nobody else can get the secret message.

3.7   Summary of Encryption

We have seen what encryption is and the difference between public key and symmetric cryptography. Public-key algorithms such as RSA yield computational security that can be breached given sufficient time and computing resources. With RSA, security is weaker for a text message encoded in ASCII than if the message is a random binary string. This is also true for DES and for other symmetric encryption systems. The reason for this reduction in security is attributed to the fact that consecutive characters in a text message are dependent upon eachother.

The only mathematical way to assess the security of symmetric systems is through information theory, which is discussed later in this book. The security depends on the uncertainty pertaining to the key and the uncertainty pertaining to the message. Roughly speaking, the longer the key the more secure the message. One reason for discussing historical ciphers, such as the Vigenère cipher, in this book is to furnish examples of how this works.

With public key algorithms, only one message can fit with a given cipher text, and the keys have to be made longer and longer to withstand brute-force attacks. What the proper length should be is a matter of conjecture and it is one of the "hot-button" issues in modern cryptography. On one hand, a financial institution using public key algorithms may not be in a hurry to report that its system has been broken. On the other hand, a successful intruder may not want to report success.

With symmetric systems, one can (at least in theory!) quantify the security, which can be measured in **Shannon bits**. In certain situations, it can be measured exactly; in other situations, it can be estimated experimentally e.g., with text messages encoded in binary. In general, it may be the case that many messages will fit with a given cipher text so that, in the end, the determination of the message may still be a matter of guesswork. Nowadays, RSA keys should be several hundred decimal digits in length. In bits, they should be at least 1024 bits long.

We should also point out that in some situations, whether dealing with symmetric or public key algorithms, it may be easier or faster to try to guess the message than to guess the key and then to get the message. Furthermore, there is a strong probabilistic component running through all of cryptography, exacerbated by the possibility of transmission errors.

We have not touched on several practical issues here such as message compression, transmission errors, and checking for key equality. These will be dealt with in Parts II and III of the book when the appropriate machinery has been built up.

We have also not discussed standards and some of the standards bodies such as NIST (National Institute of Standards in Technology), IEEE (Institute of Electrical and Electronic Engineers) in the U.S. and ISO (International Organization for Standardization). We will touch on this again in Chapter 8.

3.8 SSL (Secure Socket Layer)

The Secure Socket Layer (SSL) protocol is very widely used in modern internet application. It was developed by the Netscape corporation for secure internet transactions. Most people will have encountered SSL when sending credit card information or viewing bank statements on the internet. These types of applications usually use HTTPS (Hyper-Text Transfer Protocol - Secure) instead of the regular HTTP.

SSL uses a combination of symmetric encryption (such as DES from Chapter 5) and asymmetric encryption (such as RSA from this chapter). Asymmetric cryptography is generally slower and is used for authentication and distributing a shared secret. Symmetric encryption is quicker but needs a shared secret which the asymmetric algorithm provides. The symmetric encryption is used for the actual encrypted transaction.

Before a secure SSL transaction occurs, the server (eBay, Amazon, PayPal, etc.) must generate a public key / private key pair, and then get this pair signed by an authority. This signed information is called the **certificate**. The person (or organization) that signs the certificate is known as the **certificate authority (CA)**. Anyone may sign a certificate, but every machine maintains a list of trusted CAs, and if the server's certificate is not signed a warning is usually produced.

The signing process is merely the CA using its private key to encrypt some known data. The client may then use the CA's public key to decrypt this data and compare it to the known value. If they match, then it must have been the CA who actually signed the certificate.

The SSL handshake protocol proceeds as follows:

1. The client contacts the server, announcing its SSL version and supported protocols.

2. The server responds with its SSL version and supported protocols as well as its certificate.

3. The client checks if the server's certificate was signed by a trusted authority. The client also checks the certificate's designated address against the server's real address to guard again man-in-the-middle attacks (see Chapter 7). Lastly the client checks if the certificate has expired. If all checks pass, the algorithm continues, otherwise warnings are generated.

4. With all the data generated in the handshake so far, the client creates a **premaster secret**. The client encrypts this secret with the server's public key and transmits it.

5. The server decrypts the premaster secret using its private key, and then performs a series of steps to generate the **master secret**. The client also performs these steps and will have the same master secret.

6. Both the client and the server generate the **session keys** from the shared master secret. These session keys are symmetric keys that are used to encrypt and decrypt all the traffic that is sent from this point forward. They also may verify that the messages weren't altered in transit.

When the client and server share the same secret key (in this case they are called session keys), they may use a fast symmetric algorithm to communicate. When the two parties are finished communicating, then the session keys are forgotten, and have to be regenerated for further transactions.

3.9 PGP and GPG

Pretty Good Privacy (or **PGP**) is a computer program which provides cryptographic privacy and authentication. It is a patented technology created by Phil Zimmerman and owned by the PGP Corporation. PGP may encrypt any type of data but it is most commonly used for email.

GPG stands for **GNU Privacy Guard** which is a free (open source) version of the non-free proprietary PGP. Both of these systems are very similar in their uses and operation, but differ in the algorithms that they use. The both use RSA, DSA, or El Gamal for asymmetric encryption but PGP uses patented symmetric encryption algorithms, while GPG uses public domain (free) algorithms

Both programs may be used for encrypting / decrypting email and signing / authenticating messages.

Before using PGP or GPG, a user generates their own public key and private key pair. Then, the public key must be published so that others may access it. This may be done by putting it out on a website, sending out mass emails announcing the public key, or placing it on a **key server** and associating it to an email address.

Encrypting / Decrypting   To send an encrypted message, the user decides on a symmetric algorithm and then the computer generates a random key for use with this message. The message is encrypted using this key, and the key is encrypted using the intended recipient's public key. Both the message and the encrypted key are sent by email to the intended recipient. If the email is intercepted in this form, the eavesdropper shouldn't be able to read the contents, because they don't posses the proper private key to decode the session key, nor do they posses the session key to decode the message.

Upon receipt of an encrypted email, the user's computer will use his/her private key to decrypt the enclosed session key, and then will use the session key to decode the message. In many modern email programs that support PGP or GPG, this is done without the user knowing. That way, the user isn't inconvenienced by the added security, but will still enjoy the benefits of encrypted email.

Note that symmetric encryption is used for the actual message and asymmetric encryption is used for the key exchange. This is because symmetric cryptography is about 4000 times faster than asymmetric. That means that sending a large email with large attachments would take quite some time to encrypt if you only used RSA.

Signing / Authenticating   Just as the encryption algorithm for PGP and GPG are very similar to SSL, so is the authentication mechanism. A hash of the email message is encrypted with the user's private key, and then appended to the end of the message. Then when the email is received, the user's computer may decrypt this message with the sender's

Number Length (bits)	Machines	Memory
430	1	trivial
760	215,000	4 Gb
1020	342,000,000	170 Gb
1620	1.6 x 1015	120 Tb

Figure 3.4: Source: RSA Security Inc.

public key and check that the hash corresponds to the hash of the current message. This procedure may serve two purposes. It authenticates the original message sender (the person in possession of the private key used to encrypt the message hash), and it almost[5] guarantees that the message wasn't altered since its signing.

With the advent of PGP and GPG, secure email is finally available to anyone that needs or wants it.

3.10   RSA Challenge

RSA Security Inc. is a company that was formed to commercialize the RSA protocol. As we have learned in this chapter, RSA is based on the presumed difficulty of factoring large composite numbers. In order to prove that factoring numbers is difficult and to keep abreast of the current ability to factor, RSA Security Inc. has offered cash rewards to people who find the factors of large composite numbers. These large composite numbers are known to be the product of two large prime numbers.

At the time of the writing of this book, the largest number to be factored was 576 binary digits or 174 decimal digits long. $10 000 US was awarded to the person who discovered the two prime factors. Prizes are offered for incrementally harder numbers, up to a $ 200 000 prize for a 2048 bit (binary digit) number.

Using current algorithms and techniques, and using Pentium 500 Mhz machines, the following table estimates the resources that must be spent in order to factor the given numbers in a year.

As may be seen, longer key lengths seem to dramatically increase the resources needed to crack the RSA algorithm. Unfortunately, these estimates are based on current methods and knowledge. It has not been proved that there does not exist a polynomial algorithm (see Chapter 19) for factoring or breaking the RSA cryptosystem.

---

[5]The word 'almost' is used because altering a message while maintaining the hash is theoretically possible, but as of the writing of this book, there have been no English messages that hash to the same md5 sum. There is a $10 000 prize offered to anyone who finds two values that have the same md5 hash

3.11 Problems

Notation In some of the problems/solutions below we used the $Rem[u, v]$ notation introduced in this chapter. Recall that $Rem[u, v]$ just means the remainder when u is divided by v. For example, $Rem[37, 16] = 5$. Later on, in Chapter 19 we will also use the equivalent u (mod v) or u mod v notation.

Several of these problems need background material and may be skipped on a first reading.

1. *A fundamental issue.* If the cryptographic keys can be given to **A** and **B** over a private secure channel, why can't all messages between **A**, **B** be transmitted over this same private channel so the keys can be dispensed with?

2. *An essential property.* In a cryptographic system suppose messages M_1, M_2 are encrypted, resulting in cipher texts C_1, C_2. If $C_1 = C_2$, must $M_1 = M_2$?

3. In the RSA algorithm suppose that $1 < M_1 < N$, $1 < M_2 < N$ and $Rem[M_1^e, N] = Rem[M_2^e, N]$. Must $M_1 = M_2$?

The next two questions pertain to DES

4. *Narrowing down the key in DES.* Why is it the case that for a text message, the key search space can be reduced from approximately 2^{56} to 2^{40} after examining the first block of 8 bytes? (We assume that each text letter is encoded to a binary string of 8 bits, and that there are 64 ASCII characters.)

5. A learned reviewer of this book commented as follows. "Systems such as DES, Triple DES and Rijndael appear to be secure but can certainly be broken by exhaustive search (i.e., computational complexity is key) and for all we know could be broken very easily if somebody comes up with a clever idea". Is the reviewer correct?

Repeated squaring

6. Using the repeated squaring method, find the remainder when 5^{51} is divided by 97.

7. Using the repeated squaring method, find the remainder when 11^{22} is divided by 167.

RSA encryption

8. Using the RSA algorithm, given the cipher text $C = Rem[M^e, N]$ with gcd($e, (p-1)(q-1)) = 1$, can there be more than one decryption index d such that $Rem[C^d, N] = M$?

9. In the RSA algorithm we assume that p, q are large primes which must be unequal. Why is it that p, q must be unequal?

10. Show that for security reasons in the RSA algorithm, p and q should not be chosen too close together.

11. In the RSA algorithm, why must we choose e to be relatively prime to $(p-1)(q-1)$?

12. In the RSA algorithm show that e must be odd.

13. In the RSA algorithm the restriction is sometimes made — even in textbooks — that $\gcd(M, N) = 1$. Is this restriction necessary?

14. Suppose an eavesdropper Eve knows $N = pq$ and also $\varphi(N) = (p-1)(q-1)$.[6] Show that Eve can then find p and q.

Calculations using RSA (The Euclidean Algorithm of Chapter 19 can be used)

15. Find an RSA decryption exponent d given that $p = 41$, $q = 37$ and $e = 7$, using both methods as described in the text.

16. Repeat the previous problem with $p = 17$, $q = 19$ and $e = 5$.

17. Find an RSA decryption exponent d given that $p = 47, q = 59$ and $e = 17$.

18. Let $p = 29, q = 67$ so $N = 1943$ and $(p-1)(q-1) = 1848$. Let $e = 701$. Suppose $M = 23$. Find the cipher text C.

19. Let p, q, e be as in the previous question. Suppose $C = 1926$. What is M?

20. Find all possible values of e for $N = 55$. What is a compact formula for this quantity in general?

Sending text messages with RSA

21. Suppose Alice wishes to send a text message M to Bob using the RSA algorithm. Bob's public key is the pair $[N, e] = [2867, 17]$. Note that $2867 = (47)(61)$. Alice encodes the 26 letters of the English alphabet by putting $A = 00, B = 01, \ldots, J = 09, K = 10, \ldots, T = 19, \ldots, Z = 25$. Alice transmits the message in blocks. Each block corresponds to two letters which are encoded into their numerical equivalents. For example, the pair D, E becomes the block $[0405]$, which then gets enciphered to $C = Rem[405^{17}, 2867]$, since the block corresponds to the decimal number 405. Now suppose Bob receives the cipher text 0300. What was the message transmitted by Alice?

---

[6]$\varphi(N)$ is **Euler's Phi Function** See Chapter 19 for further details.

22. In the above problem, why can't we put more than 2 letters in a block?

23. Are the text messages that are sent in this way secure?

Elementary attacks, pitfalls, and incorrect implementations of RSA

Small message, small exponent

24. Show that if the message M is a small integer and the enciphering index is a small integer then RSA is not secure.

25. For $[N, e] = [30967, 3]$, decrypt $C = 29791$ assuming that M is "small."

26. Can you think of a real world example of enciphering a small integer where the attack of Question 25 might cause difficulties?

27. Using the fact that $C = Rem[M^e, N]$, how can RSA be attacked if e is small? (This is similar to the above attack.)

28. Decipher $C = 37$ given that $[N, e] = [51, 3]$, using the method of the previous problem.

Semantic Security and RSA

29. RSA leaks information in various ways. For example, if C_1, C_2 are the cipher texts for messages M_1, M_2 respectively, then show that $C_1 C_2$ is the cipher text for $M_1 M_2$ (we are working here by taking remainders upon division by N, i.e. we are working mod N).

    Another way in which RSA leaks information is through the Jacobi symbol, since the Jacobi symbol of the cipher text C equals that of the message M. This reveals one bit of M. A cryptosystem has **semantic security** if no information is leaked.

Broadcast Attack

30. Given an enciphering index $e = 3$ show how a plain text message M can be recovered if it is enciphered and sent to 3 different entities having pairwise relatively prime moduli $N_1, N_2,$ and N_3. This is most easily solved using the Chinese Remainder Theorem of Chapter 19.

31. Use the broadcast attack to find M when it is enciphered to 80 using $[N_1, e_1] = [319, 3]$; 235 using $[N_2, e_2] = [299, 3]$; and 121 using $[N_3, e_3] = [323, 3]$

Common Modulus Attack

32. Let Alice's public key be $[N_1, e_1]$ and let Bob's public key be $[N_2, e_2]$, with $N_1 = N_2$. Show that Bob can recover all messages sent to Alice.

Cycling Attack

33. Given $C = M^e$, an eavesdropper can compute $C^e = (M^e)^e = M^{e^2}, (C^e)^e = C^{e^2} = M^{e^3}$, etc. How can the eavesdropper obtain the message M by using this idea?

34. For $[N, e] = [187, 3]$, decrypt $C = 173$ using the cycling attack.

3.12   Solutions

1. The secure channel may not always be open. The channel may not always remain secure.

   a) **A** and **B** may not have control over when they want to communicate in secret. Thus the key may have to be in place well before the exchange.

   b) The same key will work for several different messages. For example, major banks use the same key for about 6 months.

   c) The message may be very long, yet a short key may provide adequate security.

2. Yes, we must have $M_1 = M_2$. First of all, we have $C_1 = e(M_1)$ and $C_2 = e(M_2)$ where e is the enciphering algorithm. Applying the decryption algorithm d we have $d(C_1) = M_1$ and $d(C_2) = M_2$, since $d(C_1) = d(e(M_1)) = M_1$. Similarly $d(C_2) = M_2$. Since $C_1 = C_2$, we have $d(C_1) = d(C_2)$; so $M_1 = M_2$.

3. This is a special case of 1.2. The reason is that if M is any message with $1 < M < N$ then the enciphering transformation e transforms M to $e(M)$ with $e(M) = Rem[M^e, N] = C$. As we will see in Chapter 19, the decryption algorithm d transforms C to $d(C) = Rem[C^d, N] = M$.

4. We are assuming that the plain text (= message) consists of letters, numbers and punctuation. So each 8 bits (i.e., each byte) of cipher text comes from a letter, number or punctuation mark. Including the upper and lower-case alphabet we end up with about 70 characters, and we will approximate this by 64 characters. Thus the 8-bit plain text which originally could have been any one of $2^8 = 256$ strings has been narrowed down to 64 possibilities. Thus only $\frac{64}{256} = \frac{1}{4}$ of the potential messages are realistic possibilities. Then for an 8 byte (=64 bit) block -assuming the bytes are *independent*- we narrow down the messages and end up with only about $(\frac{1}{4})(\frac{1}{4})\dots(\frac{1}{4}) = \frac{1}{4^8} = \frac{1}{2^{16}}$ of the potential messages as being realistic possibilities. Since the size of the key-space of DES is 2^{56} then, a priori, there are 2^{56} possibilities

for a message that gives rise to a given cipher text of 64 bits. However, because of our assumption on the nature of the messages only about $\frac{1}{2^{16}}$ of the potential messages are realistic possibilities. Since $\frac{1}{2^{16}}(2^{56}) = 2^{40}$ the result follows.

5. Fortunately, the reviewer is in error. To see this, let us examine the case of DES. There are 2^{56} possible keys associated with DES then, without making any assumptions on the nature of messages there are 2^{56} possible messages associated with any given cipher text. The "uncertainty" of the message space given the cipher text (i.e., the "entropy" -see Part II-) is 56. No amount of key searching, cleverness, etc., can remove this uncertainty. Of course, if additional conditions are imposed on the messages (see problem 4) then this can be exploited as explained there.

6. 69.

7. 7.

8. Yes, as we have seen in the first example discussed in the text, there can be more than one decryption index d. Here is another example. Suppose that **A** is transmitting a message to **B** whose public key is $[N, e] = [1541, 5]$. Now we have $N = 1541 = (23)(67) = pq$ with $p = 23, q = 67$. Then $(p-1)(q-1) = (22)(66) = 1452$. Now $Rem[(5)581, 1452] = 1$. So we can take $d = 581$. Then, for any cipher text $C = Rem[M^e, N] = Rem[M^5, 1541]$ it follows that $Rem[C^d, N] = Rem[C^{581}, 1541]$. However, instead of working with $(p-1)(q-1)$ we can work with any integer that is divisible by both $p-1$ and $q-1$. Such a number is 66. Note that $\gcd(e, 66) = \gcd(5, 66) = 1$. Thus, as will be shown in Chapter 19 in the general case, if d_1 is such that $Rem[(5)d_1, 66] = 1$ then d_1 also serves as a decryption index here. Now $Rem[(5)(53), 66] = 1$. So $d_1 = 53$.

Let us summarize. We have two decryption indices, namely 581 and 53, for the same values of p and q. For any message M in the form of an integer between 1 and $N = 1540$, we have $C = Rem[M^5, 1541]$, but also $Rem[C^{581}, 1541] = Rem[C^{53}, 1541] = M$.

9. Suppose a user announces his public key as $[N, e] = [pq, e] = [p^2, e]$. Then N is easily factored. Just take the square root of N to get p.

10. The idea goes back to Fermat, involving Fermat's "Difference of Squares Method", as follows. Suppose n factors as $n = pq$ with $p > q$. Recall that p, q are odd, so $p + q$ and $p - q$ are even. Then $p > \sqrt{n}$ and $q < \sqrt{n}$. Then $n = (\frac{p+q}{2})^2 - (\frac{p-q}{2})^2 = u^2 - v^2$ where u, v are integers with $u > v$. Thus $n + v^2 = u^2$ or $u^2 - n = v^2$. If p is close to q then u is bigger than, but close to, $\sqrt{n}$. So to factor n, which is our goal, we only need to try a few integers u until we find a u such that $u^2 - n$ is equal to a square integer denoted by v^2. For example, take $n = 2867$ as in problem 22. Here, $\sqrt{n} = 53.54$ So

start with $u = 54$ and immediately we get $54^2 - n = 49$, which is a square. Thus $54^2 - n = 7^2, n = 54^2 - 7^2 = (54 - 7)(54 + 7)$ and we have factored $n = (47)(61)$.

11. One explanation is as follows. The official procedure for calculating the decryption index is to find d such that $Rem[ed, (p - 1)(q - 1)] = 1$. (In other language, d is the multiplicative inverse of e modulo $(p - 1)(q - 1)$). This can be carried out, as we shall see in Chapter 19, only if e is relatively prime to $(p - 1)(q - 1)$, i.e., only if $\gcd(e, (p - 1)(q - 1)) = 1$. For example, if $p = 5, q = 11, e = 2$ then it is impossible to find d so that the remainder of $2d$ on division by 40 is 1. To see that this is the case we would need to find d so that $2d - 40x = 1$ for some x. But the left side is even and the right side is odd. So, no such d exists. What happens if we break the rules and choose an enciphering index e such that $\gcd(e, (p - 1)(q - 1)) > 1$? Take $M = 2, p = 3, q = 5, e = 2$. Then $c = Rem[M^2, 15] = Rem[2^2, 15] = 4$. However, we also get $Rem[7^2, 15] = 4, Rem[13^2, 15] = 4$ and $Rem[8^2, 15] = 4$. In other words, four different messages, namely 2, 7, 8, 13, all encipher to the same cipher text, namely 4. But we saw in problem 1.2 that in a proper cryptographic system, two different messages cannot encipher to the same cipher text.

12. If e is even, then 2 divides e. For security, p and q must be large, certainly bigger than 2. Thus p and q being primes bigger than 2 are odd numbers. So $p - 1$ (and $q - 1$) are even. Then 2 divides e as well as $(p - 1)(q - 1)$. This contradicts the assumption that $\gcd(e, (p - 1)(q - 1)) = 1$.

13. No, the algorithm works without that requirement. However, an attacker might test if $g = \gcd(M, N) > 1$, and if so, then $g = p$ or q. To see this, note that $g < N$, since $M < N$. Since g divides $N = pq$, and p is prime, this forces that g divides p or g divides q. Then, since $g > 1$ and both p and q are prime, this implies that $g = p$ or $g = q$. Thus, an attacker can easily factor N when $\gcd(N, M) > 1$.

14. We have $N - \varphi(N) + 1 = (pq) - (p-1)(q-1) + 1 = p + q$. Thus Eve knows $p + q$ and pq. Now $(p + q)^2 - 4pq = (p - q)^2$ so Eve knows $(p - q)^2$. Eve can assume that p is bigger than q. So Eve calculates the positive square root of $(p - q)^2$ which is $p - q$. Since Eve knows both $p + q$ and $p - q$ then, by adding, she gets p and then q. For example, suppose Eve knows $N = 323$ and $\varphi(N) = 288$. Then $323 - 288 + 1 = 36 = p + q$. Now $36^2 - 4(323) = 1296 - 1292 = 4 = (p - q)^2$, so $p - q = 2$. Since $p + q = 36$ we get $p = 19$ and $q = 17$.

15. Solution (a). We find d such that $Rem[7d, (40)(36)] = 1$. This gives $d = 823$.

    Solution (b). Let $t = 360$. Then 40 divides t and 36 divides t. We find d such that $Rem[7d, t] = 1$. This gives $d = 103$.

16. Solution (a). We find d such that $Rem[5d, (16)(18)] = 1$. This gives $d = 173$.

    Solution (b). Let $t = 144$. Then 16 divides t and 18 divides t. We find d such that $Rem[5d, t] = 1$. This gives $d = 29$.

17. If $p = 47, q = 59, e = 17$ we get, using either method, that $d = 157$.

18. $C = 458$.

19. We can take $d = 29$, giving $M = 12$.

20. N factors as $p = 5, q = 11$, so we find all e for which there is a d so that $Rem[ed, (p-1)(q-1)] = 1$. That is, we find all integers relatively prime to, and less than, $(p-1)(q-1) = 40$. There are 16 (excluding the number 1); they are: 3, 5, 7, 9, 11, 13, 17, 19, 21, 23, 27, 29, 31, 33, 37, 39. In general, this quantity is called **Euler's Phi function**, so $\varphi(40) = 16$. This is further discussed in Chapter 19.

21. We can take $d = 2273$ so $M = 0408$ corresponding to the pair of letters EI.

22. If we put 3 letters in a block we might have a block as large as 252525, which is larger than $N = 2867$.

23. If we encode with, say 2 letters in a block we know that there are only $26^2 = 676$ possible messages even though the largest message is encoded as 2525. So, to increase the security (i.e. to make a brute-force attack more difficult) we need a large number of letters in a block and so a large modulus N. More economical methods of encoding are possible (see Mollin [Mol02] and [Mol00]).

24. If M^e is less than N, the message M is immediately obtained by getting the eth root of the cipher text C. For example, if $e = 3$, we need only calculate the cube root of C. The reason is that $Rem[M^e, N] = M^e = C$ so $M = C^{1/e}$.

25. $M = 29791^{\frac{1}{3}} = 31$.

26. Suppose **A** chooses a binary string of length 56 as a proposed session key K for the DES algorithm and transmits this key to **B** using RSA. When we represent K as an integer it will be less than 2^{56}. Then the cipher text C is less than 2^{56e}. If e is small, say $e = 3$, then this may be much less than $N = pq$ when p and q are large. Then, by calculating the eth root of C an eavesdropper can recover the key K.

27. From the stated fact, it follows that $M = C^{\frac{1}{e}}$ or $(C+N)^{\frac{1}{e}}$ or $(C+2N)^{\frac{1}{e}}$ or $(C+rN)^{\frac{1}{e}}$ for some r. If r is small, then RSA can be attacked.

28. For $r = 6$, we find $(C + rN) = 343$ so $M = 343^{\frac{1}{3}} = 7$ (One can verify from theory, or directly, that $Rem[M^e, N] = Rem[7^3, N] = 37 = C$).

29. We have $C_1 = Rem[M_1^e, N]$ and $C_2 = Rem[M_2^e, N]$. Therefore, $C_1 = M_1^e + \lambda_1 N$, $C_2 = M_2^e + \lambda_2 N$, $M_1 < N, M_2 < N$. Then $C_1 C_2 = (M_1 M_2)^e + \mu N$. Thus $Rem[C_1 C_2, N] = Rem[(M_1 M_2)^e, N]$.

30. This follows from the Chinese Remainder Theorem discussed in Chapter 19. The main point is this. We have $Rem[M^3, N_1] = C_1, Rem[M^3, N_2] = C_2, Rem[M^3, N_3] = C_3$. By the Chinese Remainder Theorem there is a unique x less than $N_1 N_2 N_3$ such that $Rem[x, N_1] = C_1, Rem[x, N_2] = C_2, Rem[x, N_3] = C_3$. Now M^3 also satisfies these 3 equations. Moreover since $M < N_1, N_2, N_3$ we get $M^3 < N_1 N_2 N_3$. Thus $x = M^3$. The cube root of x then gives M. Note that this attack can be generalized to any small value of e.

31. First we check that N_1, N_2 and N_3 are pairwise relatively prime. They factor as $N_1 = 11 \times 29$, $N_2 = 13 \times 23$ and $N_3 = 17 \times 19$, and are thus pairwise relatively prime. Now, x as described in the previous solution is 74088, and so the answer is $M = 74088^{\frac{1}{3}} = 42$.

32. As pointed out by Buchmann [Buc00], given d one can factor N. Assuming this, suppose Bob intercepts a cipher text intended for Alice. By the result above, Bob can factor N using his private key d_B. Once N is factored, Bob can use the Euclidean Algorithm to recover Alice's private key d_A from her public key e_A. Note that this question is irrelevant if one allows that private keys are the triples $[p, q, d]$ rather than merely d.

33. Since e is relatively prime to $(p-1)(q-1)$, there are integers t, λ so that $e^t = 1 + \lambda(p-1)(q-1)$. This follows from a generalization to Fermat's Theorem. Using this fact, we have $C^{e^t} = C^{1+\lambda(p-1)(q-1)} = C(C^{\lambda(p-1)(q-1)})$. Now $Rem[C^{p-1}, p] = Rem[C^{q-1}, q] = 1$, (by Fermat's Theorem) and so it follows from the Chinese Remainder Theorem that $Rem[C^{e^t}, N] = Rem[C(C^{\lambda(p-1)(q-1)}), pq] = C$. Then, given that $C = Rem[M^e, N]$, we conclude that $Rem[C^{e^{t-1}}, N] = M$. Thus, if t is small, RSA can be attacked. Note that in using Fermat's Theorem, we have supposed that $Rem[C, p] \neq 0$ and $Rem[C, q] \neq 0$, but, we in fact can only be guaranteed that one of $Rem[C, p] \neq 0$ or $Rem[C, q] \neq 0$ is true, although it is possible that both are true. However, the result still holds in this case, and the proof is similar.

34. For $t = 4$, $Rem[C^{e^t}, N] = C$, so $M = Rem[C^{e^{t-1}}, N] = 24$.

Chapter 4

The Fundamentals of Modern Cryptography

Goals, Discussion We present a summary of most of the important procedures and ideas in modern cryptography. The important question of authentication is pursued further in Chapter 8.

New, Noteworthy We discuss the material in such a way that mathematical prerequisites are minimized. For example, the RSA material discussed in Chapter 3 will give more than enough background for an understanding of digital signatures. On the question of the design of modern block ciphers, several authors will suggest that Shannon's diffusion principle comes down just to a permutation. However, quoting from the master, we point out that there is much more to diffusion than a mere permutation.

Issues from complexity theory, relating to computational security, are pursued in Chatper 19

4.1 Encryption Revisited

Recall from Chapter 3 that cryptography is the art or science of keeping messages secret. Briefly, **A** (for Alice) wants to send a secret message M to **B** (for Bob). The message M is also called the plain text. The message might be a text message or a binary string, such as $\{1\,1\,0\,1\,1\,0\}$, etc. As with all computer communications, the text message will first be converted to binary. One way of doing this is with the ASCII conversion code described in the Appendix.

A scrambles (encrypts, enciphers) M with an encryption key, often called the **session key**, into unintelligible cipher text C. Then **A** transmits C **in the open** — that is, over a public channel — to **B**.

When the cipher text (= scrambled message) reaches **B**, it can be descrambled by **B** using the decryption key. Thus **B** can recover the original message M. The security of the message rests on two assumptions.

1. Only **B**, or an authorized representative, has the descrambling (= decryption) key.

2. The message M cannot be recovered without the decryption key.

The procedure or algorithm for encryption is called a *cipher*. Several different kinds of ciphers — Caesar, Vigenère, affine — have been described in Chapter 2 on classical ciphers.

Cryptanalysis is the science of trying to "break the system", or trying to capture the message M from the cipher text C by first getting the key (or even by getting the message M directly). An attempted cryptanalysis is called an **attack**. An attack which involves trying all possible keys is called a **brute-force attack**. As we saw in Chapter 3, this involves trying only about half the key-space. A **cryptosystem** consists of an encryption and decryption algorithm, together with all possible messages (= plain texts), cipher texts and keys. Traditionally, a fundamental — if theoretical — principle in cryptography is known as **Kerchoff's principle**. It asserts that the security of a cryptosystem must depend only on the key. In other words, in assessing the security of a cryptosystem, one must assume that a possible intruder or eavesdropper has complete knowledge of the cryptosystem and its implementation. Of course, in practice, as little as possible is revealed.

Although we do not make extensive use of mathematical notation, we can summarize the encryption procedure in the following equation:

$$d(e(M)) = M \qquad (4.1)$$

where e is the encryption (enciphering) transformation which scrambles the message M to give the cipher text C, so that $e(M) = C$. Further, d is the decryption or deciphering algorithm!algorithm, so that $d(C) = d(e(M)) = M$.

An important mathematical consequence of Formula (4.1) is Formula (4.2):

$$e(d(M)) = M \qquad (4.2)$$

Formula (4.1) says that we get the original message M back if we first encrypt the message M and then decrypt the encrypted message (= the cipher text).

In formula (4.2), the order is reversed. In other words, if we first perform the decryption algorithm on the message itself (not on the cipher text, as in (4.1)) and then perform the encryption algorithm on the result, we also end up with the message M. We are tacitly assuming here that the message set equals the set of cipher texts (see Chapter 15), so that e and d are defined on the same set.

Formula (4.2) is the basis for **digital signatures** as we will see later. The quantity $d(M)$ is called the signature. Note that, in (4.1) and (4.2), we should, strictly speaking, write $e_K(M)$ instead of $e(M)$ because the enciphering transformation e will depend on a key K. Similar remarks pertain to the deciphering transformation d.

As explained in Chapter 3, the enciphering and deciphering keys are equal in the case of symmetric cryptography; for public key systems they differ. In the notation of Chapter 3, $e(M)$ is simply $\text{Rem}[M^e, N]$ for the RSA system, i.e. the remainder of M^e upon division by N.

The main public key cryptosystem is RSA, but elliptic curve cryptography is also used. For symmetric cryptography the main algorithms are **AES, DES, Triple DES** and (a modification of) the **Vernam Cipher** or **one-time pad**. Frequently, the lengthy random key that is needed for the one-time pad is approximated by the output of a suitable feedback shift register. Also, we will be discussing another important algorithm which straddles the border between symmetric and public key exchanges, called the **Diffie–Hellman key-exchange**. Nowadays, in industry, a kind of **hybrid system** is used. This means that, using RSA for example, **A** transmits a message K to **B**.

Then **A** and **B** use K as a session key for a symmetric encryption using DES or AES.

One difficulty with transmitting messages with RSA is that it is quite slow, relatively speaking, whereas AES and DES are much faster. However, for **A** and **B** to use AES or DES, they must first be in possession of a common secret key.

In any cryptographic system, various other related issues besides encryption must be addressed. These issues include the following:

1. **Authentication**. How does **B** (or **A**) know that the entity with whom **B** (or **A**) is communicating is really **A** (or **B**)?

2. **Message Integrity**. How does **B** know that the message received from **A** has not been tampered with?

3. **Digital Signatures/Non-Repudiation**. How do **A**, **B** arrange to obtain legally binding reassurances, similar to a signature, from each other?

In this chapter we discuss some of these concerns. Authentication issues are also pursued in Chapter 8.

4.2 Block Ciphers, Shannon's Confusion and Diffusion

In some modern applications of symmetric cryptography, such as GSM (Global System for Mobile communications) phones, a **stream cipher** encryption algorithm is used. The message, represented as a stream of binary digits (bits), is encrypted, bit by bit. One advantage of stream ciphers is that an error will not affect future bits, i.e. the error is not propagated. One of the main examples of a stream cipher comes from a **linear feedback shift register** (LFSR), which is discussed in detail in Chapter 16.

In other applications of symmetric cryptography, **block cipher** encryption algorithms are used. The message is broken up into substrings (called blocks) of a fixed length $n > 1$ and encrypted block by block. The integer n is called the **block-length**. In the case of DES, $n = 64$, i.e. the block-length is 64 bits long and the key-length is 56. Thus, for each block, the input is 64 bits, the output is 64 bits and the key-length is 56, so there are 2^{56} possible keys.

At a basic level, DES is simply a combination of the two fundamental techniques for construction of ciphers advocated by Shannon in 1949, namely, **confusion** and **diffusion**.

Confusion   This technique tends to block the cryptanalyst from obtaining statistical patterns and redundancies in the cipher text arising from the plain text. Thus, the statistical dependency of the cipher text on the plain text is obfuscated. The easiest way to cause confusion is through the use of substitutions. In the case of a binary string, we substitute various ones and zeros by zeros and ones respectively, according to a pre-determined formula.

Diffusion   This technique dissipates the redundancy of the plain text by spreading it over the cipher text. Redundancy will be precisely explained in Part II. For the moment, we can think of it informally as statistical patterns. Diffusion implies that, if we change just one letter or character in the plain text, we cause a big change in the cipher text. Thus, we will need a large amount of cipher text to capture redundancy in the plain text.

Several authors have suggested that, according to Shannon, diffusion simply means a permutation or rearrangement of the characters in the message string. However, this is not quite correct because a permutation will still preserve the frequency of characters. Shannon's recipe for diffusion also involves acting on a string with a diffusing function.

In the words of the master himself, Shannon [Sha49]: "In the method of diffusion, the statistical structure of M which leads to its redundancy is 'dissipated' into long range statistics, i.e., into statistical structure involving long combinations of letters in the cryptogram. The effect is that the enemy must intercept a tremendous amount of material to tie down this structure, since the structure is evident only in blocks of very small individual probability. Furthermore, even when he has sufficient material, the analytical work required is much greater, since the redundancy has been diffused over a large number of individual statistics."

An example of diffusion of statistics is operating on a message $M = m_1, m_2, m_3 \dots$ with an "averaging" operation, for example

$$y_n = \sum_{i=1}^{s} m_{n+i} \pmod{26}, \tag{4.3}$$

adding s successive letters of the message to get a letter y_n. One can show that the redundancy of the y sequence is the same as that of the m sequence, but the structure has been dissipated. Thus, the letter frequencies in the y sequence will be more nearly equal than in the m sequence; the diagram frequencies also more nearly equal, etc. Indeed any reversible operation, which produces one letter out for each letter in and does not have an infinite "memory", has an output with the same redundancy as the input. The statistics can never be eliminated without compression, but they can be "spread out."

We mention here that the historical ciphers discussed in Chapter 2, such as the Caesar and Vigenère ciphers, do not have the properties of confusion and diffusion. On the other hand, DES and AES use confusion and diffusion to good effect.

In general, the cipher texts in DES and AES will reveal some information about the message, the amount depending on the nature of the message. The extreme case, mentioned in Chapter 3, is when knowledge of the cipher text reveals no information whatsoever about the message, as is the case with the Vernam cipher (= one-time pad). This extreme case is called **"perfect secrecy"** or **"perfect security"**. We will have much to say about this both in the next section and in Parts 2 and 3 of the book.

4.3 Perfect Secrecy, Stream Ciphers, One-Time Pad

We recall that in modern digital communications all numbers and symbols are converted to binary strings. All symbols consist of zeros and ones and are known as bits (= binary digits). The conversion to binary is made using some encoding table, such as the ASCII code (see the Appendix).

To better explain how a one-time pad works, we use binary arithmetic in which we add according to the following four basic rules.

$$0 + 0 = 0, \quad 0 + 1 = 1, \quad 1 + 0 = 1 \quad \text{and} \quad 1 + 1 = 0. \tag{4.4}$$

An important property of binary arithmetic is that *subtraction is the same as addition*, i.e. $a - b = a + b$. For example, $0 - 1 = 0 + 1 = 1$, $1 - 1 = 1 + 1 = 0$, etc.

Note that these operations correspond to the well-known XOR (= exclusive OR) Boolean function. Moreover, the usual rules of arithmetic are satisfied, namely:

1. $a + a = 0$, no matter whether a is 0 or 1.

2. $a + b = b + a$, for all binary integers a, b.

3. $(a + b) + c = a + (b + c)$ for all binary integers a, b, c.

Multiplication in binary is also easy, since $1^2 = 1$, $0^2 = 0$ and $(1)(0) = (0)(1) = 0$. Thus, $x^2 = x$ for all binary variables x.

Let us suppose now that **A** wishes to send a secret message M to **B** in the form of a binary string of length n. We also assume that **A**, **B** are in possession of a secret key K, known only to them, which is also in the form of a random binary string of length n, *the same length as the message M.*

For example, assume that $n = 6$ and that K is the string $\{110010\}$. Suppose that the message M is $\{100011\}$. The enciphering algorithm is remarkably simple: add the two strings, bit by bit, using binary addition. Thus, we get:

$$\begin{array}{lcccccc}
\text{Plain text } M: & 1 & 0 & 0 & 0 & 1 & 1 \\
\text{Key } K: & 1 & 1 & 0 & 0 & 1 & 0 \\
\text{Cipher text } C: & 0 & 1 & 0 & 0 & 0 & 1
\end{array}$$

In symbols we can write $C = M + K$ (see Figure 4.1).

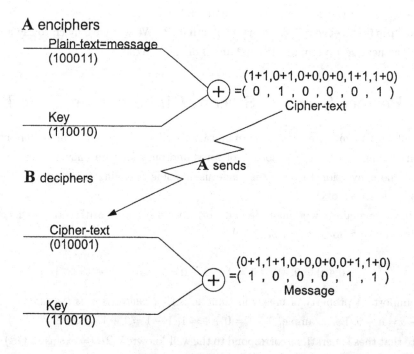

Figure 4.1: One-Time Pad

A transmits the cipher text C to **B**. When **B** receives C, he must undo the enciphering algorithm that **A** carried out on M. Since **A** added the key K to the message M to get the cipher text C, **B** must subtract the key K from the cipher text C. Since addition and subtraction are the same in binary, **B** just adds the key K to C, without "carries" of course, to recover M.

Thus, **B** deciphers C by adding K to C to obtain:

$$C + K = (M + K) + K = M + (K + K) = M + 0 = M \qquad (4.5)$$

In other words, **B** recovers the original message M that was transmitted to **B** by **A**, as required.

This encryption procedure is known as the Vernam cipher or one-time pad since it is recommended that the key be used just once, for added security. This is discussed in more detail in Part II.

What does unconditional (= perfect) security mean? Why does the one-time pad yield unconditional security?

Answering these questions in a rigorous mathematical way requires the explanation of several concepts and formulae from the discipline of information theory, following the pioneering work of Claude E. Shannon.

It is not difficult to explain the main ideas.

An encryption algorithm gives unconditional or perfect security (= perfect secrecy), if knowledge of the cipher text gives no additional knowledge of the plain text. In the one-time pad, we can think of each 0 or 1 in K (i.e., each element of K) as having been obtained in a completely random way. For example, one can imagine a coin as having been tossed for each element of the key, with heads giving a 0 and tails dictating that a 1 was chosen. Thus, the key K will have no known patterns in it.

In the example above, the plain text, key, and cipher text were, respectively, $M = \{100011\}$, $K = \{110010\}$ and $C = \{010001\}$. If an eavesdropper named Eve recovers C, no inference on the contents of the message can be drawn by Eve. For all Eve knows, the real plain text might well be the message $M_1 = \{011110\}$ and the real key might be the key $K_1 = \{001111\}$, since when we encipher M_1 using the key K_1, we get the cipher text $C = \{010001\}$.

Indeed, for any binary string S of length 6, there is a unique key K of length 6 such that S, enciphered by binary addition using K, gives the plain text C. The total number of binary strings S of length 6 is easily calculated. There are 2 choices, zero or one, for the first digit, 2 for the second digit, ... and 2 for the sixth digit, giving $2 \times 2 \times \ldots \times 2 = 2^6 = 64$ total choices for S.

Thus, Eve can only guess which of these 64 strings is the real message, as the cipher text gives no information on the message (= plain text).

In the case where the message is a binary string of length n, the number of possible messages is 2^n and Eve can only guess which one of these 2^n possibilities is the correct one.

This indicates why the one-time pad is perfectly secure. Indeed, the probability that Eve makes the correct guess is $1/2^n$, which is vanishingly small as n gets large. *Moreover, even if Eve made the right guess, she would never know that she had done so.*

One-time pads form the most secure encryption algorithm. Nowadays, they are mainly used for ultra-secure low-bandwidth channels. To quote Schneier: [Sch96]

"Many soviet-spy messages to agents were encrypted using one-time pads. These messages are still secure today and will remain that way forever."

We emphasize that for many applications the keys for one-time pads can only be used once and then discarded, as the name implies. If not, the security can be compromised.

It should also be pointed out that the general definition of perfect secrecy does not necessarily imply that the probability of getting the correct message is vanishingly small (as is the case for the one-time pad of length n, with n large). However, perfect secrecy means that the cipher text gives no extra information about the original message that was not available already.

Note that the Vernam cipher can be modified to work for any alphabet, not just the binary alphabet. Thus, **A** enciphers by adding a key element and **B** deciphers by subtracting the key element. For example, we mention that the Vigenère cipher (see Chapter 2) where the key is as long as the message, is essentially equivalent to a one-time pad, but over an arbitrary alphabet, not just the binary alphabet.

A famous result of Shannon asserts that, roughly speaking, the key must be as long as the message for perfect secrecy. This is discussed in Part II.

Instead of using a random key in the Vernam cipher, we can modify it by using a **pseudo-random key**, which can be generated from a **linear feedback shift register** (LFSR). This is discussed in more detail in Chapter 16.

4.4   Hash Functions

Hash functions are of fundamental importance in the design and implementation of cryptosystems. They arose in computer science where they are used for carrying out insertions, deletions, and searches of tables. However, there has been some concerns recently about their security as mentioned in our "Update" section in the preface.

There are many "off the shelf" hash functions that are readily available. However, in some situations it can be advantageous to "custom design" hash functions using some theory, e.g. the theory of linear codes. We present a nice application in Chapter 24.

In many situations in data processing and communications alike, one has to condense a long binary string M representing a large message to a shorter binary string M_1 of a fixed length. We think of M_1 as providing a digest or "snapshot" of M. The formula which changes M to M_1 is called a **hash function**. The shorter string M_1 is called the **hash** of the longer string M. M is the *input* and M_1 is the *output*.

As a very simple example, if M is a binary string $\{x_1 x_2 x_3\}$ of length 3 we could define M_1 to be the string of length 1 given by $M_1 = \{x_1 + x_2 + x_3\}$, addition being in binary.

Thus if $M = \{011\}$ then $M_1 = \{0 + 1 + 1\} = \{0\}$.

An essential feature of hash functions is that the output $(= M_1)$ can easily be calculated from the input M — but not the other way around! A desirable feature of hash functions is that each bit of the output depends — in a complicated way — on all the bits in the input. This is a recasting of Shannon's confusion and diffusion for hash functions rather than encryption transformations.

In general, the output of a hash function has a shorter length than the input.

From this it follows that **collisions** necessarily occur (a collision being two different inputs that hash to the same output).

In order for a hash function to be useful for encryption it must be **secure**. This means that:

(a) Given an output M_1 it is not computationally feasible (=intractable) to find an input M such that $f(M) = M_1$.

(b) It is computationally not feasible to find collisions i.e. to find messages $x_1 \neq x_2$ such that $f(x_1) = f(x_2)$.

Condition (a) implies that there will be no easy way of calculating any input which hashes to a given output.

Condition (b) is tied in with various kinds of attacks as described in Chapter 7

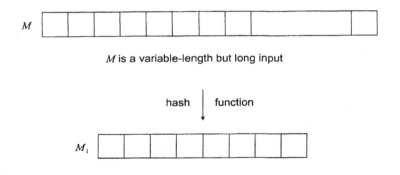

M is a variable-length but long input

hash | function

M_1 is a fixed-length (shorter)output

As we will see later, secure hash functions are fundamental in digital signatures. Another application of hash functions concerns password-protection schemes such as that used in Unix. Characters of the password are converted to binary (using the ASCII encoding for example). To save storage space on the computer each password M is hashed to an output M_1 using a secure hash function f. When somebody logs in, their password is hashed, and a check is made with the valid list of stored hash values. If the hash checks out, then access is granted; otherwise it is denied.

There are several commercially available "off the shelf" hash functions such as MD4 and MD5 developed by R. Rivest using random permutations of bytes and various nonlinear compression functions. These give 128-bit hash values.

Together with the NSA, NIST has designed the so-called SHA (Secure Hash Algorithm) which has a 160-bit hash value.

Later on in Part III of this book, we will show how to customize hash functions, suitable for the purpose at hand, using the theory of block designs and error-correcting codes. With such hash functions the security level can be calculated with greater precision than the "off the shelf" hash functions mentioned above.

Hash functions can be constructed in many ways. For example, we can use standard symmetric encryption algorithms such as DES for their construction as follows. Let M be the input. For simplicity, assume that M is a binary string whose length is divisible by 56. (Otherwise, we can pad M with zeros or ones). Divide M into 56-bit substrings namely $K_1, K_2, K_3, \ldots, K_t$. Start with L_0 equal to some string of length 64 (L_0 could be a piece of secret information). Calculate L_1 by applying the DES algorithm (see Chapter 5), with key K_1 to L_0. Similarly, calculate $L_2, L_3, \ldots, L_t$ by using the keys $K_2, K_3, \ldots, K_t$ applied to $L_1, L_2, \ldots, L_{t-1}$. The final 64-bit output L_t can be used as the hash value.

If we run this algorithm again with a different L_0 we get another 64-bit output unrelated to the first 64-bit output which can be combined with the previous output to give a 128-bit hash value.

Note that the hash above is easy to calculate, using standard software. The output seems to depend on every bit of the message and it is not computationally feasible to find two messages which give the same hash value.

Another very important general application of hash functions is in the area of key-verification. For example, using public key cryptography such as RSA, we assume that **A** has sent a secret message M to **B**. How do **A**, **B** verify that they both have the same message M? Because of the length of the keys in RSA, transmission errors are always a possibility. One method in use, is this: **A** and **B** publicly compare some number of corresponding bits, say 50 bits. If these 50 bits coincide, **A** and **B** can conclude with reasonably high probability, that they both have the same message M. The idea is that the 50 bits provide a hash function: then **A** and **B** conclude that if the hash functions are equal then the corresponding messages are equal. We will have much more to say about this situation in Part III.

Hash functions are also widely used to check message integrity as described in the next section.

4.5 Message Integrity Using Symmetric Cryptography

Here the evil *Eve* herself has designs on Romeo (see Figure 4.2) and, by altering the message, wishes to put Juliet out of the picture. In general one is concerned with the question of whether the message itself has been transmitted without alteration. This is the quest for message integrity.

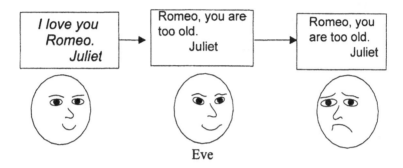

Figure 4.2: How does Romeo know that Juliet's message has not been altered?

The procedure used to simultaneously ensure authentication and data integrity using symmetric encryption is simple (Figure 4.3). Suppose m is the message that **A** wants to send to **B**. **A** and **B** will work with a publicly known family of hash functions. Thus, this general family of hash function is also known to Eve. However, suppose the specific hash function in the family to be used by **A** and **B** is the kth hash function, where k is a secret number or key is known only to **A** and **B**. So **A**, **B** work with a *keyed hash function*.

A calculates the hash $m^* = f_k(m)$ of the message m, appends this to m and transmits the package to **B**. It is entirely possible now that the pair (m, m^*) transmitted by **A** will be altered by Eve to the pair (m_1, m_2).

Now **B** must act. He simulates the procedure followed by **A**. Thus, he hashes the first half of the received message, namely, m_1 with the kth hash function (used by **A**) and checks whether the result is equal to m_2.

If they are not equal, **B** knows for sure that tampering has occurred, or that transmission errors may have occurred.

However, if they are equal, **B** can be quite certain that the message has not been tampered with *and* that the sender is **A**. After all, who else but **A** has the secret key k (apart from **B**)?

The more secure the hash function the more certain B can be that

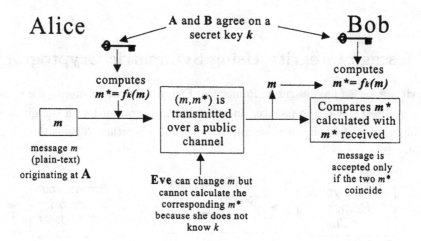

Figure 4.3: Authentication and message integrity check to ensure that nobody tampered with the message from **A** to **B**.

1. The message came from **A** *and*

2. This message from **A** has not been altered.

Thus, in this way, we can use symmetric encryption both for authentication and for message integrity.

4.6   General Public Key Cryptosystems

In this situation one assumes that each entity **B** in a network has a *pair of keys* namely:

1. a public key E_B, known to all members of the network and,

2. a private key D_B, known only to **B**.

It is assumed that the enciphering algorithm E_B can be carried out easily and quickly. The keys E_B (E for enciphering) and D_B (D for deciphering) have the property that, for all messages m, $D_B(E_B(m)) = m$. The keys are created by a public key authority (PKA).

If **A** wants to send a message m to **B** (which will usually be a session key for a future communication), the procedure is as follows.

1. **A** looks up the public key E_B for **B**, or gets it from the PKA.

2. Using E_B, **A** enciphers the message m, yielding the cipher text $E_B(m)$. This cipher text is transmitted to **B**.

3. **B** applies the secret key D_B to the cipher text and recovers the original message since $D_B(E_B(m)) = m$.

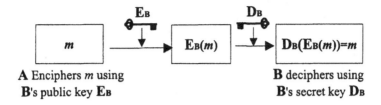

A Enciphers m using
B's public key $\mathbf{E_B}$

B deciphers using
B's secret key $\mathbf{D_B}$

Figure 4.4: Public key encryption principle.

This system using Public Key Cryptography for generating session keys between **A** and **B** will never have perfect (Shannon) security. We must settle instead for **computational security**. This means that the security rests on the assumption that, given the cipher text $E_B(m)$, it is infeasible to calculate m, in a reasonable time at least, and thereby break the system. It has been argued that, from the theoretical point of view, public key algorithms have **no real security at all**! This is because, given sufficiently large amounts of time and computational resources, an adversary is **certain** to break the system.

The main public key algorithms in use, are RSA and ECC (= elliptic curve cryptography which is briefly described below and detailed in Chapter 6). The McEliece public key cryptosystem, using error-correcting codes, is briefly described below and is detailed in Part III.

It has been shown by P. Shor, that the factoring of integers can be carried out easily using a quantum computer. Thus the physical construction of a *quantum computer* would kill off the RSA system but not necessarily the McEliece system or ECC.

First we briefly discuss ECC. We focus on the key-exchange obtained from an **elliptic curve (= non singular cubic curve)**. Details are provided in Chapter 6.

In the Diffie–Hellman key-exchange described below we work with the p numbers, $\{0, 1, 2, \ldots, p-1\}$. In the elliptic curve case we have n points on an elliptic curve **T** (which is a special kind of cubic curve). It turns out that any two points **P** and **Q** on **T** can be geometrically added to obtain a point **R** on **T**. This geometric addition obeys all the usual rules of arithmetic and so we have a set of n numbers corresponding to the n points on **T**. One can then construct a key-exchange, i.e., a common key for **A** and **B** using a method that is analogous to the Diffie–Hellman (DH) key-exchange. The analogous **discrete-log problem** is much more complex than DH. However the overhead for implementing ECC is much larger than for DH.

Recent attacks on ECC were successful in breaking messages encrypted with a 109-bit key. This is consistent with the assumption from Mollin [Mol00] that key-lengths starting at 300 bits for ECC are needed for adequate security.

We turn now to a brief description of the **McEliece cryptosystem**.

This ingenious public key system, discovered by McEliece in 1978, is of interest for a number of reasons. One of these is that, just like the other public key systems, it provides only computational security, so that it can be broken by any adversary with sufficient time and computing resources. However, unlike the RSA system, the McEliece system cannot be directly broken by a quantum computer.

The McEliece system uses the theory of error-correcting codes. Rather than giving full mathematical details, we will explain the basic idea.

Each user B has a public key in the form of a binary code G_1 (which is usually linear). This code is just a list of allowable binary strings called code words, all of the same length.

Mathematically, if the code is linear, it can be described concisely by a binary matrix with k rows and n columns which we also denote by G_1.

Suppose user A wishes to send a secret message m — such as a session key for a future communication with B — to B.

First, A obfuscates the message m by introducing a random error vector R. R is a binary vector having at most $t - 1$ ones in it where t is some predetermined positive integer. Then A encodes m using G_1 and adds R to the result, yielding the cipher text $c = mG_1 + R$.

The public key for B, as mentioned, is G_1. G_1 is obtained from the private key G of B by scrambling G. Adversary Eve, knowing G_1 and the cipher text, cannot recover the message m, but B can, by using G.

G is a very special code: it is designed to correct up to t errors. Thus, G has the property that if a string is not equal to a code word in G but is only off by a few digits, say at most $t - 1$ digits, then G can match that string to the unique code word in G which differs from that string in at most $t - 1$ positions. Then B is able to get rid of the random error part of the cipher text and recover the message m.

One disadvantage of the system is that, offering only computational security, it is again vulnerable to Moore's law[1]. To offer any kind of security the key-length must be very large.

4.7 Electronic Signatures

Some attributes of the usual handwritten signature on a document signed by party A are as follows:

a) Only A may produce this signature.

b) Anyone can check or verify that the signature belongs to A.

---

[1] Moore's law is an observation made by Gordon Moore in 1965, which predicts that processing power will double every 18 months for constant cost.

We want to achieve these objectives electronically. One difference with respect to the handwritten signature is this: The handwritten signature is always the same, regardless of the message. On the other hand, the digital or electronic signature varies with the message. In the digital case it will turn out that nobody can alter the message being signed without everyone involved seeing that the message has been tampered with: forgery is much more difficult!

Recall that any user **A** in a public key network has a public key $e = e_A$ and a private key $d = d_A$. When **A** wants to sign a digital message m and send it to **B**, the following steps are carried out. We assume here that e and d act on the same set. (The El Gamal signature scheme discussed below and other signature schemes are not dependent on this assumption.)

1. **A** enciphers m with her private key d and sends the message m along with the enciphered message to B. Thus, A transmits the pair $(m, d(m))$ to B.

2. **B** applies the public key of **A**, namely e, to the second half of the received pair, and verifies that the result is the first half of the received pair. In other words, **B** verifies that $e(d(m)) = m$. **B** then concludes that the message m has been signed by **A** (see Figure 4.5)

To reduce overhead, what frequently happens is that **A** calculates a hashed version, m_1 say, of the message m to **B**. The hash m_1 is obtained from m by applying a publicly known hash function f to m so that $f(m) = m_1$. Then A transmits the pair $(m, d(m_1))$ to **B**. When **B** receives the cipher text pair, **B** carries out the following steps.

a) **B** applies the public key e that belongs to **A** (obtained from the PKA) to the second half of the received pair.

b) **B** applies the publicly known hash function f to the first part of the message.

c) **B** verifies that the answers from steps a) and b) are equal. In symbols, **B** verifies that $e(d(m_1))$ equals $f(m) = m_1$.

If this checks out, **B** concludes with very high probability that **A** has digitally signed the message m.

One can also construct digital signature schemes based on the Diffie–Hellman algorithm, which will be discussed in Section 4.8.

Digital Signatures Using Symmetric Encryption . The procedure detailed in the section on "authentication and data integrity using symmetric cryptography" can also be considered a digital signature scheme.

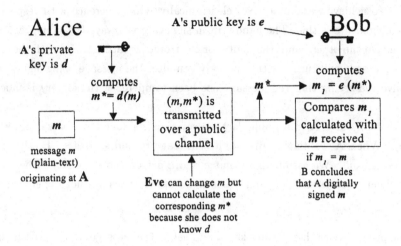

Figure 4.5: Digital Signature scheme using Public Key Cryptography.

Theoretically, using public key algorithms for digital signatures is better for settling disputes between **A** and **B**. In the symmetric case, **A** and **B** both have the same key which might cause difficulties in the case of a dispute between **A** and **B**. It could conceivably be argued that the message that **A** reportedly signed was in fact a message created by **B** who then used the common key to produce the "signature"!

4.8   The Diffie–Hellman Key Exchange

This is one of the most mathematically elegant algorithms in cryptography. Communicating parties **A** and **B** end up generating a common secret key, so there is a connection with symmetric encryption. On the other hand, the method of generating the common key is quite similar to the RSA algorithm and indeed is said to have inspired the RSA algorithm. The security of DH, like the security of RSA, is merely computational.

The DH key exchange may proceed in the following way.

Participants **A**, **B** wish to generate a common secret key. First, a suitable prime p is publicly chosen and then a generator g for p. Here, a **generator** g (which always exists!) has the property that if we take all powers of g from 1 to $p-1$ and calculate their remainders when we divide by p, we obtain all possible numbers $1, 2, 3, \ldots, p - 1$ in some order (see Chapter 19). Recall that $Rem[u, v]$ means the remainder when u is divided by v. *Here, in this section, and in the problems, $v = p$ and $Rem[u, p]$ will be simply denoted by $Rem(u)$.*

Procedure. **A**, **B** choose secret numbers a, b and transmit $u = Rem(g^a)$, $v = Rem(g^b)$ to **B**, **A**, respectively.

B receives u and calculates $Rem(u^b) = K_1$.

A receives v and calculates $Rem(v^a) = K_2$.

Now $K_1 = Rem(g^{ab}) = Rem(g^{ba}) = K_2$ and **A**, **B** are in possession of a common secret key $K_1 = K_2 = K$, since $g^{ab} = g^{ba}$.

An example with a small prime p . $p = 11$, $g = 2$, $a = 4$, $b = 3$. Then:

$$u = Rem(g^a) = Rem(2^4) = 5$$
$$v = Rem(g^b) = Rem(2^3) = 8$$
$$K_1 = Rem(u^b) = Rem(5^3) = 4$$
$$K_2 = Rem(v^a) = Rem(8^4) = 4$$

The common secret key possessed by **A**, **B** is 4. In calculating, we may use the shortcuts that were introduced in Chapter 3.

The security of the Diffie–Hellman (DH) key-exchange rests on the assumption that the DH problem described now cannot be solved in a reasonable amount of time, i.e. it is intractable.

Diffie–Hellman Problem . Given a prime p, $u = Rem(g^a)$ and $v = Rem(g^b)$ find $w = Rem(u^b) = Rem(v^a)$.

A (potentially) more general problem is the discrete log problem.

(We remark that in the DH problem it suffices to consider the cases when $0 \le a < p - 1$ and $0 \le b < p - 1$.)

Discrete Log Problem . Given a prime p and $Rem(g^x)$ where x is one of the numbers $\{0, 1, 2, \ldots, p - 2\}$, find x.

It is called the discrete log problem because $\log(g^x) = x$ when g is chosen as the logarithmic base.

A solution to the discrete log problem (i.e., finding an algorithm for calculating x in a reasonable amount of time) would imply a solution to the Diffie–Hellman problem. The converse statement is not known to be true, although there is experimental evidence pointing in that direction.

We should point out that, for security, one wants p to be well-behaved meaning that $p - 1$ has large factors. The ideal case is when $p - 1 = 2q$ where q itself is prime. For example, take $p = 11$ so $q = 5$. In the ideal case, p has the greatest possible number of different generators (for its size) so that it is easy to find a generator. However, such primes

p, known as **Sophie–Germain** primes, are conjectured to be rare. In any event, only a finite number are known to exist.

Using the Diffie–Hellman idea, it is possible to construct a public key cryptosystem called the **El Gamal Cryptosystem**.

El Gamal Cryptosystem : As before we are given a prime p and a generator g. Each participant **B** has, as a private key, a secret integer b (which can be assumed to lie between 1 and $p - 1$) and a public key $\beta = Rem(g^b)$. Suppose **A** wants to send **B** a secret message m, which is in the form of a positive integer less than p. Let the integer a be the private key for **A**. **A** has, for a public key, $\alpha = Rem(g^a)$. **A** also computes the key $k = Rem((g^b)^a)$, obtained by getting the remainder upon raising β, the public key of **B**, to the power a, and dividing by p (As in the DH key-exchange, **B** can also find k by raising α to the power of b and dividing by p to get the remainder).

Finally, **A** transmits the cipher text $C = Rem(km)$ to **B** (as well as α). From α, **B** can calculate k. Since p is a prime, **B** can calculate k^{-1} where $Rem(kk^{-1}) = 1$. Then **B** calculates $m = Rem(k^{-1}(km))$ This is the El Gamal Cryptosystem.

Remark 4.1 *Instead of taking the cipher text $C = Rem(km)$, we could also choose the cipher text $C_0 = f_k(m)$ where f_k is any keyed symmetric algorithm such as DES or AES.*

The RSA digital signature protocol is relatively easy because e and d are both defined on the same set, namely $\{0, 1, 2, \ldots, m - 1\}$. For a more complicated digital signature example, we present the El Gamal Digital Signature Scheme.

The El Gamal Digital Signature Scheme : **A** wants to send a signed message M to **B**. **A** begins with a large prime p and a generator g (a primitive root) for that prime. Then **A** chooses an integer a such that $1 \leq a \leq p - 2$. The public key of **A** is (p, g, g^a) and the private key of **A** is a. Since p and g are publicly known, it is important that it be infeasible to calculate the solution x to $g^x \equiv g^a \pmod{p}$. This requires that p be large. When signing a message M, **A** chooses an integer k relatively prime to $p - 1$ such that $1 \leq k \leq p - 2$. Then **A** transmits a signature in the form of the pair $(Rem(g^k), s)$ where

$$s \equiv k^{-1}(h - ag^k) \pmod{p - 1} \tag{4.6}$$

$$(\text{In other words, } Rem[k^{-1}(h - ag^k), p - 1] = s)$$

and h, the hash of the message M, is some integer such that $1 \leq h \leq p - 2$. (The hash function maps strings of 0s and 1s onto the integer h in the proper range).

The signature is verified in the following way: **B** checks the conditions

1. The left integer in the signature pair, namely $Rem(g^k)$, lies in the interval $[1, p - 1]$.

2. The congruence $g^{ag^k} g^{ks} \equiv g^h \bmod p$ is satisfied. In other words, we check that $Rem[g^{ag^k} g^{ks}] = Rem[g^h]$.

If both conditions hold, then **B** accepts the signature as coming from **A**. The reason for this is as follows. Substituting Equation 4.6 into the second condition, we get

$$g^{ag^k} g^{k(k^{-1}(h-ag^k))} \equiv g^h \pmod{p} \tag{4.7}$$

This means that if **A** computed s, then condition 2 will be satisfied. Conversely, if condition 2 is satisfied, then (since g is a primitive root of p) the exponents must give the same remainder on division by $p - 1$. Hence,

$$ag^k + ks \equiv h \pmod{p-1} \tag{4.8}$$

i.e., $Rem[ag^k + ks, p - 1] = Rem[h, p - 1]$

which is a restatement of (4.6). So the computation carried out by **A** is the only way to satisfy the second condition.

For a simple example, let $p = 11$, $g = 2$, $a = 4$ and $k = 7$. Then **A** has the private key $a = 4$ and the public key $(p, g, g^a) = (11, 2, 5)$ since $g^a = 2^4 = 16$ and $Rem(16) = 5$. Let the hash of the message M be $h = 1$. $Rem[k^{-1}(h - ag^k), p - 1] = Rem[3(1 - (4)(7)), 10] = Rem[9, 10]$ Then $s = 9$. Thus, **A**'s signature is $(7, 9)$. Condition 2 is then satisfied for this signature.

4.9 Quantum Encryption

Quantum cryptography was introduced by Wiener and developed also by Bennett, Bessete, Brassard, Salvail, Smolin and others. It is based on a plausible (if unproved) special case of the uncertainty principle of quantum physics having to do with the measurement of photon polarization. The work attempted to generate a secret key by using principles of quantum mechanics. We refer to Mollin [Mol00] and Nielsen and Chung [NC00].

Roughly, the idea is this: As photons travel in a plane, they vibrate in some direction, up and down, left and right, etc. When a large group of photons vibrate in the same direction, we say they are polarized.

If a pulse is polarized in one direction, say horizontal-vertical, and we measure it in that direction, the correct count for the number of photons is obtained.

On the other hand, if we measure with a polarization detector whose axis is on the diagonal (at an angle of 45° to the vertical) we get a random result for the count.

Here is the protocol to generate a secret key.

1. **A** sends **B** a string of photon signals. Each photon is polarized in one of four directions

namely left-diagonal, right-diagonal, horizontal and vertical. We denote the horizontal or vertical polarizations by the term "rectilinear". Denote the left or right diagonal polarizations by the term "diagonal".

2. **B** has a polarization detector with two settings, rectilinear or diagonal. In effect, both settings cannot be used at the same time by Heisenberg's uncertainty principle. Thus **B** successively chooses one setting or the other at random, for measurement purposes.

3. **B** tells **A**, over an unsecure channel, which of the two settings he has chosen in every instance.

4. **A** tells **B** which settings were correct, i.e. the polarization method that **A** used for each photon (rectilinear or diagonal) but not how the photon was polarized. For example, if the rectilinear polarization is used, **A** does not reveal whether a horizontal or vertical direction was favored. In effect, **A** does not announce, over the unsecured channel, which of 0 or 1 has been transmitted. However, **B** will have measured this accurately if **B** has chosen the rectilinear setting on his polarization detector.

5. **A** and **B** only keep those polarizations that were correctly measured. For rectilinear polarization assign 1 for vertical, 0 for horizontal. For diagonal, assign 1 for left-diagonal and 0 for right-diagonal.

This generates a string of bits corresponding to the correct measurements by **B**. This bit string in principle is random and can be used as a one-time pad. Moreover, in theory, this one-time pad is absolutely unbreakable!

Note that, on the average, **B** will have correctly set the detector for about 50% of the measurements. So, in theory, if **A** transmits n photons, **A** and **B** will generate a secret key of length $\frac{n}{2}$ on the average.

However, this is not utopia. The most serious limitation of this scheme is the difficulty of transmitting photons accurately at a considerable distance without a loss of information due to the interaction between the photons and the medium.

We have not brought the eavesdropper Eve positioned between **A** and **B** into account either. Eve can listen in and her eavesdropping can introduce errors in the pulse. Thus **A** and **B** may end up with possibly different bit strings.

Indeed Brassard, Bennett and other colleagues worked on mathematically repairing the damage using the beginnings of techniques in **Privacy Amplification** and **Bit Reconciliation**. This initially seemed to work.

However, the final blow came when it was discovered that the whole protocol was flawed. As Brassard himself later admitted the devices that were used to polarize the photons made noticeably different noises depending on the type of polarization chosen, thus security was compromised. Nonetheless the experiment became very well known and thus made its way into the literature. Furthermore, this ground-breaking attempt in the quantum domain, even if impractical, has paved the way for much subsequent research.

The protocol is such that if Eve is eavesdropping, this can be detected with high probability by **A** and **B**. However, in the case of eavesdropping the usual protocol calls for **A** and **B** to start all over again! We will say more on this in Part III.

4.10   Key Management and Kerberos

So far we have discussed several procedures using symmetric encryption. For this to work, the parties **A** and **B** who are communicating with each other use a particular encryption algorithm that is specified by a secret key, known only to **A** and **B**. Frequently the key is used for just one particular communication session; it is time stamped and is called a session key.

However, we now come to the problem: *how does one get the key to **A**, **B** securely?* As mentioned earlier, this is a fundamental problem. In Chapter 3, we referred to it as the Catch-22 of cryptography.

Now we briefly describe various methods that can be used to provide **A** and **B** with a secret key k.

(A) **Off Line**. In this case session keys are exchanged beforehand through a secure private channel such as a courier. This method is still used nowadays for certain financial transactions. Also, various encryption companies utilize this method by supplying sets of session keys to their clients at regular intervals.

(B) **Central Server (Kerberos)**. Each party **A** in a given network has established a secret authentication key T_A with the trusted central server T (for Trust). These passwords are often generated off-line. The basic procedure is very straight-forward.

1. **A** contacts **T** and requests a session key to communicate with **B**.

2. **T** generates a random session key k (usually a binary string), encrypts k (which is now a message from **T** to **A**) with the key T_A — using a symmetric encryption algorithm that **A**, **T** have previously agreed upon — and transmits the encrypted message to **A** along with a time-stamp.

3. **A** decrypts the message from **T**, obtaining the session key k.

4. **T** encrypts k (treated as a message) with the key T_B and transmits this encrypted message to **B** along with a time-stamp. **T** informs **B** that **A** wishes to communicate with **B**.

5. **B** decrypts the message from **T**, obtaining the session key k.

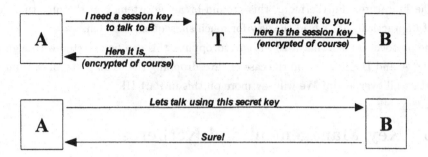

Figure 4.6: The basic Kerberos scheme for trusted server authentication.

6. **A** and **B** communicate using the session key k.

Note that the role of the trusted central server in Kerberos is logically very similar to the role of the Certificate Authorities in the Public Key Infrastructure (PKI).

(C) **Pseudo-Random Keys**. This is usually accomplished by means of a linear feed-back shift register (LFSR), which is easily implemented in hardware. Using a shared secret, **A** and **B** generated a common pseudo-random sequence. A common (non-repeating) portion of this pseudo-random sequence can be used by **A** and **B** as a session key for symmetric encryption or as a one-time pad.

The security of the system can be improved by ensuring that the period of the sequence is very long. Also, instead of using a linear function f, one may use a nonlinear shift register.

More general than LFSR's from the mathematical point of view are the so-called **Cellular Automata**. Introduced in the 1950's by Ulam and von Neumann, the theory of Cellular Automata provides a framework for the understanding of computing processes in digital machines.

Several uses of Cellular Automata in cryptography have been proposed.

(D) **Quantum Encryption**. Using properties of quantum physics, such as the Heisenberg uncertainty principle, one can construct a session key as described earlier.

(E) **Provable Information-Theory Security by Public Discussion**.

This is a very new technology. One can achieve unconditional security by using classical physics and information theory. Such systems are only now coming on the market. The methods also apply to quantum encryption. Further details are in Part III.

(F) **Public Key Cryptosystems**. This has already been described in Section 4.7.

(G) **The Diffie–Hellman Key-Exchange**. This has already been described in Section 4.8.

4.11 A Summary of DES

Symmetric key algorithms use the same key to encrypt and decrypt data. Given the key, both operations are easily performed. As an example of a modern symmetric key algorithm, we briefly describe the *Data Encryption Standard* (DES). The Data Encryption Algorithm[2] (DEA) was one of the most widely used encryption algorithms. It was designed to meet a series of criteria set by the National Bureau of Standards (NBS, now known as NIST for National Institute of Standards and Technology) in 1973. It was developed by a group of researchers at IBM and published in 1975. DES is a *block cipher* that takes 64-bit blocks of message text and outputs them as 64-bit cipher text blocks using a 56 bit key. In general terms, the algorithm works on each block over several iterations (or rounds) by splitting, after a permutation of the bits, the input string into two 32 bits strings (Left and Right). The right string is left untouched and switched to the left side, while the left string is modified by the following procedure:

1. Expansion permutation — This permutes and expands the 32 initial bits to 48 bits. The process is designed in such a way as to maximize the dependence of every resulting bit on each bit of the input and every bit of the key. This process introduces diffusion and, in the technical literature, is usually referred to as the **avalanche effect**.

2. The resulting string is XOR'ed with the compressed key. The compressed key is obtained by choosing subsets of the key in a cyclical fashion that is governed by the "round number" and a fixed table. This function gives a 48 bit key from the 56 bit original key.

3. S-box permutation. The 48 bits resulting from the previous operation are substituted according to substitution boxes (S-boxes). There are 8 S-boxes (or tables) that replace a 6 bit input with a 4 bit output (confusion).

4. P-box permutation. This performs a fixed permutation on the 32 bits resulting from the previous step (yet more diffusion).

The modified right string is then XOR'ed with the original left string and the resulting string replaces the original right string.

The process is repeated 16 times (16 rounds) and the resultant cipher text is permuted by using the inverse of the initial permutation. Note that the key applied to each round is different; these are modified in a recursive way.

---

[2] DEA by ANSI and DEA-1 by the ISO.

To decrypt, the same algorithm is applied to the cipher text by using the modified keys in reverse order, i.e., if the modified keys for encryption are $K_1, K_2, \ldots, K_{16}$, the key for decryption will be $K_{16}, K_{15}, \ldots, K_1$.

For further details on DES and AES we refer to Chapter 5.

4.12   Problems

1. Suppose **A**, **B** are in possession of the secret key $K = \{1010110\}$ and **A** has enciphered the message M (using K for a Vernam cipher). If **B** receives the cipher text C where $C = \{0110111\}$, what is M?

2. If, in Question 1, an eavesdropper **E** intercepts the communication and listens in to C, can **E** then narrow down possibilities for M?

3. Let A be a binary matrix with k rows and n columns. Let M be a message in the form of a binary string of length n. Show how to use A to construct a hash function such that the hash of M has length k.

4. Let M be the message 1101 and let A be the matrix $\begin{pmatrix} 1 & 0 & 1 & 1 \\ 0 & 1 & 0 & 0 \\ 1 & 1 & 0 & 1 \end{pmatrix}$.   Use A to construct the hash of M as suggested in the previous problem.

5. **B** receives the following message $m = (1101, 011)$, supposedly sent by his friend. In this case, the right part of the received message is an enciphered hash of the left part, using the matrix A given in the previous problem. The secret key $K = 001$ used to encipher the hash is a Vernam cipher key known only to **B** and his friend. Should **B** accept the message as a valid communication from his friend?

6. Find all generators associated with the prime 11.

7. Find all generators of the prime 7.

8. Find $Rem(5^{51})$ where $p = 97$.

9. The following problem is an easy version of the discrete log problem: Find x if $p = 11$ and $Rem(2^x) = 9$.

4.13   Solutions

1. $M = \{1100001\}$. In general, $K+C = K+(M+K) = K+(K+M) = (K+K)+M = M$. So, to get M, **B** simply adds K to C, where all operations are in binary.

2. No. Given C then any message M can be made to fit with C: we just need to solve for K. We want $M + K = C$: this gives $K = C + M$ and with this choice of K it follows that M fits with C. In other words, enciphering M with K gives C. In this case as far as an eavesdropper is concerned the number of possibilities for M is the number of binary strings of length 7, which is $2^7 = 128$, and the eavesdropper can only guess what M might be.

3. Make M into a column matrix and compute the matrix product AM. Suppose for example that $m = 4$, $k = 3$ and

$$A = \begin{pmatrix} 1 & 1 & 1 & 1 \\ 0 & 1 & 1 & 1 \\ 1 & 0 & 0 & 1 \end{pmatrix} \tag{4.9}$$

and $M = (1010)$. Then the column matrix form of M is .

$$M = \begin{pmatrix} 1 \\ 0 \\ 1 \\ 0 \end{pmatrix} \tag{4.10}$$

and the matrix product

$$AM = \begin{pmatrix} 1 & 1 & 1 & 1 \\ 0 & 1 & 1 & 1 \\ 1 & 0 & 0 & 1 \end{pmatrix} \begin{pmatrix} 1 \\ 0 \\ 1 \\ 0 \end{pmatrix} = \begin{pmatrix} 1 & + & 0 & + & 1 & + & 0 \\ 0 & + & 0 & + & 1 & + & 0 \\ 1 & + & 0 & + & 0 & + & 0 \end{pmatrix} = \begin{pmatrix} 0 \\ 1 \\ 1 \end{pmatrix} \tag{4.11}$$

Therefore, the hash value of $M = (1010)$ is (011).

4. The matrix product AM (where M has been turned into a column matrix) gives the hash: (011).

5. **B** adds K to the right half of the received message to get (010). He also computes the hash of the left part of the message by multiplying by the matrix A on the left. The matrix product is the column matrix containing (011). Since (011) is not the same as (010), **B** should be doubtful about the source of the communication. (An error in transmission could account for the discrepancy. In cases where the message and key are much larger, this is much less likely to be a possible explanation.)

6. The possible remainders when a positive integer is divided by 11 are 0, 1, 2, 3, 4, 5, 6, 7, 8, 9 and 10. These integers are also known as the integers mod 11. Now 0 is not

a generator since 0 raised to any power just gives 0. Similarly, 1 is not a generator. Let us try 2. So we take successive powers of 2 and get the remainder when we divide by 11. This gives us $2, 4, 8, 5, 10, 9, 7, 3, 6, 1$. Thus, 2 is a generator. Note that 9 is not a generator (even though 9 is the additive inverse of 2) since when we get successive powers of 9 and get the remainder upon division by 11 we just get $\{9, 4, 3, 5, 1\}$. The other generators, apart from 2, are 6, 7 and 8.

7. 3, 5.

8. 69.

9. $x = 6$.

Chapter 5

DES, AES and Operating Modes

Goals, Discussion We present the basic ideas for two fundamental modern block encryption symmetric algorithms, namely DES, (and Triple DES) and AES. These algorithms make fundamental use of Shannon's principles of **confusion** and **diffusion**. We also discuss the three basic methods of applying the DES coding to a long text. These methods can be used for any block-encryption symmetric algorithms and are completely general.

New, Noteworthy We present several problems on the algorithms, and we discuss how many Unix implementations rely on DES for password security.

5.1 The Data Encryption Standard Code

The DES code was constructed to work well with computer equipment of the late 1970s and therefore restricts its operations to a maximum of 64 bits. It uses both confusion and diffusion to achieve the encryption. The large number of rounds, 16, was thought to provide adequate protection against possible attacks. However, it can now be attacked by linear cryptanalysis; apparently this weakness was not known to the designers. Since the mid-1990s a number of other new methods for attacking codes such as DES have been developed.

DES was initially developed by IBM to protect computer data. In 1974, IBM submitted its original version of DES to NBS (the American National Bureau of Standards) for adoption as a publicly available coding system. According to Alan Konheim, an IBM employee, the NSA (National Security Agency) changed the S boxes submitted by IBM. This raised suspicion that NSA had built a "trapdoor" into the code so that it could easily decode

DES-encrypted messages. This possibility was investigated by the US Senate Committee on Intelligence, whose summary claimed that DES had no known mathematical or statistical weaknesses. Nevertheless, there still remained some doubt because the committee's report was never made public. However, there are no known trapdoors and it seems likely that they would have been discovered or leaked by now. In 1981, the American National Standards Institute approved DES as a private-sector standard, and it has been widely used since then, especially for banking.

Using the Key for the Code

The key for a DES code consists of 64 bits, but every eighth bit is chosen to satisfy a parity check condition on the bytes of the key and is ignored when using the key. This leaves a key of 56 bits. A 48-bit subkey is generated from the 56-bit key for each of the 16 rounds by carrying out the following steps:

1. The 56-bit key is divided into two 28-bit halves. The left half is the first 28 bits of the key, and the right half is the last 28 bits of the key.

2. To determine the subkey K_j for the jth round, each half is shifted circularly left by v_j bits, where $v_j = 2$ except for $j = 1, 2, 9$, or 16, in which case $v_j = 1$. These shifts are cumulative. This means that, for example, for the third subkey K_3, the bits of each half of the original key have been circularly shifted $4 = 1 + 1 + 2$ bits to the left.

3. The compression permutation is then used to select 48 of the 56 bits.

The Compression Permutation

If the 56 bits (after shifting) are $k_1 k_2 \ldots k_{56}$, then the output of the compression permutation is the following 48 bits.

$$k_{14} k_{17} k_{11} k_{24} k_1 k_5 k_3 k_{28} k_{15} k_6 k_{21} k_{10}$$
$$k_{23} k_{19} k_{12} k_4 k_{26} k_8 k_{16} k_7 k_{27} k_{20} k_{13} k_2$$
$$k_{41} k_{52} k_{31} k_{37} k_{47} k_{55} k_{30} k_{40} k_{51} k_{45} k_{33} k_{48}$$
$$k_{44} k_{49} k_{39} k_{56} k_{34} k_{53} k_{46} k_{42} k_{50} k_{36} k_{29} k_{32}$$

Because of the shifts, a different set of 48 bits will be used in each subkey. Each of the original 56 bits is used in approximately 14 of the 16 subkeys, but they are not used exactly the same number of times. This inherent randomness serves to increase the diffusion of the encryption.

Using the S-Boxes for the Code

There are eight S-Boxes for the DES code; each is a 4×16 matrix. These are given below. The purpose of these S-boxes is to make the code nonlinear; this is the major reason why it is difficult to decode text encrypted with the DES algorithm.

$S_1 =$

	0	1	2	3	4	5	6	7	8	9	10	11	12	13	14	15
0	14	4	13	1	2	15	11	8	3	10	6	12	5	9	0	7
1	0	15	7	4	14	2	13	1	10	6	12	11	9	5	3	8
2	4	1	14	8	13	6	2	11	15	12	9	7	3	10	5	0
3	15	12	8	2	4	9	1	7	5	11	3	14	10	0	6	13

$S_2 =$

	0	1	2	3	4	5	6	7	8	9	10	11	12	13	14	15
0	15	1	8	14	6	11	3	4	9	7	2	13	12	0	5	10
1	3	13	4	7	15	2	8	14	12	0	1	10	6	9	11	5
2	0	14	7	11	10	4	13	1	5	8	12	6	9	3	2	15
3	13	8	10	1	3	15	4	2	11	6	7	12	0	5	14	9

$S_3 =$

	0	1	2	3	4	5	6	7	8	9	10	11	12	13	14	15
0	10	0	9	14	6	3	15	5	1	13	12	7	11	4	2	8
1	13	7	0	9	3	4	6	10	2	8	5	14	12	11	15	1
2	13	6	4	9	8	15	3	0	11	1	2	12	5	10	14	7
3	1	10	13	0	6	9	8	7	4	15	14	3	11	5	2	12

$S_4 =$

	0	1	2	3	4	5	6	7	8	9	10	11	12	13	14	15
0	7	13	14	3	0	6	9	10	1	2	8	5	11	12	4	15
1	13	8	11	5	6	15	0	3	4	7	2	12	1	10	14	9
2	10	6	9	0	12	11	7	13	15	1	3	14	5	2	8	4
3	3	15	0	6	10	1	13	8	9	4	5	11	12	7	2	14

$S_5 =$

	0	1	2	3	4	5	6	7	8	9	10	11	12	13	14	15
0	2	12	4	1	7	10	11	6	8	5	3	15	13	0	14	9
1	14	11	2	12	4	7	13	1	5	0	15	10	3	9	8	6
2	4	2	1	11	10	13	7	8	15	9	12	5	6	3	0	14
3	11	8	12	7	1	14	2	13	6	15	0	9	10	4	5	3

$S_6 =$

	0	1	2	3	4	5	6	7	8	9	10	11	12	13	14	15
0	12	1	10	15	9	2	6	8	0	13	3	4	14	7	5	11
1	10	15	4	2	7	12	9	5	6	1	13	14	0	11	3	8
2	9	14	15	5	2	8	12	3	7	0	4	10	1	13	11	6
3	4	3	2	12	9	5	15	10	11	14	1	7	6	0	8	13

$S_7 =$

	0	1	2	3	4	5	6	7	8	9	10	11	12	13	14	15
0	4	11	2	14	15	0	8	13	3	12	9	7	5	10	6	1
1	13	0	11	7	4	9	1	10	14	3	5	12	2	15	8	6
2	1	4	11	13	12	3	7	14	10	15	6	8	0	5	9	2
3	6	11	13	8	1	4	10	7	9	5	0	15	14	2	3	12

$S_8 =$

	0	1	2	3	4	5	6	7	8	9	10	11	12	13	14	15
0	13	2	8	4	6	15	11	1	10	9	3	14	5	0	12	7
1	1	15	13	8	10	3	7	4	12	5	6	11	0	14	9	2
2	7	11	4	1	9	12	14	2	0	6	10	13	15	3	5	8
3	2	1	14	7	4	10	8	13	15	12	9	0	3	5	6	11

The S-Boxes are used to change an input block of 48 bits $(c_1 c_2 \ldots c_{48})$ into an output block of 32 bits $(d_1 d_2 \ldots d_{32})$. The input bits are broken up into 8 sub-blocks of 6 bits each, and the output of 32 bits is broken up into 8 sub-blocks of 4 bits each. Thus the second sub-block of the input would consist of the bits $(c_7 c_8 \ldots c_{12})$. The first and last bits in the jth sub-block determine the row used in the S-Box S_j and the middle four bits in the jth sub-block determine the column used in S_j. For example, if $(c_7 c_8 \ldots c_{12}) = 100110$, then the row used in S_2 is 10 in binary or row 2 and the column used in S_2 is 0011 in binary or column 3. The number given in row 2 and column 3 of S_2 is 11, which in binary, has the expression 1011. Hence, the bits $(d_5 d_6 d_7 d_8)$ of the output block are 1011. For another example, if $(c_{25} c_{26} \ldots c_{30}) = 010110$, then the row used in S_5 is 00 in binary or row 0 and the column used in S_5 is 1011 in binary or column 11 . The number given in the row 0 and column 11 of S_5 is 15, which in binary, has the expression 1111. Therefore, the bits $(d_{17} d_{18} d_{19} d_{20})$ of the output block are 1111. So each of the eight S-boxes determines one sub-block of the output block.

The Initial Permutation

The input consists of a block of 64 bits $a_1 a_2 \ldots a_{64}$. These bits are permuted so that their new order is

$$a_{58} a_{50} a_{42} a_{34} a_{26} a_{18} a_{10} a_2 a_{60} a_{52} a_{44} a_{36} a_{28} a_{20} a_{12} a_4$$

$$a_{62} a_{54} a_{46} a_{38} a_{30} a_{22} a_{14} a_6 a_{64} a_{56} a_{48} a_{40} a_{32} a_{24} a_{16} a_8$$

$$a_{57} a_{49} a_{41} a_{33} a_{25} a_{17} a_9 a_1 a_{59} a_{51} a_{43} a_{35} a_{27} a_{19} a_{11} a_3$$

$$a_{61} a_{53} a_{45} a_{37} a_{29} a_{21} a_{13} a_5 a_{63} a_{55} a_{47} a_{39} a_{31} a_{23} a_{15} a_7$$

This initial permutation has no effect on the security of DES. The historical reason for this permutation seems to be related to the process of loading plain text into a DES chip in byte-sized chunks.

The Final Permutation

The input consists of a block of 64 bits $a_1 a_2 \ldots a_{64}$. These bits are permuted so that their new order is

$$a_{40} a_8 a_{48} a_{16} a_{56} a_{24} a_{64} a_{32} a_{39} a_7 a_{47} a_{15} a_{55} a_{23} a_{63} a_{31}$$

$$a_{38} a_6 a_{46} a_{14} a_{54} a_{22} a_{62} a_{30} a_{37} a_5 a_{45} a_{13} a_{53} a_{21} a_{61} a_{29}$$

$$a_{36} a_4 a_{44} a_{12} a_{52} a_{20} a_{60} a_{28} a_{35} a_3 a_{43} a_{11} a_{51} a_{19} a_{59} a_{27}$$

$$a_{34} a_2 a_{42} a_{10} a_{50} a_{18} a_{58} a_{26} a_{33} a_1 a_{41} a_9 a_{49} a_{17} a_{57} a_{25}$$

This final permutation is the inverse of the initial permutation, and hence could have been computed mathematically from the initial permutation. Its main purpose is to help make decryption essentially the same process as encryption, while ensuring that the data can be efficiently loaded onto the appropriate chip.

The Expansion Permutation

The expansion permutation uses a 32-bit block as input and produces a 48-bit block as output. If the input block is $b_1 b_2 \ldots b_{32}$, then the output block is given by

$$b_{32} b_1 b_2 b_3 b_4 b_5 b_4 b_5 b_6 b_7 b_8 b_9$$
$$b_8 b_9 b_{10} b_{11} b_{12} b_{13} b_{12} b_{13} b_{14} b_{15} b_{10} b_{17}$$
$$b_{16} b_{17} b_{18} b_{19} b_{20} b_{21} b_{20} b_{21} b_{22} b_{23} b_{24} b_{25}$$
$$b_{24} b_{25} b_{26} b_{27} b_{28} b_{29} b_{28} b_{29} b_{30} b_{31} b_{32} b_1$$

The main purpose of the expansion permutation is to allow some bits to affect two different substitutions when using the S-boxes. This increases the rate of reaching the condition that every bit of the coded text depends on every bit of the input text and is called an **avalanche effect**.

One Round of DES

At the beginning of round j the coded text block is expressed as $L_{j-1} R_{j-1}$, where both the left and the right halves consist of 32 bits. At the end of the round, the new left half is $L_j = R_{j-1}$, and the new right half, R_j, is computed in the following way.

1. R_{j-1} is used as input for the expansion permutation, producing an output E_j of 48 bits.

2. The subkey for the jth round K_j is added bitwise to E_j. This is denoted by $E_j \oplus K_j$ where $\oplus$ is the *xor* operation on bits.

3. The resulting 48 bits are changed into the 32-bit output R_j by using the eight S-boxes for the code.

Overview of DES Encryption

A block of 64 bits of text is the basic input to the algorithm. The following steps are applied.

1. The initial permutation is applied to the 64 bits.

2. After the initial permutation, the block is broken into the left half L_0 and the right half R_0, which are both 32 bits long. By using the subkeys $K_1, K_2, \ldots, K_{16}$ successively over 16 rounds, we reach the output $L_{16} R_{16}$.

3. The final permutation is applied to the output $R_{16}L_{16}$. Note that the order of the sub-blocks has been changed here to help make decryption simpler.

Decryption of DES-Coded Text

Decryption is essentially the same as encryption. The steps of the DES algorithm have been carefully chosen so that this is possible. The encryption subkeys must be used in the opposite order in the decryption process, namely, $K_{16}, K_{15}, \ldots, K_2, K_1$. The subkeys can be generated in the opposite order by starting with the key for the code and shifting the bits of each half circularly to the right by the correct number of bits. In the first case, no shift is needed to obtain K_{16} because the bits of each half of the key have been circularly shifted 28 bits to the left when K_{16} is computed. The subkey K_j is obtained from the subkey K_{j+1} by a circular shift to the right of w_j bits where $w_j = 2$ except when $j = 15, 8$, or 1, in which cases $w_j = 1$. The details of both subkey creation and the use of the S-boxes and permutations are the same as for encryption.

Modes of Operation

There are several ways of applying the DES coding to a long text. The simplest is called the **Electronic Code Book (ECB) mode** and consists of breaking the long text into blocks of 64 bits that are then encoded, block-by-block, using the same key for each block. Therefore, if the plain text blocks are $B_1, B_2, \ldots, B_m$, then the coded text will be $C_1, C_2, \ldots, C_m$, where C_j is the DES code for B_j. The problem with this mode is that some plain text blocks will be repeats of other plain text blocks. If a large amount of the coded text is analyzed, it may be possible to determine the meaning of some coded blocks (from context) without having any knowledge of the key itself. A dictionary for the coded text could then be gradually built up and used to break the cipher.

A more secure mode of operation is called the **Cipher Feedback mode (CFB)**. In this case, the plain text is broken up into blocks of 8 bits (or one byte), $A_1, A_2, \ldots, A_m$. Some predetermined 64 bits of plain text, B_1, are encoded with DES to produce the code block F_1. The next code block F_2 is obtained by applying DES to the last (rightmost) 56 bits of B_1 followed by the 8 bits $A_1 \oplus G_1$ where G_1 is the first (leftmost) eight bits of F_1. Now F_3 is obtained by applying DES to the last 48 bits of B_1 followed by the 8 bits $A_1 \oplus G_1$, followed by the 8 bits $A_2 \oplus G_2$, where G_2 is the first eight bits of F_2. In general, G_j is the first eight bits of the jth coded block F_j. Continuing in this manner, we eventually compute F_{m+1} by applying DES to the block that consists of $A_{m-7} \oplus G_{m-7}$ followed by $A_{m-6} \oplus G_{m-6}$, followed by ... followed by $A_m \oplus G_m$. In this way every block of 8 bits is eventually coded into one of the output text blocks. One of the advantages of this approach (over some other modes of operation that also involve feedback) is that errors in transmission only affect part of the decoded message. For example, if the first code block, F_1, contains an error, this will affect the interpretation of only the first nine bytes $A_1, \ldots, A_9$. The remaining bytes will be correctly decoded provided that there are no other errors in transmission. In other words, each error will only propogate through 9 bytes.

Another mode of operation is called the **Output Feedback mode**. Some predetermined 64 bits of plain text, F_0, are encoded with DES to produce the code block F_1. This is repeated indefinitely so that F_j is the DES code for F_{j-1}. If the plain text blocks of 64 bits are $B_1, B_2, \ldots, B_m$, then the coded text will be $C_1, C_2, \ldots, C_m$, where C_j is $B_j \oplus F_j$.

5.2 Triple DES

There are some other modifications of standard DES that have been tried to improve upon the security of DES. **Double DES** is the procedure that first encrypts the text with one key K_1 and then encrypts the coded text with a different key K_2 to produce the final coded text. However, there is a "meet in the middle" attack (see Chapter 7) that makes this only slightly more secure than standard DES. This attack does not work for **Triple DES**. Triple DES has security approximately equal to the security offered by the the original DES algorithm with a key of 112 bits. A variant of Triple DES that is sometimes used is the following. The 64-bit code block B is DES-encrypted with the key K_1. The coded text C is then DES-decrypted with a different key K_2 to produce the block D. Then D is DES-encoded with a third key K_3 to produce the final coded text block E. (Note that if $K_1 = K_2 = K_3$, then this reduces to standard DES.) Of course, Triple DES procedures require triple the computing time to encode or decode text.

Security of DES-Encoded Text

When DES was introduced in the 1970s, computers were much slower than they are today. This meant two things: It took longer to encode text with complicated algorithms, and it took longer to break computer-coded text. DES was a reasonable code when introduced given the sophistication of computers at the time. The first IBM PC (a 4.7-MHz 8088) could encode or decode approximately 370 DES blocks per second, whereas a 66-MHz 80486 could encode approximately 43,000 DES blocks per second. To put things in perspective, a Pentium 4 processor operating at over 2 GHz can encode over one million DES blocks per second. Clearly, the growth in computer power over the last 25 years has been phenomenal.

By the 1990s, computers were sophisticated enough to attack DES using a brute-force attack, a seemingly impossible task only a few decades earlier. In 1997 the RSA Data Security company challenged the world to crack a DES-coded message and offered \$10,000 to the first person who could correctly decode the message. The prize was claimed a mere five months later by Rocke Verser, a programmer and software consultant who utilized the combined computing power of thousands of computers to accomplish this feat. In 1998 a similar challenge was issued, and this time the prize was claimed in only 39 days. What a difference a year makes in computing! It should be clear that in today's world, DES offers only a limited degree of security. Unfortunately, not everyone has adopted a more secure code, such as AES, discussed in the next section.

5.3   DES and Unix

Several implementations of the Unix operating system use DES as a method of storing passwords. It is undesirable to store plain text passwords on the hard drive because anyone with privileged access to the file system (or physical access to the drive) could read the passwords.

So, when a user establishes a password, the computer encodes the password into bits using ASCII (see Appendix). If the password is less than 8 characters, the operating system pads it with "null" characters to ensure it is of appropriate length. Then, because the first bit of any standard ASCII code is 0, the password program discards the first bit for each character and the remaining block consists of 56 bits, just enough for a DES key.

Then the password program uses this key for an encoding very similar to DES to encode a string of all 0s. The designers of the password program changed the algorithm slightly to prevent hardware implementations of DES crackers from obtaining the set of passwords. It then stores this encrypted version of the password on the hard drive. Whenever a user attempts to log in, the computer repeats this process with the password supplied by the user, and if the final result matches what is stored on the hard drive, the user is granted access.

One other important detail about the Unix password method is that there are 4096 different ways to encipher the same password, where each encoding is a slight variation of the original DES encoding method. Before the password is encoded, the computer chooses one of these methods at random, enciphers the password, and records which method it used (called the "**salt**"), so that the next time the user attempts to log in, the password program knows which method to use again. This is done to discourage people from making large tables for all possible passwords, and then doing a simple lookup when they get access to an adversary's encrypted password.

Because of the advanced computer power that is now available, this method is slowly being phased out in favor of newer hashing techniques like MD5 or SHA1. Clearly, the cryptographic techniques must stay ahead of brute-force attack in order to maintain security.

5.4   The Advanced Encryption Standard Code

The Advanced Encryption Standard (AES) code (also known as the Rijndael Code after its co-inventors Joan Daemen and Vincent Rijmen) is based on properties of the finite field of 256 elements (refer to Chapter 19 for mathematical details). Each byte can be associated with a unique element of this field. Because field elements can be multiplied and added, this association makes it possible to add and multiply bytes. Moreover, in a field, each byte has a multiplicative inverse. These algebraic properties make it possible to encode entire bytes at a time by using matrix operations that are nonlinear. The AES code was designed for use with keys of length 128, 192, or 256 bits. For simplicity, the discussion here is restricted to the case in which the key has a length of 128 bits. In this case, every round starts with the input of 16 bytes (or 128 bits) and has an output of 16 bytes (or 128 bits). There are ten rounds to promote diffusion of the bits, and each round has its own round key derived from the original round key.

The Field of 256 Elements

Definition 5.1 *A set S of algebraic elements together with a binary operation $\oplus$ defined on the elements of S is* **associative** *if $(a \oplus b) \oplus c = a \oplus (b \oplus c)$ for all elements a, b, and c in the set S.*

Definition 5.2 *A set S of algebraic elements together with a binary operation $\oplus$ defined on the elements of S is* **commutative** *if $a \oplus b = b \oplus a$ for all elements a and b in the set S.*

Recall that matrix multiplication is not commutative. The fact that $(9-5)-2 = 4-2 = 2$ whereas $9 - (5 - 2) = 9 - 3 = 6$ shows that the subtraction of integers is not associative.

Definition 5.3 *A* **field** *$(F, +, *)$ is a set F together with two binary operations $+$ and $*$ defined on F so that the following axioms are all satisfied:*

1. *The binary operation $+$ is associative and commutative on F.*

2. *There is an identity element for $+$ in F, which is denoted by 0 and called the additive identity.*

3. *Every element $a \in the field F$ has an inverse for $+$ in F, which is denoted by $-a$ and called the additive inverse of a.*

4. *The binary operation $*$ is associative and commutative on F.*

5. *The binary operation $*$ is distributive over $+$ in F. This means that $a * (b + c) = a * b + a * c$ for all elements a, b, and c in the field F.*

6. *There is a nonzero identity element for $*$ in F, which is denoted by 1 and is called the multiplicative identity.*

7. *Every element $a \in the field F$ except 0 has an inverse for $*$ in F, which is denoted by $1/a$ and is called the multiplicative inverse of a.*

For example, the set Z of integers is not a field since the number 3 has no multiplicative inverse *in the set* Z, even though it does have one outside of Z. The smallest field that contains all the integers is the field of rational numbers.

All finite fields of n elements are isomorphic, so there is essentially only one field of 256 elements. The field of two elements, denoted by $F_2 = \{0, 1\}$, is constructed from the ring of integers Z, by setting all multiples of 2 equal to 0. Hence F_2 is the factor ring $Z/\langle 2 \rangle$, in which both addition and multiplication are modulo 2. To get the field F_{256}, start with the polynomial ring $F_2[x]$ and use the irreducible polynomial $g(x) = x^8 + x^4 + x^3 + x + 1$. (Other irreducible polynomials of degree 8 generate isomorphic fields of 256 elements.) The factor ring $F_2[x]/\langle g(x) \rangle$ is a field because $g(x)$ is irreducible in $F_2[x]$. The elements in this field may be considered to be all expressions of the form: $\sum_{j=0}^{7} c_j \alpha^j$ where the c_j are elements from $F_2 = \{0, 1\}$ and α is considered to satisfy the polynomial $g(x) = 0$. There are 256 elements of this type. Addition in F_{256} is just addition modulo 2 of polynomials in α, and multiplication is both modulo 2 and modulo the polynomial $g(\alpha)$. For example, $(\alpha^7 + 1)(\alpha^3 + 1) = \alpha^7 + \alpha^6 + \alpha^5 + \alpha^2 + 1$ because $\alpha^{10} = \alpha^2 \alpha^8 = \alpha^2(\alpha^4 + \alpha^3 + \alpha + 1) = \alpha^6 + \alpha^5 + \alpha^3 + \alpha^2$.

A byte consists of eight bits $(c_7 c_6 \ldots c_0)$. The following one-to-one map from a byte to an element of F_{256} is central to the AES code construction: $(c_7 c_6 \ldots c_0) \rightarrow \sum_{j=0}^{7} c_j \alpha^j$. This shows that every byte can be associated with exactly one element of F_{256}. This map allows the multiplication of bytes, because the elements of F_{256} can be multiplied together. The alternative one-to-one map $(c_7 c_6 \ldots c_0) \rightarrow \sum_{j=0}^{7} c_j 2^j$ maps every byte onto a unique integer in the range [0,255]. This map will be used to express the S-Box for the code. Thus, there are three interchangeable ways to express a byte, (1) as a series of 8 bits, (2) as an element of the field F_{256}, and (3) as an integer in the range [0,255].

Using the S-Box for the Code The S-Box for the AES code is given by the following 16×16 matrix:

	0	1	2	3	4	5	6	7	8	9	10	11	12	13	14	15
0	99	124	119	123	242	107	111	197	48	1	103	43	254	215	171	118
1	202	130	201	125	250	89	71	240	173	212	162	175	156	164	114	192
2	183	253	147	38	54	63	247	204	52	165	229	241	113	216	49	21
3	4	199	35	195	24	150	5	154	7	18	128	226	235	39	178	117
4	9	131	44	26	27	110	90	160	82	59	214	179	41	227	47	132
5	83	209	0	237	32	252	177	91	106	203	190	57	74	76	88	207
6	208	239	170	251	67	77	51	133	69	249	2	127	80	60	159	168
7	81	163	64	143	146	157	56	245	188	182	218	33	16	255	243	210
8	205	12	19	236	95	151	68	23	196	167	126	61	100	93	25	115
9	96	129	79	220	34	42	144	136	70	238	184	20	222	94	11	219
10	224	50	58	10	73	6	36	92	194	211	172	98	145	149	228	121
11	231	200	55	109	141	213	78	169	108	86	244	234	101	122	174	8
12	186	120	37	46	28	166	180	198	232	221	116	31	75	189	139	138
13	112	62	181	102	72	3	246	14	97	53	87	185	134	193	29	158
14	225	248	152	17	105	217	142	148	155	30	135	233	206	85	40	223
15	140	161	137	13	191	230	66	104	65	153	45	15	176	84	187	22

The S-Box is used to replace a byte with a coded byte in the following way: If the input byte is $(c_7 c_6 \ldots c_0)$, then the number in row $\sum_{j=4}^{7} c_j 2^{j-4}$ and in column $\sum_{j=0}^{3} c_j 2^j$ of the S-Box is the integer representation of the new byte. For example, if the input byte is 10010101, then the row is $9 = 2^3 + 2^0$ and the column is $5 = 2^2 + 2^0$. Since both the rows and the columns start their numbering with 0, the corresponding output byte is found in the 10th row and 6th column and has integer representation $42 = 32 + 8 + 2$ and, therefore, the bits of the output byte are 00101010.

Representing the Input Data

The input consists of 128 bits (or 16 bytes). If the input in byte form is

$$A_{0,0}, A_{1,0}, A_{2,0}, A_{3,0}, A_{0,1}, \ldots, A_{0,3}, A_{1,3}, A_{2,3}, A_{3,3}$$

then these bytes are arranged in a 4×4 matrix in the following way:

$$A = \begin{pmatrix} A_{0,0} & A_{0,1} & A_{0,2} & A_{0,3} \\ A_{1,0} & A_{1,1} & A_{1,2} & A_{1,3} \\ A_{2,0} & A_{2,1} & A_{2,2} & A_{2,3} \\ A_{3,0} & A_{3,1} & A_{3,2} & A_{3,3} \end{pmatrix} \tag{5.1}$$

The ByteSub Transformation

The first step in a round of the code is called the ByteSub Transformation (BS), which is a nonlinear transformation and is therefore resistant to linear and differential attacks. In this step each of the 16 bytes in the matrix above is replaced with a new byte using the S-Box and the procedure described above. The result is a new matrix

$$B = \begin{pmatrix} B_{0,0} & B_{0,1} & B_{0,2} & B_{0,3} \\ B_{1,0} & B_{1,1} & B_{1,2} & B_{1,3} \\ B_{2,0} & B_{2,1} & B_{2,2} & B_{2,3} \\ B_{3,0} & B_{3,1} & B_{3,2} & B_{3,3} \end{pmatrix} \tag{5.2}$$

where $B_{i,j}$ is the output byte that arises from using $A_{i,j}$ as the input byte. Thus if $A_{3,2}$ is the byte 10010101 then $B_{3,2}$ is the byte 00101010.

The ShiftRow Transformation

The second step is called the ShiftRow Transformation (SR). This linear step causes diffusion of the bits over multiple rounds. Row j of the matrix is shifted cyclically to the left by j offsets (recall that the rows are numbered starting with 0).

So the new matrix is

$$C = \begin{pmatrix} B_{0,0} & B_{0,1} & B_{0,2} & B_{0,3} \\ B_{1,1} & B_{1,2} & B_{1,3} & B_{1,0} \\ B_{2,2} & B_{2,3} & B_{2,0} & B_{2,1} \\ B_{3,3} & B_{3,0} & B_{3,1} & B_{3,2} \end{pmatrix} \tag{5.3}$$

The MixColumn Transformation

The third step is called MixColumn Transformation (MC). This step creates high diffusion between the columns over multiple rounds of the code. In this step, the bytes in the matrix C above are written as elements of the field F_{256} and multiplied on the left by a matrix M of elements from F_{256} as indicated below.

$$MC = \begin{pmatrix} \alpha & \alpha+1 & 1 & 1 \\ 1 & \alpha & \alpha+1 & 1 \\ 1 & 1 & \alpha & \alpha+1 \\ \alpha+1 & 1 & 1 & \alpha \end{pmatrix} \begin{pmatrix} B_{0,0} & B_{0,1} & B_{0,2} & B_{0,3} \\ B_{1,1} & B_{1,2} & B_{1,3} & B_{1,0} \\ B_{2,2} & B_{2,3} & B_{2,0} & B_{2,1} \\ B_{3,3} & B_{3,0} & B_{3,1} & B_{3,2} \end{pmatrix} \tag{5.4}$$

For example, if the elements in the first column of the matrix on the right have bit form

$$\begin{pmatrix} 10000001 \\ 00000000 \\ 00001001 \\ 00101010 \end{pmatrix} \tag{5.5}$$

then the entry in the left uppermost position of the product matrix is $\alpha(\alpha^7 + 1) + (\alpha + 1)(0) + (1)(\alpha^3 + 1) + (1)(\alpha^5 + \alpha^3 + \alpha) = \alpha^8 + \alpha^5 + 1 = \alpha^5 + \alpha^4 + \alpha^3 + \alpha$. The bit form of this byte would be 00111010.

Creating the W-Matrix That Contains the Keys for the Code

Initially, the key is a 128-bit number that is written as 16 bytes. These bytes are placed in the first four columns (columns w_0, w_1, w_2 and w_3) of the W-matrix, which is 4×44. The other columns of the W-matrix are generated recursively from the first four columns by the following procedure.

Let column $j + 1$ of the W-matrix be denoted by w_j and assume that $j \geq 4$. Provided that j is not a multiple of 4, then $w_j = w_{j-4} + w_{j-1}$ (addition may be considered to take place in the field F_{256} or by using the XOR operation on the bit representation of the bytes). If j is a multiple of 4, then the computation is more complicated and should be regarded as a series of steps: (1) Every byte in column w_{j-1} is used to find a new byte in the S-Box by using the ByteSub Transformation. (2) The order of the new bytes is cyclically changed by moving the top byte to the bottom and moving every other byte up one place to create the column vector n_{j-1}. (3) The new column vector in the W-matrix is $w_j = w_{j-4} + n_{j-1} + v_j$, where v_j is the column vector consisting of the bytes

$$
\begin{pmatrix}
\alpha^{(j-4)/4} \\
0 \\
0 \\
0
\end{pmatrix}
\tag{5.6}
$$

For example, if $j = 16$, then $\alpha^{(j-4)/4} = \alpha^3$, which has bit form 00001000. Suppose that

$$
w_{12} = \begin{pmatrix}
10110010 \\
01100010 \\
01101001 \\
01001000
\end{pmatrix}
\quad \text{and} \quad
w_{15} = \begin{pmatrix}
10011110 \\
11110000 \\
01100110 \\
10111111
\end{pmatrix}
\tag{5.7}
$$

Then, using the S-Box, we find that

$$
n_{15} = \begin{pmatrix}
00000100 \\
00001011 \\
10001100 \\
00110011
\end{pmatrix}
\tag{5.8}
$$

since $11 = 8 + 2 + 1$, $140 = 128 + 8 + 4$, and $51 = 32 + 16 + 2 + 1$. Therefore, $w_{16} = w_{12} + n_{15} + v_{16} =$

$$
\begin{pmatrix}
10110010 \\
01100010 \\
01101001 \\
01001000
\end{pmatrix}
+
\begin{pmatrix}
00000100 \\
00001011 \\
10001100 \\
00110011
\end{pmatrix}
+
\begin{pmatrix}
00001000 \\
00000000 \\
00000000 \\
00000000
\end{pmatrix}
=
\begin{pmatrix}
10111110 \\
01101001 \\
11100101 \\
01111011
\end{pmatrix}
\tag{5.9}
$$

RoundKey Addition

The final step is referred to as the RoundKey Addition (ARK) step. At the end of the

jth round, the product matrix (MC) computed during the MixColumn Transformation is added (modulo 2) to the 4×4 matrix E_j obtained by taking columns w_{4j}, w_{4j+1}, w_{4j+2}, and w_{4j+3} from the W-matrix.

Overview of the Rijndael Encryption

Step (1) The W-matrix is computed from the key. The first four columns of this matrix (the original keyword) is added to the input data consisting of 16 bytes arranged in a 4×4 matrix.

Step (2) Nine rounds of BS, SR, MC, and ARK are processed, using the appropriate columns of the W-matrix for each application of ARK.

Step (3) The final round consists of BS, SR and ARK, using the last round key (the final four columns of the W-matrix).

Notice that MC is not applied in the final round. This is to help with the decryption of the code word.

Decryption of the AES Code

The decryption of the AES code is based on the fact that each of the steps BS, SR, MC, and ARK are invertible. The inverse of the BS step is based on the fact that the map from bytes onto integers in [0,255] is one-to-one and, hence, there is another lookup table called InverseByteSub (IBS) that inverts BS. The inverse of SR consists of shifting the rows to the right the same number of times that they were shifted to the left by SR. This operation is denoted by ISR. Moreover, the order of the operations BS and SR can be interchanged without changing the result. (In other words, these two operations commute with each other.) This means that the order of IBS and ISR can also be interchanged.

The inverse of MC is based on the fact that the 4×4 matrix M used in the MC operation is invertible over the field F_{256}. In fact, the inverse of M is the matrix

$$M^{-1} = \begin{pmatrix} \alpha^3 + \alpha^2 + \alpha & \alpha^3 + \alpha + 1 & \alpha^3 + \alpha^2 + 1 & \alpha^3 + 1 \\ \alpha^3 + 1 & \alpha^3 + \alpha^2 + \alpha & \alpha^3 + \alpha + 1 & \alpha^3 + \alpha^2 + 1 \\ \alpha^3 + \alpha^2 + 1 & \alpha^3 + 1 & \alpha^3 + \alpha^2 + \alpha & \alpha^3 + \alpha + 1 \\ \alpha^3 + \alpha + 1 & \alpha^3 + \alpha^2 + 1 & \alpha^3 + 1 & \alpha^3 + \alpha^2 + \alpha \end{pmatrix} \quad (5.10)$$

Therefore, the inverse of MC, denoted by IMC, consists of multiplying the 4×4 matrix of 16 coded bytes by the matrix M^{-1} on the left. Because ARK is simply addition modulo 2, it is its own inverse. However, ARK does not commute with MC. MC followed by ARK has the form $MC + E_j = H$, where C is the matrix obtained at the end of the SR step and E_j is the key for the jth round. Solving this for the matrix C, we multiply by M^{-1} on the left to obtain $C + M^{-1}E_j = M^{-1}H$. So, $C = M^{-1}H + M^{-1}E_j$ (since in modulo 2, $M^{-1}H - M^{-1}E_j = M^{-1}H + M^{-1}E_j$) , which shows that the path from H back to C can be accomplished by first applying IMC to H and then adding the new key $M^{-1}E_j$. Because this key is directly computed from the key E_j (using the fixed matrix M^{-1}), we denote the operation of adding the key $M^{-1}E_j$ as IARK.

Overview of Decryption of AES

Step (1) ARK using the last round key.

Step (2) Nine rounds of IBS, ISR, IMC, and IARK using the round keys in the opposite order, i.e. key 9 down to key 1.

Step (3) A final round consisting of IBS, ISR, and ARK using the original round key (the given key word).

Thus decryption has essentially the same format as encryption.

5.5 Problems

1. A key is a **weak key** if the subkey used in each of the 16 cycles is the same. Find the four weak keys for the DES algorithm.

2. Use the appropriate DES S-Box to find the output code sub-block if the input code sub-block is 101101 and this sub-block is

   (a) The second sub-block of input code

   (b) The seventh sub-block of input code

3. Suppose that $E_4 \oplus K_4$ is the 48 bits 11001011 00110001 10101101 11011011 01100010 10001110. Find the first 8 bits of R_4.

4. Suppose that the subkey K_6 is 10110101 01101100 10001010 00011100 10010110 01110101 and suppose that E_6 is 01101101 11010011 11011101 01110011 01110101 11011101. Find $E_6 \oplus K_6$.

5. Suppose that the 56-bit key for a DES algorithm is
   11011001 01001101 01100101 00010011 01010110 01100110 00110100.
   Find K_1, the subkey for the first round.

6. Suppose that $R_1 = 10011100011010111101001010100011$. Find the first four bits of R_2 if the subkey for the second round is
   $K_2 = 0111010101110010111001010100010100010001101010110$.

7. Check that the matrix given for M^{-1} above is the correct inverse of the matrix M by finding the product of M and M^{-1}

8. Use the AES S-Box to find the byte that is changed into the output byte 10010111 by the ByteSub Transformation.

9. Compute the product of the bytes 10010001 and 00100010 in F_{256}.

10. Suppose that the matrix at the end of ShiftRow Transformation is

$$
C = \begin{pmatrix}
00000001 & 00000000 & 00000001 & 00000010 \\
00000000 & 00000100 & 00000010 & 00000001 \\
00000010 & 00000000 & 00000000 & 00000001 \\
00000100 & 00000001 & 00000001 & 00000001
\end{pmatrix}
\tag{5.11}
$$

Find the matrix that is computed from this matrix by the MixColumn Transformation.

11. Suppose that the first four columns of the W-matrix are

$$
\begin{pmatrix}
00110001 & 01011010 & 01101001 & 00000010 \\
01100010 & 00110100 & 01110010 & 00000001 \\
10110010 & 00110010 & 11100011 & 00000001 \\
01000100 & 00110001 & 10110011 & 00000000
\end{pmatrix}
\tag{5.12}
$$

Compute the fifth column of the W-matrix.

12. Let b be a byte in bit form and let $\hat{b}$ be $b + 11111111$ (the complement of b). For a fixed given key, if AES encrypts b to g, does AES encrypt $\hat{b}$ to $\hat{g}$?

5.6 Solutions

1. Since each half of the 56-bit key word is manipulated independently of the other half, the subkeys will be the same provided that each half consists of 28 identical bits. Therefore, the possible weak keys are 56 0s, 56 1s, 28 0s followed by 28 1s and 28 1s followed by 28 0s.

2. The first and last digits of 101101 are 11, which indicates that we should look in row three (the bottom row) of the S-Box. The middle four digits are 0110, which is the binary representation of 6. This means that we should look in column 6 (the seventh column) of the S-Box.

   (a) In S_2 the number in column 6 and row 3 is 4, and hence the output is 0100.

   (b) In S_7 the number in column 6 and row 3 is 10, and hence the output is 1010

3. The first six bits of the input code are 110010. This means that we need to find the number in row 2 and column 9 of the S-Box S_1. This number is 12, which has the binary representation 1100. The second six bits of the input code are 110011. This means that we need to find the number in row 3 and column 9 of the S-Box S_2. This number is 6, which has the binary representation 0110. Hence, the first eight bits of R_4 are 11000110.

4. $E_6 \oplus K_6$ is obtained by the rule that the bit in the jth place is 0 if the bits in the jth places of E_6 and K_6 are the same and is 1 if the bits are different. The result is
11011000 10111111 01010111 01101111 11100011 10101000.

5. First, each half of the key is shifted circularly one bit to the left. This gives the 56-bit word
10110010 10011010 11001010 00100110 10101100 11001100 01101001
We now apply the compression permutation to obtain K_1:
01001010 10101011 00011010 10110010 11100001 01111000

6. The expansion permutation is applied to R_1, giving the 48-bit code word E_2
11001111 10000011 00110111 11101100 01001101 00000111
Because we are only interested in the first four bits of R_2, we now need to compute only the first six bits of $E_2 \oplus K_2$. These are 101110. Now, using S-Box S_1, we find that we need to use the row labeled 2 and the column labeled 7. The number 11 found inside the matrix S_1 indicates that the first four bits of R_2 are 1011.

7. $M * M^{-1} = I$, where I is the identity matrix.

8. 10010111 has integer representation $128 + 16 + 4 + 2 + 1 = 151$. This number occurs in row 8 and column 5 of the S-Box. Hence the input byte is 10000101.

9. Representing the bytes as elements of the field, we find that we need to compute the product of $\alpha^7 + \alpha^5 + 1$ and $\alpha^5 + \alpha$. Multiplication gives $\alpha^{12} + \alpha^8 + \alpha^{10} + \alpha^6 + \alpha^5 + \alpha$. However, $\alpha^8 = \alpha^4 + \alpha^3 + \alpha + 1$ and, hence, $\alpha^{10} = \alpha^6 + \alpha^5 + \alpha^3 + \alpha^2$ and $\alpha^{12} = \alpha^8 + \alpha^7 + \alpha^5 + \alpha^4 = \alpha^7 + \alpha^5 + \alpha^3 + 1$. Therefore, the product is $\alpha^7 + \alpha^5 + \alpha^4 + \alpha^3 + \alpha^2 + \alpha + 1$. This byte has bit form 10111111.

10. $MC = \begin{pmatrix} \alpha & \alpha+1 & 1 & 1 \\ 1 & \alpha & \alpha+1 & 1 \\ 1 & 1 & \alpha & \alpha+1 \\ \alpha+1 & 1 & 1 & \alpha \end{pmatrix} \begin{pmatrix} 1 & 0 & 1 & \alpha \\ 0 & \alpha^2 & \alpha & 1 \\ \alpha & 0 & 0 & 1 \\ \alpha^2 & 1 & 1 & 1 \end{pmatrix}$

The product is $MC = \begin{pmatrix} \alpha^2 & \alpha^3+\alpha^2+1 & \alpha^2+1 & \alpha^2+\alpha+1 \\ \alpha+1 & \alpha^3+1 & \alpha^2 & \alpha \\ \alpha^3+1 & \alpha^2+\alpha+1 & 0 & \alpha \\ \alpha^3+1 & \alpha^2+\alpha & \alpha+1 & \alpha^2 \end{pmatrix}$

11. The new column,

$$w_4 = w_0 + n_3 + v_4 = \begin{pmatrix} 00110001 \\ 01100010 \\ 10110010 \\ 01000100 \end{pmatrix} + \begin{pmatrix} 01100011 \\ 01110110 \\ 01111011 \\ 01111011 \end{pmatrix} + \begin{pmatrix} 00000001 \\ 00000000 \\ 00000000 \\ 00000000 \end{pmatrix} = \begin{pmatrix} 01010011 \\ 00010100 \\ 11001001 \\ 00111111 \end{pmatrix}$$

$$(5.13)$$

12. Let $b = 00000000$. Then $\hat{b} = 11111111$. When we come to the ByteSub Transformation we find that b is replaced by 01100011 and $\hat{b}$ is replaced by 00010110. These are not complements of each other. The S-Box is constructed in a way that does not preserve complementation.

Chapter 6

Elliptic Curve Cryptography (ECC)

Goals, Discussion   We want to give a geometrical overview of elliptic curves as well as some cryptographic applications.

Elliptic curves were introduced into cryptography in the 1980s. They can be used as a key exchange, analogous to the Diffie–Hellman key exchange. They can also be used for public key encryption and digital signatures analogous to the El Gamal algorithms, based on Diffie–Hellman, which are discussed in Chapter 4.

New, Noteworthy   We present some historical background and also some of the geometry of elliptic curves. In particular we present a detailed derivation of the algebraic formula for addition on the curve based on the geometrical definition. We also give a detailed discussion of the idea of a nonsingular cubic curve, i.e. an elliptic curve. Related work in finite geometries due to the author and others is presented. (In fact, one of the results here will be mentioned again in Part III in connection with MDS codes). We also briefly describe the brilliant insight due to Frey, which was a major catalyst for the solution to Fermat's famous last theorem. Frey showed that the Fermat problem—which can be phrased as a question on curves of *unknown degree*—could be solved if the solution to a question on *cubic* curves could be found. This led to the famous result by A. Wiles that the Fermat equation $x^n + y^n = z^n$ has no nontrivial solution in the integers when $n > 2$.

6.1   Abelian Integrals, Fields, Groups

The subject of elliptic curves is indeed a venerable one. In fact, it has been asserted that more mathematical papers have been written on topics related to elliptic curves (= nonsingular

113

cubic curves) than on the rest of mathematics put together!

Historically, elliptic curves arose in connection with doubly periodic functions. For the motivation, recall from calculus that $\int_0^x \frac{1}{\sqrt{1-t^2}} \, dt$ is a function of x, say $f(x)$, that arises in the computation of the arc length of a circle. In the familiar notation, we have $y = f(x) = \arcsin x$. Then $x = \sin y$, so the inverse function is the sine function. So, when "inverting the integral" we end up with the sine function—which has some fundamental properties. In particular, although not an algebraic function, it is periodic of period 2π, i.e., $\sin(x + 2\pi) = \sin x$.

It was Gauss, in 1797, who studied the integral $\int_0^x \frac{1}{\sqrt{1-t^4}} \, dt$ in connection with the arc length of the leminiscate or the infinity symbol. (whose equation in polar coordinates is $r^2 = a^2 \cos 2\theta$). According to Whittaker and Watson [WW46] this gave rise to one of the earliest new functions defined by the inversion of an integral. Long before Gauss, Count Fagnano had investigated similar integrals, but not their inversion.

The study of elliptic integrals and elliptic functions was developed in the eighteenth and nineteenth centuries, giving rise to doubly periodic functions in general and the Weierstrass function $W(z)$ in particular. It was observed that W and W' satisfy a cubic relationship and parameterize a cubic curve that, topologically, can be made to correspond to the period lattice and thus to a torus. Moreover, the "group law" on the cubic played a major role in Abel's theory of Abelian Integrals: See Shafarevich [Sha77] for an account of this, along with Clemens [Cle80].

Algebraically, nonsingular cubic curves are an order of magnitude more complicated than second-degree curves such as circles. For example, the unit circle $x^2 + y^2 = 1$ in the Euclidean plane has a "rational parameterization." This is because if we put $x = \frac{1-t^2}{1+t^2}$, $y = \frac{2t}{1+t^2}$ and let t vary over all real numbers we get all points on the circle. Contrast that with cubic nonsingular curves (= elliptic curves) which have no such rational parameterization (easy!). Of course, there exist rational cubic curves (such as the familiar calculus curve given by $y = x^3$).

In the twentieth century, cubic curves were intensively studied mainly on the algebraic side, related to theta functions, number theory, and algebraic geometry. Few people have not heard of the solution of the three-hundred and fifty year-old problem known as "Fermat's last theorem" by Andrew Wiles. The proof of Wiles built on the work of Frey and Ribet, who were able to reduce the question to a problem on cubic curves. However, many deep questions in number theory related to cubic curves still remain unsolved. We mention in particular the so-called Birch–Swimmerton Dyer conjectures. Some brief remarks connecting the Fermat problem and cubic curves are presented at the end of this chapter.

Fields As mentioned in Chapter 19, a field is an algebraic system with two operations, addition and multiplication, obeying all the usual rules so we can add, subtract, multiply and divide. Addition and multiplication are connected by the distributive law, namely,

$a(b + c) = ab + ac$. The rational numbers (comprised of all fractions $\frac{u}{v}$ with $v \neq 0$ where u, v are integers) form a familiar example, as do the real numbers (calculus again!). The reals can be extended to the complex numbers.

The above fields are infinite, but there also exist finite fields. The number of elements in the field is called the **order** of the field. The work of Galois shows that the order of a finite field F must be equal to a power of a prime p. In symbols, $|F| = q = p^n$. (In honor of Galois, a finite field of order q is known as a Galois field and is written $GF(q)$.) Conversely, if we take any prime power such as 25 ($=5^2$) there is a field F with exactly 25 elements. In fact, as shown by Galois, there is essentially only one such field. Sitting inside F is the field Z_5 consisting of the remainders $\{0, 1, 2, 3, 4\}$ when we divide by 5. Thus $4 + 4 = 3$, $(4)(3) = 2$, etc.

Note especially that in the field F of order 25 we have $1 + 1 + 1 + 1 + 1 = 0$. In general the **characteristic** of a field is the smallest number of ones that add up to zero. For example, if $|F| = p^n$ then the characteristic of F is p. If no sum of ones equals zero the field (such as the reals or the complexes or the rationals) is said to have **characteristic zero**.

To construct a field of order 5^n we use Z_5 together with an equation of degree n that does not factor over Z_5; see Chapter 19.

Groups A group is a nonempty set with just one operation, unlike a field, which has two operations. The operation in a group is usually called addition or multiplication. For example, in Diffie–Hellman, the set is $\{1, 2, 3, \ldots p - 1\}$ for some prime p and the operation is multiplication. Another example of a group is given by $\{0, 1, 2, \ldots, p - 1\}$ where we add 2 elements and get the remainder on division by p. In a field, the nonzero elements form a group under multiplication and all of the elements form a group under addition. In this chapter, we construct a group using the points on an elliptic curve.

6.2 Curves, Cryptography

Here we develop a key exchange with elliptic curves. As Hendrik Lenstra showed, elliptic curves can also be used to factor integers (see [Len87]).

Start off in the plane $AG(2, F)$ over F, where F is any field. Let us define this plane, denoted by π. It has points and lines. The points of π are all possible ordered pairs (a, b) with a, b in F. The lines are all sets of points satisfying linear equations of the type $y = mx + b$ or $x = c$, where m, b, c are fixed in F. There are no surprises here: When F is the reals, π is just the Euclidean plane, following Descartes.

In what follows now we assume, in order to get a nonsingular curve, that *the characteristic of F is neither two nor three*. (We explain in detail the reason for this in the

Figure 6.1: From left to right: Peter van Emde Boas, L. Lovasz, Hendrik W. Lenstra and Arjen K. Lenstra. Photo from Bonn in 1982, after the publication of the LLL algorithm [LLL82]. This celebrated algorithm is used in many branches of cryptography such as factoring. It was also used to break the knapsack cryptosystem

problems at the end of this chapter.) By an **elliptic curve** E we mean all ordered pairs (x, y) satisfying the following equation:

$$y^2 = x^3 + ax + b. \tag{6.1}$$

In addition, we impose the following condition:

$$4a^3 + 27b^2 \neq 0 \tag{6.2}$$

to make the curve nonsingular (see below). Finally, apart from the solutions in (6.1), which in $AG(2, F)$ and are known as the **affine points**, we decree that E also contains the infinite point P_∞. We think of this point as the "point of intersection" of all vertical lines in the plane: They meet at infinity (at P_∞). For example, think of P_∞ of lying at both ends, extended infinitely, of the y-axis. This can all be explained rigorously with projective or homogeneous coordinates, well-known to students of engineering and computer graphics. Later on, we give a brief discussion of homogeneous coordinates. In homogeneous coordinates P_∞ does lie on the curve and we have

$$P_\infty = (0, 1, 0).$$

For example, if we take the curve $y^2 = x^3 - 4x$ over the real numbers we get Figure 6.2.

From equation (6.1) we see that if (x, y) is on the curve then so also is $(x, -y)$. In other words, the curve is symmetric about the x-axis. *If we reflect the top half in the x-axis we get the bottom half of the curve.* Note that equation (6.1) is a cubic equation because of the x^3 term. *This implies that any line meets the curve in at most 3 points.*

6.3 Nonsingularity

Condition in equation (6.2) can be explained in two different ways. Geometrically, it implies that each point of the curve is a "simple" point , and so has a unique tangent at each point. Algebraically, we have that at no point P in the plane are the three partial derivatives zero. The reason why we must assume that the characteristic of $F \neq 2, 3$ is to ensure nonsingularity. All this is discussed in the problems at the end of the chapter. There are algebraic complications: When checking whether the partial derivatives are zero we may have to work over algebraic extension fields.

6.4 The Hasse Theorem, and an Example

Given a curve E such as in equation (6.1), where we are working with a finite field $F = GF(q)$, how many points including P_∞ are there on E? The number of elements in F is denoted by q where q is necessarily a prime power, say $q = p^n$. Denote the number of points on the curve by N. We have the following famous result due to Hasse.

Theorem 6.1 N *lies between* $q + 1 - 2\sqrt{q}$ *and* $q + 1 + 2\sqrt{q}$. *In symbols we have* $q + 1 - 2\sqrt{q} \leq N \leq q + 1 + 2\sqrt{q}$.

There is also the following result in Waterhouse [Wat69].

Theorem 6.2 *If* $q = p$, *a prime with* $p > 3$, *and if* N *is any integer such that* $p + 1 - 2\sqrt{p} < N < p + 1 + 2\sqrt{p}$, *then there exists an elliptic curve satisfying equations (6.1), (6.2) and having exactly* N *points.*

To get a very rough idea of why Theorem 6.1 works, suppose we have a curve as in equation (6.1). There are q possible values for x. Now in any field of characteristic other than 2, about half the elements are squares, i.e. half the elements have square roots. (This is similar to what we expect from the real numbers where "about half" the elements, namely the positive elements, have square roots.) If $x^3 + ax + b$ is not zero and has a square root y it will also have a square root $-y$. For fields of characteristic different from 2, $y \neq -y$. So, roughly, the total number of points should be about $2 \left(\frac{1}{2} q \right) = q$.

As mentioned earlier (for the real numbers), an elliptic curve over a finite field has no rational parameterization, unlike a rational cubic curve. However, subject to minimal conditions, any cubic over any field can be obtained as the projection of a quartic, i.e. a curve of degree 4, in 3 dimensions. Such a quartic can be regarded as the intersection of two quadrics in 3 dimensions. Then by space dualizing (point to plane, line to line) we get two different quadrics and a different quartic. This then easily leads to the following result, in Bruen and Hirschfeld [BH88, Theorem 8.4].

Theorem 6.3 *For every rational cubic or elliptic curve C with exactly N points over $GF(q)$ there is a curve C_1 with N_1 points over $GF(q)$ such that $N + N_1 = 2(q + 1)$.*

The proof of Theorem 6.3, by algebraic methods and involving two projectively distinct curves with the same absolute invariant, is in Cicchese ([Cic65] and [Cic71]).

In fact, Theorem 6.2 was extended by Rück in 1987 (see [Ruc87]), as follows.

Theorem 6.4 *For every N with $q + 1 - 2\sqrt{q} \le N \le q + 1 + 2\sqrt{q}$ there exists an elliptic curve E with exactly N points. Moreover, the group of E is cyclic.*

We will explain the meaning of cyclic again below, but we have met the idea already. It simply means that the group has a generator as in the Diffie–Hellman protocol when the group is the multiplicative group of integers $(\bmod p)$.

6.5   More Examples

We work with the field $GF(7)$, also known as Z_7. The elements of Z_7 are $0, 1, 2, 3, 4, 5, 6$. These are just the remainder when we divide any integer by 7. To add or multiply, we add or multiply in the usual way and take the remainder on division by 7. Thus $4 + 5 = 2$, $(3)(5) = 1$. The additive inverse of 5, for example, is 2 since $5 + 2 = 0$. Thus $5 = -2$ and $-5 = 2$. The multiplicative inverse of 5 is 3 since $(5)(3) = 1$, so $\frac{1}{5} = 3$.

Which elements of $GF(7)$ are squares, i.e. which elements have square roots?

We have $0^2 = 0$, $1^2 = 1$, $2^2 = 4$, $3^2 = 2$, $4^2 = 2$, $5^2 = 4$, $6^2 = 1$. So the squares are $0, 1, 2, 4$. Note that the calculations above can be shortened since $5 = -2$ implies that $5^2 = (-2)^2 = 2^2$. For a general prime p, $p > 2$, we only need to calculate $0^2, 1^2, 2^2, \ldots, \left(\frac{p-1}{2}\right)^2$, from which we get the rest.

Now consider the curve E given by

$$y^2 = x^3 + 2 \tag{6.3}$$

so that $a = 0$, $b = 2$, in equation (6.1). As $(27)(2^2)$ is not divisible by 7 it is not zero in the field $GF(7)$, so equation (6.2) is satisfied and we have a nonsingular cubic curve C. We can now make up the following table.

Values of x	0		1	2	3	4	5	6
Values of x^3	0		1	1	6	1	6	6
Values of $x^3 + 2$	2		3	3	1	3	1	1
Values of y from (6.3)	± 3				± 1		± 1	± 1
Affine points on E	$(0,3),(0,4)$				$(3,1),(3,6)$		$(5,1),(5,6)$	$(6,1),(6,6)$

Thus, including the infinite point P_∞, the number of points on E is exactly $9 = q + 2$.

From the point of view of cryptography, the crucial fact is that the points on any elliptic curve actually form a group. For example, if we add P_∞ to $(3,1)$ in this group, we get $(3,1)$. If we add $(5,1)$ to $(3,1)$ we get $(6,6) = (-1,-1)$. This is explained below.

6.6 The Group Law on Elliptic Curves

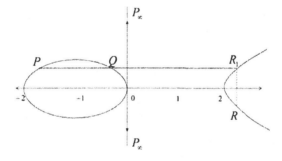

Figure 6.2: The Group Law

Theorem 6.5 *The set of points on any elliptic curve over any field F forms an additive abelian group G. If $F = GF(q)$ then $q + 1 - 2\sqrt{q} \leq |G| \leq q + 1 + 2\sqrt{q}$.*

The first part of this is one of the show pieces of classical algebraic geometry. A formal proof can be constructed from the theorem of the "nine associated points" to wit: Given 9 points that form the intersection of 2 cubic curves, any other cubic curve passing through 8 of the 9 points automatically passes through the ninth.

The definition of addition is very simple. First of all, P_∞ is the identity element so $P_\infty + P = P$ for all points P on the curve including $P = P_\infty$.

Next let $P = (x_1, y_1)$ and $Q = (x_2, y_2)$ be any 2 affine points. To find $P \mid Q$, perform the following 2 steps (see Fig 6.2).

(i) Let the line PQ meet the curve again in R_1.

(ii) Reflect R_1 in the x-axis to get $R = -R_1$.

Then define $P + Q$ to be the point R.

A few remarks are in order.

1. The reflection of P_∞ in the x-axis is defined to be P_∞ : As P_∞ is on either end of the y-axis, this makes sense. So $-P_\infty = P_\infty$, and we can define $P + Q$ even if $R_1 = P_\infty = R$.

2. The definition of addition can be extended to any 2 points on the curve, not just to 2 affine points.

3. If $P = Q$ then the line PQ is defined to be the (unique!) tangent to the curve at $P = Q$.

4. From the definition it is clear that R is on the curve, so that given P and Q we have that $P + Q$ is one of the points on the curve. Thus we have "*closure under addition.*" It is clear that $P + Q = Q + P$, so the group is *abelian*. The point P_∞ acts as the identity element. Each element (=point) P in the group has an inverse, namely $-P$, its reflection in the x-axis.

    In general, the axioms for a general group involve closure, an identity element, inverses, and associativity. For an abelian group we need that $P + Q = Q + P$. (This will imply that, when adding any number of points, the order does not matter.) From the above definition, all axioms for a group (in fact, an abelian group) are satisfied save for the associative law. This law says that $(P + Q) + R = P + (Q + R)$ and is the big stumbling block. As mentioned earlier, it can be proved with the theorem of the 9 associated points.

5. The addition on the cubic is not the same as the usual vector addition whereby, if $P = (x_1, y_1)$ and $Q = (x_2, y_2)$ then $P + Q = (x_1 + x_2, y_1 + y_2)$. Also, to get a formula for the sum of a point with itself on a cubic, one needs to calculate the tangent at a point on the curve $y^2 = x^3 + ax + b$. This can be done (calculus again) by noting that, from implicit differentiation, $2y \frac{dy}{dx} = 3x^2 + a$.

The geometrical definition of addition above gives rise to the following algebraic result.

Theorem 6.6 *Let $P = (x_1, y_1)$, $Q = (x_2, y_2)$. Let $(x_1, y_1) + (x_2, y_2) = P + Q = R = (x_3, y_3)$. We have two major cases.*

(1) $x_1 = x_2$. Then if $y_1 \neq y_2$, we have $P + Q = R = P_\infty$. However, if $y_1 = y_2$ (i.e $P = Q$), then

$$
\begin{aligned}
x_3 &= \gamma^2 - x_1 - x_2 \\
y_3 &= \gamma(x_1 - x_3) - y_1
\end{aligned}
$$

and $\gamma = 3x_1^2 + \frac{a}{2y_1}$.

(2) $x_1 \neq x_2$. Then x_3, y_3 are as described above, except that in this case $\gamma = \frac{y_2 - y_1}{x_2 - x_1}$, so that γ is the slope of the line joining the points.

Proof. Suppose $x_1 = x_2$ and $y_1 = y_2$ so $P = Q$. The equation of the tangent at P is obtained by using implicit differentiation on the curve $y^2 = x^3 + ax + b$. We get $2y\frac{dy}{dx} = 3x^2 + a$. Thus $\frac{dy}{dx} = \frac{3x^2 + a}{2y}$. Thus the slope of the tangent at (x_1, y_1) is $\frac{3x_1^2 + a}{2y_1}$. If $y_1 = 0$ then the tangent line is vertical and $P + Q = P_\infty$. Suppose $y_1 \neq 0$. The equation of the tangent line at P is as follows:

$$
y - y_1 = \frac{3x_1^2 + a}{2y_1}(x - x_1),
$$

i.e., $y = y_1 + \gamma x - \gamma x_1$, where $\gamma = \frac{3x_1^2 + a}{2y_1}$. Substituting into the cubic we get

$$
(y_1 + \gamma x - \gamma x_1)^2 = x^3 + ax + b.
$$

Write this out as a cubic in x of the form $x^3 + Ux^2 + Vx + W = 0$. The roots are x_1, x_1, and x_3. Thus we have

$$
x^3 + Ux^2 + Vx + W = (x - x_1)^2(x - x_3).
$$

So $-\gamma^2 = U = -(x_1 + x_2 + x_3)$ on comparing the coefficients of x^2. Thus $x_3 = \gamma^2 - x_1 - x_2$. Now $y_3 = y_1 + \gamma x_3 - \gamma x_1$ by substituting in the equation of the tangent line. Thus $y_3 = \gamma(x_3 - x_1) + y_1$. So when we join P to Q, i.e., when we take the tangent at P, we get the point $(\gamma^2 - x_1 - x_2, \gamma(x_3 - x_1) + y_1)$. To get the sum of P and Q we take the negative of the y-coordinate and use the same x-coordinate. Thus

$$
P + Q = P + P = (\gamma^2 - x_1 - x_2, \gamma(x_1 - x_3) - y_1).
$$

The case $x_1 \neq x_2$ is handled in the same way. ∎

Going back to our curve over $GF(7)$ we see that, for example, $(3, 1) + (5, 1) = (6, 1) = (-1, 1)$.

6.7   Key Exchange with Elliptic Curves

Given an elliptic curve E satisfying equations (6.1), (6.2) over $GF(q)$ we have a group G. It is convenient to assume that G is cyclic or has a generator. This means that there is a point $P = (u, v)$ on the elliptic curve such that all points on the curve are of the form P, $2P = P + P$, $3P = P + P + P, \ldots, nP = P + P + \cdots + P$ (n times).

Let A, B be communicating parties. Then A, B (Alice, Bob) proceed exactly analogous to the situation in Diffie–Hellman as follows.

A, B choose secret integers a, b, respectively. Suppose P is a given generator for the elliptic curve. Then A, B transmit aP, bP openly to each other. When A, B receive bP, aP they calculate $a(bP)$, $b(aP)$, respectively. They end up with a common secret key $(ab)P = (ba)P$. The discrete log problem in DH comes out to the following in ECC.

Given P, xP, find x.

In the above if P is a generator and the group G has order n then $nP = P_\infty$, which is the zero element of the group, giving that $(n + 1)P = P$, $(n + 2)P = 2P$ and, in general, $mP = Rem[m, n]P$.

We can also develop a cryptographic system in ECC analogous to that for DH as well as a digital signature scheme.

6.8   Elliptic Curves mod n

Instead of working with elliptic curves over a field, it is possible to work over the integers mod n, which is the set $Z_n = \{0, 1, 2, 3, \ldots, n - 1\}$. Recall from Chapter 3 that the only numbers in Z_n that have a multiplicative inverse are those that are relatively prime to n. Thus Z_n is not a field unless n is a prime. However, one can get important results such as algorithms for factoring n, called elliptic curve factorization, developed by H. Lenstra in 1986 [Len87]. When $n = pq$ is the product of 2 distinct primes, this can be used on the RSA cryptosystem developed in Chapter 3.

6.9   Encoding Plain Text

Let m be a message, represented as a binary string. To use an ECC cryptosystem, how do we make m correspond to a point on an elliptic curve E? We can initially make m be a part of the x-coordinate of a point on E by ensuring that the length of m is suitably small and by appending zeros to m. If the corresponding y-value for $x = m$ is then not a square (and there is only about half a chance that it will be) we can use a method developed by

Koblitz and append bits suitably in such a way that the y-value is now a square (with very high probability).

6.10 Security of ECC

This is analogous to the security of DH. For example, in the above notation we have the following questions.

Problem 1 *Elliptic Curve Problem. Given aP, bP, P, how does one find $(ab)P$?*

Problem 2 *Discrete Log Problem for Elliptic Curves. Given xP, P, how does one find x?*

Given a solution to problem 2 we obtain a solution to problem 1. Whether the converse holds is not clear.

It seems that the security of ECC may be a bit stronger than the security of DH (the Diffie–Hellman key exchange) and RSA. One reason is that the size of the prime (or field) used in ECC need not be as large as that in DH or RSA to guarantee comparable security. In any event, ECC is still in its infancy even though the subject of elliptic curves is a venerable one. Also, ECC has not yet undergone the kind of testing that has been applied to RSA and DH.

6.11 More Geometry of Cubic Curves

Working over an algebraically closed field (such as the complex numbers), let $F(x, y, z) = 0$ be a cubic curve in homogeneous coordinates—see section 6.13. We consider the equation

$$y_1 F_x + y_2 F_y + y_3 F_z = 0.$$

where F_x denotes the partial derivative of F with respect to x. We can interpret this equation in two ways (see [SR85]).

1. Suppose P is a given point on the curve. Then, varying y_1, y_2, y_3 we get an equation of the form

$$y_1 a + y_2 b = y_3 c = 0$$

with $a = F_x(P)$, $b = F_y(P)$, $c = F_z(P)$. Then it follows that (if the curve F is nonsingular), $(a, b, c) \neq (0, 0, 0)$ As a result, the above, in homogeneous coordinates, is the equation of a line. This line is in fact the tangent line to the curve at P.

On the other hand, think of $P = (y_1, y_2, y_3)$ as being a given point that is not necessarily on the curve. Now if F has degree n, then F_x, F_y, F_z all have degree $n - 1$. Then equation (6.11) yields a curve of degree $n - 1$ called the *polar curve* of P with respect to F. This polar curve represents the points lying on the "feet of the tangents" from P to the curve $F = 0$.

A basic fact from algebraic geometry, called Bezout's theorem, asserts that two curves of degree u, v intersect in exactly uv points . (This is valid only for algebraically closed fields, so in the smaller field we may have less.) The conclusion is that, in general, a point P off F lies on exactly $n(n - 1)$ tangents to F.

6.12   Cubic Curves and Arcs

Let Γ be an elliptic curve in the plane over $F = GF(q)$. We then have $q + 1 - 2\sqrt{q} \leq |\Gamma| \leq q + 1 + 2\sqrt{q}$, where $|\Gamma|$ denotes the number of points on Γ.

Suppose now that $|\Gamma|$ is even, so 2 divides $|\Gamma|$. Recall that P_∞ serves as the identity element for the group G associated with Γ: Denote it by 0. From our earlier definition it will follow that $P + Q + R = 0$ if and only if P, Q, R are collinear. By hypothesis, 2 divides $|G|$. From group theory there exists a subgroup H of G with $2|H| = |G|$. So we can write G as the disjoint union of H and the coset L where $|H| = |L|$. Then, from group theory we have that

(a) The sum of any 2 elements in L lies in H.

(b) The sum of any 3 elements in L lies outside H.

From (b) it follows that the sum of any 3 elements in L is never 0, since 0 is in H. Thus the set L gives rise to an **arc** with size $|L|$ i.e. a set of $|L|$ points in the plane with no 3 collinear, where $|L| = \frac{1}{2}|G|$. Such large arcs have been intensively studied in connection with Segre's theorem and a generalization of the Hasse theorem known as the Hasse–Weil theorem. We refer the reader to [Bru84] for more detail. As pointed out there, the arc L was independently discovered by A. Zirilli and by P. M. Neumann (see [Bru84]). In fact, building on the work of J. Voloch, Alderson and Bruen [AB04] show that in many cases the code associated with L is maximal.

6.13   Homogeneous Coordinates

Given any "Euclidean" or "affine" equation such as (6.1), namely, $y^2 = x^3 + ax + b$, we can convert it (by adjoining appropriate powers of z for each term) to a homogeneous equation in which each term now has degree 3. Thus the above equation becomes

$$y^2 z = x^3 + ax^2 z + bz^2. \tag{6.4}$$

On the other hand, given equation (6.4), we can get the Euclidean equation by putting $z = 1$. The point (x, y) corresponds to the point $(x, y, 1)$ in homogeneous coordinates. In fact, two triples $(u_1, v_1, w_1) \neq (0, 0, 0)$ and $(u_2, v_2, w_2) \neq (0, 0, 0)$ are said to represent the same point if and only if $u_2 = \lambda u_1$, $v_2 = \lambda v_1$, $w_2 = \lambda w_1$ with $\lambda \neq 0$, i.e. one triple is a nonzero scalar multiple of the other.

The advantage of homogeneous coordinates is that algebraically they are easier to work with, compared to the usual Euclidean coordinates. Also, the behavior at infinity is captured by putting $z = 0$. If we do this in equation (6.4) we get $x^3 = 0$, giving $x = 0$. So the point(s) at infinity of equation (6.1) are all points corresponding to the triples $(0, y, 0)$ with y in F. By "scalar multiples" we just get one point at infinity, namely, the point corresponding to the triple $(0, 1, 0)$, which is the point P_∞.

6.14 Fermat's Last Theorem, Elliptic Curves, Gerhard Frey

Fermat's last theorem is the assertion that the equation $x^n + y^n = z^n$ has no solutions when x, y, z are nonzero integers and $n > 2$ is an integer. Fermat, from Toulouse in France, stated this around 1630, but it was not proved, despite many attempts, until 1997, by A. Wiles.

In fact, it suffices to prove the equation has no solutions when $n > 2$ is an odd prime. For, let $n = pq$ where p is an odd prime. Since $x^n + y^n = z^n$, we have $(x^q)^p + (y^q)^p = (z^q)^p$ where now also x^q, y^q, and z^q are nonzero positive integers.

Suppose then that we are looking at the equation $x^p + y^p = z^p$ for p an odd prime and let $x = a, y = b, z = c$ be a solution. Thus we have $a^p + b^p = c^p$ where a, b, c are nonzero integers. Then, $a^p + b^p - c^p = 0$. Frey's insight then was to study the Frey elliptic curve with the equation given by $y^2 = x(x - a^p)(x + b^p)$. He predicted that this curve would be incompatible with the Taniyama–Shimura conjecture. This is a central conjecture about curves that states that rational elliptic curves are **modular** so that they arise from modular forms. Wiles then proved the Taniyama–Shimura conjecture for a large class of rational elliptic curves including the so-called **semistable** curves. Because Frey's elliptic curves were semistable, Fermat's result followed as a corollary.

6.15　Problems

1. For the elliptic curve E with equation $y^2 = x^3 + 2$ given in the text over the field $GF(7)$, describe some calculations for the group G associated with E. In particular, determine whether or not G is cyclic.

Nonsingular Curves.

2. We examine the curve E given in the text where E has the equation $y^2 = x^3 + ax + b$ over a field K. In homogeneous coordinates E has equation $y^2 z = x^3 + axz^2 + bz^3$. Under what conditions does there exist a point P in the plane such that $F_x(P) = F_y(P) = F_z(P) = 0$?

3. If $F_x(P) = F_y(P) = F_z(P) = 0$ for some homogeneous polynomial F, must P lie on E (defined above)?

4. Use the equation $y_1 F_x + y_2 F_y + y_3 F_z = 0$ to find the equation of the tangent line to E (defined above) at a point. Reconcile this with the calculus approach suggested in the text.

5. In general, how many tangents can be drawn to E from a point P?

6. Give an easy example of a curve that does not factor over a field F but does factor over a larger field K containing F.

An Application of Nonsingularity.

7. Show that the Fermat curve $F = 0$ given by $x^n + y^n - z^n = 0$ over any field K where the characteristic of K is zero or does not divide n is irreducible (i.e. does not factor).

8. For the elliptic curve with equation $y^2 = x^3 + 4x + 4$ over Z_{13}, what is the size of the group G?

9. If G is the group in Problem 8, what is the structure of G?

6.16　Solutions

1. As shown in the text, $|G| = 9$. We know that G has P_∞ as the neutral or zero element 0. Take any nonzero point P such as, say, $(3,1)$. To find $P + P$ we have $\gamma = \frac{3x_1^2 + a}{2y_1} = \frac{3\,3^2 + 0}{2} = \frac{27}{2} = (27)(4) = (6)(4) = (-1)(4) = 3$. Then $P + P = (x_3, y_3)$ where $x_3 = \gamma^2 - x_1 - x_2 = 9 - 3 - 3 = 3$ and $y_3 = \gamma(x_1 - x_3) - y_1 = 3(3-3) - 1 = -1$. So $(3,1) + (3,1) = (3,-1)$ and $(3,1) + (3,1) + (3,1) = (3,1) + (3,-1) = P_\infty$. Thus $(3,1)$, and in fact all points $Q \neq P_\infty$, have the property that $3Q = P_\infty = 0$. Thus there is no point P such that all points on the curve E are of the form xP, $1 \leq x \leq 9$ i.e. there is no generator. So G is not cyclic.

2. We have $y^2z = x^3 + axz^2 + bz^3$ giving

$$x^3 + axz^2 + bz^3 - y^2z = 0. \qquad (6.5)$$

$F_x = 0$ implies $3x^2 + az^2 = 0$.

$F_y = 0$ implies $-2yz = 0$.

$F_z = 0$ implies $2axz + 3bz^2 - y^2 = 0$.

Suppose $z = 0$. Then $F_x = 0$ implies $3x^2 = 0$. Assume that the characteristic of the field is not 3. Then x must equal 0. So a point satisfying $F_x = 0$ and $F_y = 0$ must look like $(0, y, 0)$. For this triple to be a point we must have $y \neq 0$. But then $F_z = 0$ is not satisfied by $(0, y, 0)$.

Conclusion. If the characteristic of K is not 3, then no point with $z = 0$ satisfies $F_x = 0$, $F_y = 0$, $F_z = 0$. However, if char $K = 3$ the point $(1, 0, 0)$ satisfies $F_x = 0$, $F_y = 0$, $F_z = 0$. Note that this point $(1, 0, 0)$ is **not** on the curve. So $z \neq 0$.

Now, suppose $z \neq 0$. Because of scalar multiples we can take $z = 1$. Our equations become

$$3x^2 + a = 0$$
$$-2y = 0$$
$$2ax + 3b - y^2 = 0$$

We have 3 cases.

char $K = 2$. We have $3x^2 + a = 0$, $3b - y^2 = 0$.

So we have solutions if $x^2 = \frac{-a}{3}$, $y^2 = 3b$. Any such solutions will give points on the curve.

char $K = 3$. Then our conditions become $a = 0$, $y = 0$, $2ax - y^2 = 0$.

That is, $a = 0$, $y = 0$. Substituting in the equation of the curve gives a singular point $(v, 0)$ on the curve where $v^3 = -b$.

char $K \neq 2, 3$.

We have $3x^2 + a = 0$, $y = 0$, $2ax + 3b = 0$. Suppose $a = 0$. Then $3b = 0$. So $4a^3 + 27b^3 = 0$. Suppose $a \neq 0$. Then $x = \frac{-3b}{2a}$. But $3x^2 + a = 0$, giving $3\left(\frac{-3b}{2a}\right)^2 + a = 0$. This implies $(3)\frac{9b^2}{4a^2} + a = 0$, $27b^2 + 4a^3 = 0$, i.e. $4a^3 + 27b^2 = 0$.

Conclusion. If char $K = 2$ or char $K = 3$ there can be points P with $F_x(P) = F_y(P) = F_z(P) = 0$ even if $4a^3 + 27b^2 \neq 0$. However, if char $K \neq 2$ and char $K \neq 3$ and $4a^3 + 27b^2 \neq 0$ there are no points P in the plane with $F_x(P) = F_y(P) = F_z(P) = 0$.

Note that there are other criteria that will guarantee that we cannot have $F_x(P) = F_y(P) = F_z(P) = 0$. For example, for char $K \neq 2, 3$ we have $3x^2 + a = 0$, so $x^2 = \frac{-a}{3}$. Thus if $\frac{-a}{3}$ has no square root then we cannot have $F_x(P) = F_y(P) = F_z(P) = 0$. However, in this case P is singular because $\frac{-a}{3}$ has a square root in the quadratic extension field.

3. The Euler identity says that

$$xF_x + yF_y + zF_z = nF. \qquad (6.6)$$

Now if $F_x(P) = F_y(P) = F_z(P) = 0$ this implies that $nF(P) = 0$. Thus if the characteristic of the field does not divide n we conclude that $F(P) = 0$ so P is on F.

4. We work with the equation $y^2 z = x^3 + axz^2 + b^3$ i.e. $x^3 + axz^2 + bz^3 - y^2 z = 0$. Suppose $P = (u, v, w)$ is on the curve. Then the equation of the tangent line at P can be calculated. We have $F_x = 3x^2 + az^2, F_y = -2yz, F_z = 2axz + 3bz^2 - y^2$. So the equation of the tangent line at P is

$$x(3u^2 + aw^2) + y(-2vw) + z(2auw + 3bw^2 - v^2) = 0. \qquad (6.7)$$

Assume $w \neq 0$. So P is not equal to P_∞. So we can assume $w = 1$. Then the point $(u, v, 1)$ satisfies the equation

$$x(3u^2 + a) + y(-2v) + (2au + 3b - v^2) = 0. \qquad (6.8)$$

The affine equation of the curve is

$$y^2 = x^3 + ax + b. \qquad (6.9)$$

From calculus, we have $2y\frac{dy}{dx} = 3x^2 + a$ so $\frac{dy}{dx} = \frac{3x^2 + a}{2y}$. The equation of the tangent at (u, v) is as follows:

$$y - v = m(x - u) = \frac{3u^2 + a}{2v}(x - u), \qquad (6.10)$$

and so

$$2vy - 2v^2 = (3u^2 + a)(x - u). \qquad (6.11)$$

This gives

$$x(3u^2 + a) - 2vy - 3u^3 - au + 2v^2 = 0. \qquad (6.12)$$

Since (u, v) is on the curve, $v^2 = u^3 + au + b$, $u^3 = v^2 - au - b$. The constant term in (6.12) becomes $-3(v^2 - au - b) - au + 2v^2 = -v^2 + 2au + 3b$. So (6.8) and (6.12) give the same answer.

5. If P is off the curve we get in general 6 tangents. If P is on the curve we get 4 tangents. (Note that $6 = n(n-1) = (3)(2)$.)

6. Let's work with the curve $x^2 + y^2 = 0$. This does not factor over the real numbers. However, over the complex numbers, which contain the reals (i.e. are an extension of the reals) we have $x^2 + y^2 = (x+iy)(x-iy)$ with $i^2 = -1$. So over the complex numbers the curve $x^2 + y^2 = 0$ factors into the product of 2 lines with slope i, $-i$.

7. Suppose F factors as GH. Then G, H are homogeneous with $\deg(G) = n_1$, $\deg(H) = n_2$ and $n_1 + n_2 = n = \deg(F)$. Now there exists a point P defined over $\bar{K}$, the algebraic closure of K, such that $G(P) = 0$ and $H(P) = 0$. It follows that $F_x(P) = G_x(P)H(P) + H(P)G_x(P)$. Thus $F_x(P) = 0$. Similarly, $F_y(P) = 0$, $F_z(P) = 0$. But F_x is nx^{n-1}, $F_y = ny^{n-1}$, $F_z = nz^{n-1}$. Let $P = (u, v, w)$. Then $F_x(P) = nu^{n-1}$. So $F_x(P) = 0$ with characteristic of K not dividing n implies $u = 0$. Similarly $F_y(P) = 0$ implies $v = 0$ and $F_z(P) = 0$ implies $w = 0$. So $P = (0, 0, 0)$ which is the one triple that does not represent a point in homogeneous coordinates. We conclude that F does not factor over a field K if the characteristic of K does not divide n.

In summary, the moral is this: *F is nonsingular implies that F is irreducible.*

8. G has exactly 15 elements.

9. G is an abelian group with the size of G being pq where $p \neq q$ are primes. From group theory G must be cyclic. Any point P on the curve not equal to P_∞ will serve as a generator. For example, the point $(1, 3)$ or the point $(-1, -5)$ is a generator.

Chapter 7

General and Mathematical Attacks in Cryptography

Goals, Discussion   Some of the classical cryptanalytic techniques were already introduced in Chapter 2 for what are often referred to as "pencil and paper" ciphers. In Chapter 3 we also discussed several attacks on various implementations of RSA. Here we introduce several techniques, mathematical and otherwise, developed for the breaking of modern ciphers and other components of crypto-systems used in real-life applications and protocols. These are in addition to the attacks on insecure implementations of RSA discussed in various problems in Chapter 3.

New, Noteworthy   In addition to giving a comprehensive overview of the most common attacks on privacy, we detail how the combination of a result described in Chapter 3 with the known attack on low decryption exponent suggest a possible new attack on RSA.

7.1   Cryptanalysis

Technically speaking, we can define **cryptanalysis** (or cryptanalytics or cryptoanalysis) as the art and science of solving unknown codes and ciphers. Cryptanalysts try to break the codes and ciphers created and used by cryptographers. This can sometimes be achieved either by obtaining the key or by directly obtaining the message. By extension, the art of exploiting weaknesses in protocols used to communicate secure information (authentication, key-management, software/hardware defects, etc.) also falls within the bounds of cryptanalysis. Practitioners often refer to the cryptanalytic process as "breaking the crypto-system".

In general, cryptanalysts are faced with the task of first determining the language being used for the communication, the general type of cipher or code being used, the specific key(s) and finally, the re-construction of the message or plain text. These determinations have to be made based on variable amounts of cipher text and related information such as the identity of sender and receiver, statistical analysis of traffic, knowledge of some specific information about the contents of the message, etc.

A famous literary example of these ideas can be found in Edgar Allan Poe's tale *"The Golden Bug"*[Poe93]. In this tale, the principal character describes, in vivid detail, the entire process of cryptanalysis of a simple substitution cipher based only on the cipher text and a scant knowledge of the sender's identity and intent. In real life, examples abound of broken secrecy systems based on knowledge of side information or weaknesses of the system rather than on the breaking of the cipher itself. Frequently in cryptanalysis we try to piece together various existing probable pieces of information in order to determine the secret key or the message.

7.2   Soft Attacks

No matter how sophisticated the attack techniques become, one must not forget that when the ultimate goal is to obtain the secret message, **coercion** or **social engineering** are often the most effective attack techniques. These attacks are based on using physical or psychological threats, robbery, bribery, embezzlement, etc. The attacks are mostly directed to human links of the data security chain. Extensive consultation with experienced professionals and security experts is a defense against these attacks. In a sense, design of the whole data security system includes the selection of hardware, software and human resources, as well as the implementation of mechanisms to check the proper functioning of these elements. However, as so many high profile cases have shown, no matter how much effort is spent on keeping information secret, human nature will always conspire against, and defeat, even the best designed secrecy systems.

The particular techniques used in physically securing information is beyond the scope of this book. We will concentrate instead on mathematical aspects of attacks and assume that physical and operating data security has implemented. The interested reader can consult the extensive literature on the subject by authors such as Schneier [Sch03] and Mitnick [MSW02].

In the problems in Chapter 3 we have already described various mathematical attacks on RSA. We proceed with other kinds of general attacks, which are so numerous that we can only give a very short summary here. Many attacks are based on Internet vulnerabilities and are described in detail in books on Internet security. Scheneier [Sch96] also presents several examples.

7.3 Brute Force Attacks

Assuming, as **Kerchoff's principle** recommends, that the algorithm used for encryption and the general context of the message are known to the cryptanalyst, the brute-force attack involves the determination of the specific key being used to encrypt a particular text. When successful, the attacker will also be able to decipher all future messages until the keys are changed. One way to determine the key is by exhaustive search of the **key-space**[1], or **brute force**. In Chapter 3 we have described some mathematical results applicable to brute-force attacks.

Brute force is a passive or **off-line attack**. The attacker (*Eve* in this case) passively eavesdrops the communication channel and records cipher text exchanges for further analysis, without interacting with either **A** or **B**.

Brute force attacks can be carried out knowing only a small portion of cipher text and the corresponding plain text (such a collection of data is commonly referred to as a **crib**). The attack consists of systematically trying all possible keys on the key-space until the key is found that enciphers the plain text into the cipher text (or vice-versa) for the particular encryption algorithm being used. To estimate the time that a successful brute-force attack will take we need to know the size of the key-space and the speed at which each key can be tested. If N_k is the number of valid keys and we can test N_s keys per second, it will take, on average $\frac{1}{2}(\frac{N_k}{N_s})$ seconds to find the proper key by brute-force: see Chapter 3. In 1996 Schneier [Sch96] published a table with the cost-time analysis of brute-force attacks. It is clearly shown that brute-force attacks on fixed algorithms using 128 bits keys were impractical then[2]. *They still are*, even factoring in the advances in computer processing speed and the decrease in prices (roughly a factor of 10 for both speed and price).

If no amount of plain text is known with certainty or the amount of cipher text is small, the brute-force attack may not even be practical; more than one guessed key may yield meaningful messages from which the attacker may not be able to decide, with certainty, the correct one. For a brute-force attack to succeed, the attacker must have sufficient information about the contents of the message to be able to recognize the right key when it is found. In the case of public key algorithms it is possible to check, with certainty, that the correct key has been guessed.

The threat that a brute-force attack poses cannot be underestimated in the real world. Most financial institutions use cipher-systems based on DES. Keys of length 56-bits, such as the one used by the standard implementation of DES, can be obtained by brute-force using computer hardware and software available since the late 1990's with a cost of about $250,000 [PM02] given suitable information concerning the type of message being trans-

---

[1]Defined as the set of all possible valid keys for the particular crypto-system.

[2]In 1996 Schneier calculated 10^{11} years using hardware that costs $10,000,000,000,000 for a brute-force attack on a key-space of 128 bits.

Figure 7.1: Man in the Middle Attack on Symmetric Key Encryption

mitted. Again we refer to Chapter 3 for some discussion on this. Indeed, to counter this possibility, most contemporary implementations of DES use a derivative known as Triple-DES (or 3-DES) which uses three different 56-bit keys instead of one. The effective key length for the combined 3-DES key is a more secure 168 bits. Double-DES has a key 112 bits in length but as we will point out, it is susceptible to birthday attacks and therefore not practical in that it does not offer much extra security.

Brute force analysis can be used in combination with other attacks as was the case for the deciphering of the Enigma. The famous **bombes** [Bau02] designed by M. Rejewski were an example of the brute-force approach working in combination with a mathematical method that provided an important reduction of the key-space.

7.4   Man-In-The-Middle Attacks

This is a generalization to previous impersonation attacks such as IP address **spoofing** and it is by far the most powerful attack in cryptography. **A** and **B** are communicating with each other, supposedly, along with *Malone* (or *Mallory* or *Middle*, the malicious active attacker). Note that *Malone* is much more powerful than the traditional passive attacker *Eve*. Not only can *Malone* listen in to the cipher text of messages between **A** and **B**: he can alter messages, delete messages or even create fictitious messages on his own by impersonating **A** to **B** and **B** to **A**.

This kind of attack has been known since antiquity and all ciphers are susceptible to it. It will fail if **A** and **B** have a proper system of authentication (such as the use of **digital signatures**) which are discussed in Chapter 4. Moreover, one can say that this type of attack is one of the main reasons why an authentication system is needed, so that **A** and **B** can be certain that they are communicating with each other. Widely used e-commerce applications, such as the SSL protocol, are susceptible to the Man-in-the-Middle attack.

In the case in which both parties are using a shared secret key scheme, M has to intercept the initial key in transit from **A** to **B** to be successful (Figure 7.1). It is always good practice to use an independent channel to transmit keys. In the case in which a public key scheme is being used (Figure 7.2), a successful **man in the middle** M replaces **A**'s public key being transmitted to **B** and **B**'s public key being transmitted to **A** with public keys generated ad-hoc, for which M has the corresponding private keys. **Public key infrastructure** (PKI)

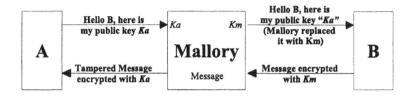

Figure 7.2: Man in the Middle Attack on Public Key Encryption

addresses this problem by creating a **Certificate Authority** (CA) that issues special files called **certificates**. These certificates can be used by **B** (or **A**) to verify the authenticity of **A**'s key (or **B**'s key). By querying the CA, **B** can independently assert whether or not the given public key is uniquely associated with the identity of **A**. This system is logically equivalent to a **Trusted Server**, which provides authentication and session keys for symmetric encryption systems,as in the **Kerberos** system.

In any case, the idea is to replace trust between **A** and **B**, who at this point might not know each other, by a third party which **A** and **B** independently trust. It has been often pointed out that shifting the trust in this way removes the problem of the man in the middle but generates new potential weaknesses such as the certification of the CA or the level of trust that can be accorded to the Trusted Server.

7.5 Known Plain Text Attacks

In this situation the adversary *Eve*, who is listening in, is assumed to possess a considerable length of message text and corresponding cipher text (i.e. a "crib") and from this seeks to find the key. This situation is a realistic one, especially for public key algorithms when the same key is used for a lengthy period.

E-mail and **e-commerce** applications are susceptible to attack in this way because the plain text for addresses, routing and headers are known or fairly easy to guess. An abundance of software is available on the Internet to stage this type of attack. Attacks using cribs were effective for the Allied forces in WWII. Since the German military even encrypted trivial data such as weather forecasts, the Allies were able to obtain very useful cribs which helped crack the Enigma machine.

7.6 Known Cipher Text Attacks

In this scenario, the adversary *Eve* only knows a piece of the cipher text and tries to obtain the plain text and the key from it. This type of attack always involves a large amount of work but is the one that requires the least input as the cipher text is almost always available to the attacker. We already discussed some particular cases such as the methods presented in Chapter 2 to attack the Vigenère cipher. Gillogly [Gil95] presented a classical example of a known cipher text attack on the Enigma cipher.

7.7   Chosen Plain Text Attacks

In this type of attack the cryptanalyst must be able to define a particular plain text, feed it into the encryption device and retrieve and analyze the resulting cipher text. This could be used to attack RSA, for example, where the public key is known.

It is not necessary for the cryptanalyst to have access to the enciphering device. In some cases it is possible to get the victim to help carry out the attack by 'planting' the chosen plain text in a normal exchange, in such a way that does not raise the victim's suspicions.

There are several well-known historical examples in which one side, directly or indirectly, revealed a purportedly 'secret text' and then eavesdropped on the communication lines of the enemy to see what the corresponding cipher text was, therefore gaining information about the enemy's cipher-system.

As mentioned, public key ciphers (in particular RSA) are vulnerable to this type of attack. In general, any system that uses the same keys to encrypt large volumes of data will be vulnerable to a chosen plain text attack. As an example, one can mount an attack of this kind to obtain the encryption keys from a wireless network using Wired Equivalent Privacy (WEP) by sending a large number of chosen plain text messages (spamming) to one user and eavesdropping on the traffic between the user and the wireless server.

7.8   Chosen Cipher Text Attacks, Digital Signatures

In this type of attack, the analyst chooses some cipher text and tries to get the victim to decrypt it in order to find the key. An alternative approach is to decrypt the cipher text with different random keys in order to get some information about the real key.

If RSA is also being used for digital signatures, *Eve* can generate a random number r and ask *Bob* to sign the message $C' = r^e C$, which looks like a random message to *Bob*. If he signs the message, *Eve* will have $C'^d = r^{ed} C^d = rM$. She needs only to divide by r to obtain the message. To avoid this situation, a **hash function** must be used when an unknown message is digitally signed.

This type of attack can be employed against popular e-mail encryption systems such as PGP, although it seems to fail if compression is used in addition to encryption.

7.9 Replay Attacks

Under certain conditions an attacker may use parts of a previous encrypted exchange between *Alice* and *Bob* to gain access to their resources.

Suppose that *Alice* sends an encrypted request to transfer funds to her bank. The key used to sign the message is only known to *Alice*. By copying and later replaying the encrypted transfer request, *Eve* can repeat the same transaction without Alice's knowledge. To be successful, *Eve* needs to guess or otherwise determine which part of the cipher text corresponds to the transfer request, copy the encrypted request, and replay it at another time, without ever having to decrypt.

In general, a *replay attack* occurs any time an intruder manages to capture a message (or part of a message) with the intention of using it at a later time. There are two variants of this attack. In the first variant, the original message is allowed to reach the destination unimpeded, but it is copied to be replayed later. In the second variant, the original message is prevented from arriving at the intended destination and is copied to be used later. See [Syv97] for further details.

To guard against such attacks, many protocols, such as Kerberos, include a time-stamp of some type or other. The protocol is designed in such a way that when **A** sends a request, **B** has a limited amount of time to acknowledge and complete the transaction. In that way when the intruder copies the encrypted exchange and later tries to replay it, **B** can decrypt and check the time stamp, refusing the transaction if the delay is larger than a predetermined amount.

7.10 Birthday Attacks

Suppose we have n people in a room. An individual's name is encoded as a binary string. Denote the corresponding variable by x. So altogether x has n distinct values, one for each person in the room. Each of the n people has a birthday which can be described by a short binary string y representing a number between 1 and 365 (or 366 in the case of a leap year).

We can now construct a hash function f which associates the string y with the string x. One can then ask the following question: what is the probability that f does not have a collision? In other words, what is the probability that no 2 people have the same birthday?

Of course if n is greater than 365 (or 366) that probability is zero: some 2 people must have the same birthday. But what if $n < 365$ (or 366)?

For the following discussion, assume that the current year is not a leap year. If we arrange the n people in some order, the total number of possible strings of length n that we obtain if we have n distinct birthdays is

$$(365)(354)(363)\cdots(365-(n-1)). \tag{7.1}$$

The total number of birthday strings of length n, distinct or not, is

$$(365)(365)\cdots(365) = (365)^n \tag{7.2}$$

Thus the probability that we end up with n birthdays such that no two are equal is (7.1) divided by (7.2) which gives

$$\left(1 - \frac{1}{365}\right)\left(1 - \frac{2}{365}\right)\cdots\left(1 - \frac{n-1}{365}\right)$$

If n is 23, the probability that we end up with no two people having the same birthday is $0.49 < 0.5$, so if we have 23 or more people it is likely that some two have the same birthday. We can generalize the argument above as follows.

Recall (calculus again) that $\ln(1-\lambda)$ is approximately equal to λ for λ small. This follows because $e^\lambda \approx 1 + \lambda$ so $e^{-\lambda} \approx 1 - \lambda$. If we generate n numbers independently, at random (i.e., with replacement) from a set $1, 2, 3, \ldots, t$ the probability that they are all distinct, denoted by $P(t, n)$, is, as above,

$$\prod_{j=0}^{n-1}\left(1 - \frac{j}{t}\right)$$

Now

$$\ln(P(t, n)) \approx \sum_{i=0}^{n-1} \ln\left(1 - \frac{j}{t}\right) \approx -\sum_{j=0}^{n-1} \frac{j}{t} = -\frac{n(n-1)}{2t}$$

If n is much smaller than $\sqrt{t}$, $P(t, n) \approx 1 - \frac{n^2}{2t}$. Thus, for birthdays, $P(365, 23) \approx 0.5$

As t gets large one can show that the expected number of draws until a collision, i.e., until two of the drawn numbers coincide, is $\sqrt{\frac{\pi t}{2}}$.

In the case of a hash function whose output is m bits, so $t = 2^m$, the average number of trials until a collision is about $2^{\frac{m}{2}} = \sqrt{t}$ which is the square root of the total number of possible outputs. Because the average number of trials until a collision is $2^{\frac{m}{2}}$ rather than 2^m, we can say informally that the output hash value should be "twice as long as it usually is" in order to withstand a brute-force collision attack.

7.11   Birthday Attack on Digital Signatures

The following attack is due to Yuval (How to Swindle Rabin, Cryptologia 3, July 1079, 187-190).

Alice wants to cheat *Bob*. She prepares 2 contracts X, Y with Y being favorable to *Bob* and X being favorable to *Alice*. *Alice* makes small harmless changes to X, such as inserting spaces, etc., and also to Y.

If each document has m lines, *Alice* can easily construct 2^m versions of X and 2^m versions of Y. If *Alice* is using a hash function with n bits and m is about $\frac{n}{2}$, the chances are good that some version of X, say X_1, hashes to the same value as some version Y_1 of Y.

Now *Alice* asks *Bob* to digitally sign Y_1 using a signing protocol whereby *Bob* only signs the hash value. Then down the road, *Alice* can replace Y_1 by X_1 — the contract favorable to *Alice* — and claim that this is what *Bob* has signed.

Birthday attacks can also be used on Diffie–Hellman (by solving the discrete log problem) and on Double DES (which is why Double DES is not used). The technique used is similar to the previous signature scheme.

7.12 Birthday Attack on the Discrete Log Problem

Given a prime p and a generator g we want to solve the **discrete log problem** given by the equation $Rem[g^x, p] = v$, i.e., $g^x \pmod{p} = v$, where we are also given v. We construct two lists (analogous to the variation on the 2 contracts in the previous discussion) called X and Y.

X consists of numbers $g^t \pmod{p}$ for about $\sqrt{p}$ values of t.

Y consists of numbers $vg^{-s} \pmod{p}$ for about $\sqrt{p}$ values of s. Note that g^{-s} is just the multiplicative inverse of $g^s \pmod{p}$.

Now, since the number of elements in X multiplied by the number of elements in Y is about p, there is a collision probability greater than $(\frac{1}{2})$, i.e., there are t, s with $g^t \pmod{p} = vg^{-s} \pmod{p}$, or, $Rem[g^t, p] = Rem[vg^{-s}, p]$. Thus $Rem[g^t - vg^{-s}, p] = 0$. Multiplying by g^s gives $Rem[g^{t+s} - v, p] = 0$, i.e., $g^{t+s} = v \pmod{p}$ and we have solved the discrete log problem.

7.13 Attacks on RSA

We continue our discussion with some specific attacks on the RSA algorithm. However, as pointed out by Boneh [Bon99], "although twenty years of research have led to a number of fascinating attacks, none of them is devastating. They mostly illustrate the dangers of improper use of RSA." One such attack is the *factoring attack*. If we can factor $N = pq$ we can break RSA. One such factoring technique which works in certain cases is called the Pollard p-1 method and is discussed in Chapter 19. Advances in computing power and in mathematical methods over the last 35 years can be gauged by the increase in the length of numbers that are practical to factor. From a 39-digit number in 1970, to a 129-digit number in 1994[3], to a 155-digit number in 1999, the progress has been faster than predicted. This last number of digits is significant because it is equivalent to a binary 512-bit long number[4], a key-length that has been used by many financial institutions for their implementations of RSA.

Another possibility that an attacker has is to re-encrypt the cipher text (anybody can do this because the public encryption exponent and the modulus are public) and iterate the encryption step on each new cipher text thus obtained, until the message is recovered. This attack is a consequence of one of the properties of the RSA algorithm, that for each M there exists a unique number k called the iteration exponent or period of M such that $C_{k+1} = C_0$, where $C_{k+1} = C_k^e \pmod{N}$ and $C_0 = M$. This attack can be efficiently staged only for relatively small p, q and e.

The RSA recommended modulus length is currently set at numbers between 640-bits to 2048-bits long. In practice, moduli in excess of 1024-bits are the norm these days. This is a hot topic of discussion for practical applications because in the case of RSA, Diffie–Hellman and ECC systems, increasing the key length increases the computational effort required to solve the underlying problem even though, the entropy of the longer key is still 0. Meeting the needs of secure communication for portable wireless systems becomes an increasingly difficult task as key-lengths grow to reach acceptable security levels. Many of the proposed attacks on RSA are effective only if RSA is not properly implemented. To increase the security offered by RSA, one option is to "pad" the message M by appending random bits to the message. While padding does add some randomness to the message, there exist attacks (such as the Coppersmith Short Pad Attack) that exploit random padding to determine M.

For further details on correct RSA implementation and RSA attacks we refer the reader to Mollin [Mol00] and Boneh [Bon99].

7.14   Attacks on RSA using Low-Exponents

To save computational effort, users may be tempted to select small encryption and decryption exponents. Because they make the modular exponentiation operation faster, Fermat primes $e = 2^1 + 1 = 3$, $e = 2^4 + 1 = 17$, or $e = 65537 = 2^{16} + 1$ are commonly chosen for the exponent e in practical implementations of RSA. Once e is fixed, it is easier to conduct the tests $\gcd(e, p - 1) = 1$ and $\gcd(e, q - 1) = 1$ while generating and testing the primes p and q, rejecting the primes that fail this test. The mathematics of RSA is such that this practice can make it vulnerable to attacks.

When the encryption exponent e is less than the number of recipients k of a given

---

[3]The challenge to factor a 129-digit number known as RSA129 was set forth in 1978 when the original RSA paper was published.

[4]RSA512

message M, an eavesdropper may recover the message as follows: *Eve* sees k cipher texts M^e (mod N_i). She knows that for all recipients to be able to uniquely decipher the message the condition $M < N_i$ must hold for $i = 1, 2, \ldots, k$. Therefore, $M^e < N_1 N_2 \cdots N_k$. If the N_i are relatively prime, *Eve* can compute M^e (mod $N_1 N_2 \cdots N_k$), where $e < k$ using the Chinese Remainder Theorem. She then has a perfect integer power over the integers, namely M^e, from which she can calculate the eth root and recover M. Alternatively, *Eve* can factor the N_i's and compute M.

To avoid these types of attacks a large encryption exponent e must be selected.

The possibility of successful attack arises when a small decryption exponent is used to reduce decryption time. Wiener [Wie90] published a detailed description of a method for recovering the decryption key when a small decryption exponent d is used. We present here the main theorem without proof.

Theorem 7.1 *Let $N = pq$ with $q < p < 2q$. Let $d < \frac{1}{3} \sqrt[4]{N}$. Given (N, e) such that $ed \equiv 1$ (mod $\phi(N)$) then an attacker can efficiently recover d.*

This means that for a 1024-bit modulus, d needs to be at least 256-bits long to be secure. There is a potentially damaging avenue here for an attack that utilizes the results presented in Chapter 3 regarding the non-uniqueness of the decryption exponent.

Partial knowledge of the decryption exponent can also be exploited. Thus, if $N = pq$ is an n-bit RSA modulus and the $n/4$ least significant bits of the decryption exponent d are known, there is an algorithm that can be used to efficiently reconstruct the whole decryption exponent and the factorization of N. We refer the reader to Boneh, Durfee and Frankel [BDF98].

7.15   Timing Attack

In 1995, Paul Kocher proposed a simple algorithm to obtain bits of an RSA key by measuring the slight time differences in the computations times for a series of decryptions [Koc96]. For each particular implementation of the algorithm there is a correlation between the time the decryption takes and whether the bits of d, the decryption exponent, are 0 or 1. If the least significant bits of the decryption key are known, by carefully measuring the average time that decrypting several (known) cipher texts takes, one can guess whether the next bit of the decryption exponent is 1 or 0. The process can be repeated to obtain one additional bit of the decryption exponent per repetition. Diffie–Hellman and Digital Signature Standard (DSS) are also susceptible to this type of attack. The attack is passive but needs to be staged on-line because the attacker needs to know the time taken by the system to decrypt each cipher text C.

7.16   Differential Cryptanalysis

This method is designed in a similar way to the chosen plain text attack. At the outset we should point out that it now appears that this attack is impractical: for details we refer to the the end of this section. Originally the method was developed to attack ciphers based on **Feistel algorithms**, such as DES. The attack consists of looking at differences produced in the cipher text by known differences in the plain text. The chosen plain text is carefully designed by the attacker to contain known differences that will result in a particular statistical distribution of the cipher text bits as they are shuffled through the successive rounds. A search is then made for the cipher texts corresponding to those patterns. With this information, it is possible to make inferences about S-boxes in DES and the bits of the key with increasing certainty, as more plain texts are used. Of course the internal details of the algorithm need to be known to give the attacker the ability to generate meaningful statistics for different patterns.

As was shown in Chapter 5, DES consists of mostly linear operations with the exception of the S-boxes, which provide all the entropy. All the operations, including the selection of entries in the S-boxes, are dependent on the bits of the key. The differential attack is designed to reveal information about which S-boxes are being used and, consequently, about parts of the key. The attack will work on ciphers designed around weak S-boxes or on watered-down versions of DES with a reduced number of rounds[5]. The method of differential cryptanalysis was known to the team that created DES in the early 1970's as the **T-attack**. To avoid revealing how these potential weaknesses of the algorithm were addressed, the team decided not to publish the design criteria for the S-boxes at that time.

The method of differential attack was rediscovered in the late 1980's by den Boer (see [dB88]) who applied it to a four-round FEAL cipher. Later it was extended to a reduced eight-round DES by Biham and Shamir [BS90]. Because of the importance of DES in industry, the issue of a theoretically possible attack generated a lot of debate in the ensuing years. Eventually, to quell the rumors that the algorithm was susceptible to attack or that it had a back door, IBM had to publicly acknowledge that in the design of the S-boxes one of the main criteria was to make the algorithm optimally resistant to differential cryptanalysis. A review article by Coppersmith [Cop94] discusses these issues to dispel concerns raised within the security community by the possibility that differential cryptanalysis works on DES. *Its main point is that because of the amount of chosen plain text needed for a successful attack on the full 16-round DES, (approximately 1.2×10^{15} bytes) the attack is impractical.* Experts in the field of security agree with Coppersmith's assessment.

---

[5]There are no reports of successful, practical attacks on the algorithm with 16 rounds as it is normally implemented.

7.17 Implementation Errors and Unforeseen States

It is very difficult to write software that is free from errors or unexpected responses to unforeseen inputs. The variety and complexity of modern hardware/software combinations make the notion of error-free systems utopian, to say the least. Readers following the science and technology section of any major news source these days would have the feeling that security holes or bugs, susceptible to use by hackers, are discovered almost weekly in popular software applications. Computer system administrators and end users find it hard to keep up with the installation of a stream of security patches released by major software vendors.

Indeed, it is a fact of life that an army of hackers is working around the clock to find the next buffer overrun on popular software applications or to learn how to fool operating systems into allowing malicious code to run undetected. Although it seems unfair, this situation is only to be expected. Many authors have already described the conflict between code-makers and code-breakers as an ongoing battle. Both sides know that any little advantage could tip the balance one way or the other.

From the cryptographical point of view, it is very important to know the implementation details for each particular algorithm, or at least to be confident that the necessary precautions had been taken at the moment of writing software. For implementations of standard algorithms such as RSA, DSS, DES, AES, etc., there is in place a validation program set up by the NIST[6] that ensures the compliance of the software to the corresponding FIPS[7] standard. From the commercial point of view it is very important to achieve certification if one hopes to sell to governments or large organizations. New (or old) cryptographic methods, not covered by existing standards, cannot be certified under FIPS.

The FIPS certification gives some reassurance of quality to users of cryptographic modules. However, it does not guarantee that the underlying operating system or the network structure is secure. There are cases of features found in widespread applications that can be used to steal information from the victim's computer. For example, *Malone* can send *Alice* an innocent-looking text file for review. Unknown to *Alice*, *Malone* added a special, legitimate instruction to embed a given file (or several files) to the document. When Alice opens the document for editing, the special, hidden field will embed the indicated files from *Alice's* computer without *Alice* ever knowing what is happening. When *Alice* returns the corrected file to *Malone* the stolen files will be embedded in it and *Malone* will have no problem retrieving information. This technique works even if encryption is used to transmit the files back and forth.

As the industry pushes programmers and software engineers for more features in hardware and software, unforeseen states and responses leave open doors for **viruses, worms**

---

[6]National Institute of Standards and Technology
[7]Federal Information Processing Standards

and **trojan horses** that can take control or extract data from the victim's computer. By July 2004, industry estimates indicate that more than 81,000 threats of this kind existed, and the number is increasing daily. Even the new generation of feature-rich cell phones are no longer immune to this type of threat (see [Cel]) as news of the development, by Russian researchers, of the first worm to spread through cell phones indicates. Defense against these types of threats has become a flourishing business, with many companies dedicated to producing software and hardware capable of stopping the attacks. The best defense is to keep software patches updated and be on the alert for new developments. All major antivirus software includes links to a network of websites that keep users updated on current security threats.

Chapter 8

Practical Issues in Modern Cryptography and Communications

Goals, Discussion In real life, mathematical concepts are not the only factors to be considered in the design of communication systems. Systems intended for commercial, military and government use interact with a variety of hardware and software applications, management systems and data security policies. The systems must also be able to handle diverse networking protocols. We briefly describe here some of the principal issues.

New, Noteworthy In this chapter, we discuss various practical issues which are of fundamental importance to the development and deployment of digital communication systems. The ideas expressed are based on our experience and industrial research.

8.1 Introduction

It is no secret that the volume of information transmitted over the Internet and the demand for connectivity have been steadily increasing over the last few years and will probably continue at a sustained rate. To stay competitive and maintain revenues, communication service providers are constantly looking at more effective ways to deliver a variety of data services. The above-mentioned growth in volume and complexity makes this task difficult. In particular, from the engineering point of view, as designs for hardware and software become more varied and complex, communication protocols must be designed more carefully to meet the two competing goals of interoperability and security. In our experience in industry, we

145

have found that in this complex and competitive environment some of the fundamental constraints are as follows:

- **Technical** The system (hardware/software/protocol) must perform the task for which it was designed under a wide variety of conditions, many of which are beyond designers' control. The complexity of modern systems often makes the task of designing them well beyond the abilities of code-developers.

- **Commercial** The business model used to commercialize new applications must make sense. One of the golden rules of business (know thy customers) is sometimes forgotten in the excitement of developing new products or services. Equally damaging is the tendency to ignore the need for official validation or certification. This can make a product unacceptable for certain market segments, no matter how good it is technically.

- **Property Rights** Patents are essential to protect and commercialize any technological advantage. There are many lawyers who offer advice on patenting issues. However, the real value of a patent resides on having a good grasp of issues related to the **prior art**, the correct mathematics and proper engineering practices, issues which are outside the field of expertise of most lawyers. It is your responsibility to make sure you have access to good technical advice on these aspects. Do not rely on lawyers or patent agents.

- **Legal** The requirement to encrypt and protect certain data has now become the law in California[1] and it is likely that other jurisdictions will follow suit. This has implications for the design of information systems, communications systems and data repositories. Also, for auditing purposes for example, information may need to be stored for several years, bringing issues of file encryption to the forefront.

The above constraints should be taken into account in the design/development cycle. We will briefly outline some specific concerns and how to address them in this chapter. In the specific area of applications of encryption, the reader can consult Graff [Gra00] and Ferguson and Schneier [FS03].

8.2   Hot Issues

Perhaps the most important issue to be addressed by any system designer is that related to **user authentication**. Based on our experience in writing patents, we offer the following advice. The strength and value of a patent depends on having a detailed knowledge of what has already been patented, i.e., the prior art. If you are writing a patent, you must be

---

[1]California Information Practices Act (SB 1386)

familiar with correct mathematical formulation and engineering practices. These issues are outside the field of expertise of most lawyers and patent agents.

Openness and accessibility are desirable characteristics for the legitimate user of a network, and a nightmare for administrators charged with the task of keeping out illegitimate users and attackers.

An adequate authentication system, if available, would solve the problem by admitting only those users that can produce the proper **credentials**. However, the implementation of an 'adequate authentication system' is easier said than done. We have briefly described several systems for the authentication of network users by utilizing keys. Here we will give more details on the practical aspects.

To give the reader a brief overview, we discuss below some of the topics that are generating intense debate and are relevant to the application of cryptography. Authentication is at the core of many of these unresolved problems. We start here with a brief summary of the subject and refer those interested in more details to the book by Smith ([Smi02]). For related issues on e-commerce, see the book by Graff ([Gra00]).

8.3 Authentication

What is authentication? In general, authentication has to do with protecting identity and privacy. For example, when withdrawing money from the bank or from an ATM it is necessary to provide evidence (account number, PIN, bank card, etc.) that one is the legal owner of the bank account. Authentication is fundamental for controlling access to data security systems. Traditionally, user authentication is based on one or more of the following:

1. Something you know, for example a password or PIN number

2. Something you have, for example, a smart card or an ATM card

3. Something that is physically connected to the user such as biometrics, voice, handwriting, etc.

What are some of the advantages and disadvantages of the various authentication systems based on 1, 2 or 3 above?

For computer and network access, **passwords** have been the main tool for verifying identity and granting access. Indeed, about 99% of all computer identification uses password authentication. There are many different kinds of password schemes and mechanisms. The potential weaknesses are well known. Also, passwords, if sent unencrypted, i.e., in plain text over a network, are vulnerable to 'sniffing attacks' (i.e. electronic interception). Password guessing has become big business. Many websites sell, for quite a modest price, password cracking software. Frequently, but not always, password cracking software is based on a brute-force attack. As we showed in Chapter 3, the password can be obtained when about half of the total number of possibilities have been tried. More mathematically sophisticated attacks are also used by these software programs. Police also make use of these commercially available programs for forensic work, whenever they need to decrypt confiscated files.

The security of a password is determined by the **Shannon entropy** of the password. Shannon entropy is a measure of the difficulty in guessing the password. This entropy is measured in **Shannon bits**. For example, a random 10-letter English text would have an estimated entropy of around 15 Shannon bits, meaning that on average we might have to try $\frac{1}{2}(2^{15}) = 2^{14} = 16384$ possibilities to guess it. In practice, the number of attempts needed would be considerably less because of side information available and redundancy (=patterns and lack of randomness). A major weakness of passwords is that the entropy is usually too small for security since users do not remember long passwords and are prone to use very well known methods for constructing their passwords. It is well-known to hackers that users commonly select passwords that include variations of the user name, make of the car they drive, name of some family member, etc. As we mentioned in Chapter 7 social engineering is one of the most powerful tools being used by hackers.

When logging on to a remote server, passwords are also subjected to **sniffing** attacks whereby the attacker eavesdrops on the traffic between the user and the server and looks for occurrences of the password. Even if an encrypted version or the hash of the password is sent to the server, the attacker can use this information to log in by replaying this portion of data. To avoid these type of attacks, modern systems use a protocol based on **challenge and response** or encrypt the traffic with a protocol called **SSL** (Secure Socket Layer).

SSL works in the following way. The server sends the user its RSA public key; and the user then generates a secret random number and encrypts this number with the server's public key. The server decrypts and uses the secret random number as a symmetric secret key for encryption. From this point on, all the traffic is encrypted and the authentication of the user can now proceed over an encrypted channel. To avoid the potential problem of an attacker setting up a fake server to obtain user information, "**Public Key Certificates**" issued by official **Certificate Authorities** are used to authenticate or validate the user to the server. See Chapter 3 for more information on SSL.

Concerning authentication based on "something you have", the most common method involves plastic cards with a magnetic stripe such as **ATM cards**, credit cards, drivers license etc. Some of the weaknesses of this method are as follows. These cards can be transferred to other individuals (by lending the card, losing it or having the card stolen), sometimes without the users knowledge. Just as it is easy to copy keys it is also very easy to copy such magnetic strips. Unscrupulous personnel have been known to copy magnetic strips with an ad-hoc modified electronic reader. Cards can also be copied easily. PIN numbers, usually in the format of four-digit numbers, are easily guessed or surreptitiously recorded by hidden cameras located on top of the point-of-sale terminals.

Other products include an active electronic circuit such as **USB biometric key**, the **SecurID** time-synchronous and event-synchronous one-time password tokens from RSA Security, the **JavaCards**, **iButtons** and the many kinds of smart cards available on the market. These active tokens use cryptographic techniques for authentication that are immune to sniffing attacks. Since they generate **one-time passwords** they are also immune to replay attacks[2]. Thus it does an attacker no good to try to replay a previous message or set of messages. However, inevitably, they have weaknesses. Some of the devices are vulnerable to off-line attacks. There is also the fundamental "lost or stolen" problem. Also, there is a security risk for a network based on the fact that the owner of the token can lend (or even sell) the token to unauthorized individuals. In some sense, these devices have the same weaknesses that PKI has. Here, PKI means "public key infrastructure", i.e. the keys, authentication protocols etc. associated with systems using public key cryptography as explained in Chapters 3 and 4. Namely one can only be confident, but not certain, that on the other side of the line, the correct user has the authentication device, be it an electronic certificate (a file) or a token. However, one can never have the same level of confidence that the authentic user is there. When transactions involve legal or financial liability, trusting a device that can be stolen or lost it is a very risky proposition.

Biometric Authentication Any characteristic uniquely associated with an individual that can be discriminated by mechanical means can be used for an authentication system. Such characteristics are commonly referred to as **biometrics**.

The idea of using biometrics as a means of identification is not new. In 1880 Dr. Henry Faulds published an article where he discussed using fingerprints for personal identification. He is also credited with the first fingerprint identification - a greasy fingerprint left on an alcohol bottle. The first documented use of a fingerprint to prevent document forgery was that by Gilbert Thompson of the U.S. Geological Survey in 1882. By 1892 Sir Francis Galton, a cousin of Charles Darwin, published his book, *Fingerprints*, in which he established the individuality[3] and permanence of fingerprints as well as the first classification system for fingerprints. In the same year the first criminal fingerprint identification was made by Juan Vucetich, an Argentine police official, who also began to systematically file the fingerprints of the general population as a mean of identification. Even today, some countries include a copy of thumbprint on **ID cards**.

Newer technologies are being introduced that can recognize a person based on differences in the iris pattern, facial shape, hand geometry and speech patterns. Authentication using

---

[2]An attack in which the eavesdropper records the traffic and later forwards the relevant information to gain access to the server

[3]According to his calculations, the odds of two individual fingerprints being the same were 1 in 64 billion.

iris scans and digitized speech is still in its infancy[4]. However, biometric authentication is taking off. Wealthy individuals now routinely use biometrics for household entry in conjunction with other security measures. Airports are increasingly making use of the technology. In the United States, several states such as California use biometric indicators such as fingerprints to authenticate individuals. Some details of the process, involving the digitizing of fingerprints are described below. The process should satisfy several major requirements. Two of these requirements are as follows.

1. A detailed statistical analysis to ensure that the number of "false positives" and "false negatives" is minimized. This means that:

   a) another person's fingerprint will not be accepted and;

   b) the valid owner's fingerprint will be accepted by the validator checking the authentication.

2. The entropy of the digital signature is sufficiently high (as measured in Shannon bits) that it is not vulnerable to attack.

We should point out that, in fact, the entropy requirements are less severe than for the smart cards and similar devices described above because, unlike such devices, which are vulnerable to various off-line attacks, biometric devices can only be attacked interactively. Also biometric devices are not vulnerable to "transference" (the "lost or stolen" syndrome). Theoretically, biometric devices are vulnerable to forgery. For example, a thumbprint could be copied to a plastic mold. However, this is much more difficult than stealing a smart card or guessing a password. We should point out that in the case when biometric readers are physically attached to the computer a viable attack can be launched whereby the attacker makes a mold of the user's fingerprint and uses it to gain access in the absence of the user. Another negative to the adoption of a biometric solution might be an individuals concern that the biometric data could fall into the wrong hands. However, with proper cryptographic techniques and security policies in place, this concern can be allayed.

The security offered by biometric devices can be combined with the security provided by time synchronous or event-synchronous tokens as described above. This is a **two-factor authentication system**, obtained by combining two well-established authentication techniques, namely biometric signatures and event-synchronization. If an unauthorized user, denoted by $\mathbf{X}$, by a stroke of fortune, guesses the biometric signature, he still has to guess the number in the events counter. Moreover, in the statistical sense, these guesses are independent and so provide mathematically independent authentication guarantees. To illustrate this point, suppose that the probability of $\mathbf{X}$ guessing the biometric signature is unrealistically high, say 0.001 i.e., $\mathbf{X}$ succeeds once in every 1000 trials. Similarly assume

---

[4]See http://news.bbc.co.uk/2/hi/technology/2584951.stm and related articles.

that the probability of **X** guessing the number in the event counter is 0.003. Then, because the two methodologies are working independently, the probability of guessing both simultaneously is only 0.000003, i.e. 3 out of 1,000,000, which is a dramatic reduction in the probability of a malfunction. This applies also to the security. If there is a 1 in 1000 chance to break one method of authentication, and a 3 in 1000 chance of breaking a second, independent method of authentication, there will then be only 3 in 1,000,000 chances of breaking a system that combines the two methods. Indeed, there is no doubt that employing biometrics dramatically increases authentication levels over devices where the ownership can be transferred. For organizations with authentication concerns for their communications and network devices we suggest that two-factor authentication should be the standard.

Analysts estimate that the biometric industry is worth between 240 million and 400 million dollars a year. From the point of view of the application of the cryptographic principles stated in this book, authentication is the most difficult part. This is definitely where money and effort are going to be spent in cryptographical systems of the future.

8.4   E-Commerce

Data encryption, authentication, message integrity and non-repudiation are essential for on-line commerce. Initial estimates for an explosive growth of e-commerce were grossly optimistic. That was in part due to the security concerns that the public expressed at the time of putting credit card information on-line. Credit card companies have countered that trend by insuring transactions, relieving the user of responsibility for on-line credit card fraud and paying for the insurance against on-line fraud.

The problem of identity theft, either by eavesdropping on legitimate transaction, exploitation of security holes in the Internet browsers and servers, viruses, trojan horses, spy-ware or the so-called **phishing**[5], is becoming a big problem. The research firm Gartner estimates that 57 million Americans in the past year received phishing e-mail messages. Gartner estimates that phishing related fraud cost banks and credit-card companies about $1.2 billion in direct losses in the last year alone. The cost to businesses of all these attacks combined, amount to many billions of dollars a year.

More secure technologies for on-line transactions must be implemented at all levels, from secure protocols to secure applications and better authentication methods, designed with security in mind. Indeed, problems such as e-mail spam cannot be solved unless the Internet protocol is modified in such a way as to force every packet of data to have a valid return address.

Higher level protocols may also need to be modified as there are reports that some

---

[5]The technique known as phishing is often based on fraudulent e-mail sent to lure people to phony websites asking for financial information.

implementations of SSL are insecure[6].

Malicious software in the form of **viruses, trojan horses** or **spy-ware** have been used by attackers to obtain financial information and personal data.

8.5   E-Government

Another emerging area of communications that is quickly expanding and for which data security is fundamental, is the so-called **e-government**. Nowadays, all the on-line exchanges between any level of government and its citizenry are lumped together under this heading.

Transactions can be of different kinds. The simplest and most frequent case is the searching for information about services or the downloading of official forms from a government website. Transactions such as renewing a driver's license, responding to jury summons and answering preliminary questions, changing an address with a government agency such as the post office, and having it disseminated automatically to all specified government agencies, will be made more convenient for both the government and the public if carried out on-line. More complex transactions such as those involving the exchange of sensitive information, (i.e., filing taxes on-line or exchanging health records) are also possible. Some other transactions, such as voting on-line, have proven very difficult to implement given the current state of the technology.

Although recent data suggest that up to 50% of internet users in the United States have visited a government website[7], still many are reluctant to complete online transactions with the government. The concerns are twofold, namely, security of the information being exchanged and the (real or perceived) threats to privacy.

A national **ID cards** system would be necessary for more complex transactions, but such an idea has proven controversial. The majority of users in the United States and Canada oppose this idea, although there is more support for such a system in many European countries.

If a system such as **PKI** is used for these purposes, a normal ID card can be replaced by a smart card with built-in public keys and corresponding electronic certificates issued by the government (which will play the role of the Certificate Authority). However, as noted in a recent GAO report[8], there are serious logistical problems for the deployment of PKI within large organizations such as the federal government. The deployment of such systems in the general population will be met with even more difficulty. In particular, the general population must be given sufficient training on key use, management, and protection issues.

---

[6]See,   for   example,:   http://news.bbc.co.uk/2/hi/technology/2785145.stm   and http://lasecwww.epfl.ch/php_code/publications/search.php?ref=Vau01.

[7]The New E-Government Equation: Ease, Engagement, Privacy & Protection, Hart-Teeter/Council for Excellence in Government, April 2003

[8]General   Accounting   Office,   Status   of   Federal   PKI   Activities   2003, http://www.gao.gov/new.items/d04157.pdf

As electronic transactions are much easier to track by automatic means, concerns about the invasion of citizen privacy may prove to be the most difficult obstacle for widespread adoption of e-government.

The only viable route for the introduction of e-government might be a system that rewards the voluntary and incremental adoption of electronic IDs by interested users.

Another hot issue closely related to the government is the possibility of casting electronic votes. Such a system would encourage more people to participate in the democratic process, be it in electing representatives or voting in referendums on important issues. It would also make vote-counting faster and more accurate. Unfortunately, to the best of our knowledge, no scheme or technology proposed so far is even remotely secure.

Protecting information about their customers, suppliers and employees is, without a doubt, in the best interest of any business. Moreover, it is likely that more and more jurisdictions will follow the precedent set by the State of California regarding the need for companies to encrypt personal data.

The **Encryption Act** has been in effect since July 2003. Under the Act, anybody dealing with the personal information of third parties such as names, addresses, Social Security Numbers, etc., must protect such data by means of encryption. It not only forces the protection of personal data in transit, (e.g. using SSL or S-HTTP) but also requires the protection of the databases containing the information. Additionally, any release of unencrypted information, either on purpose or by way of a security breach, must be immediately reported to the interested parties (the people whose data was released). The wording of the Act goes so far as to state that if you possess the personal information of even one California resident in a server located elsewhere, the information must be encrypted.

8.6 Key Lengths

There is a lot of contradictory information coming from vendors on the length of keys needed to ensure adequate security. The issue are germane for the security of portable computing devices due to their limitations in computing power. This is of particular concern to public key systems, RSA or Elliptic Curve Cryptography (see Chapters 3 and 4). Proponents of the different algorithms regularly arrange public challenges, offering monetary rewards to the first that can break a cipher obtained with a given key-length. The value of such challenges is dubious at best. There is no easy answer to the question of how long the key must be; it all depends on the particulars of each implementation. Unfortunately, in this particular aspect, the best advice is *caveat emptor*! It should be borne in mind that, on the one hand, an institution whose security has been breached may not wish to publicize this. On the other hand, the attacker also may want to keep a good thing going.

8.7   Digital Rights

The most popular web-based businesses are centered around the selling of any kind of content in digital format on-line, as there is no need for the physical exchange of goods. Content, be it text, software, music, films, TV, etc., can be downloaded by the buyer from the server and paid for through an electronic money transfer, giving the buyer the advantages of convenience and instant gratification and eliminating the cost to the seller of maintaining and transporting physical inventories.

Indeed, the publishing, music recording and motion pictures industries are expending a great deal of effort and money to develop systems that would make bookstores, record stores and video rental outlets superfluous.

One of the main obstacles to the widespread adoption of this business model is that of restricting unauthorized access and reproduction of the content once it is off-line. The problem is akin to that of authentication. Many systems employing encryption have been unsuccessfully tried so far, for example, **CSS** (Content Scrambling System) for DVD's. In most cases the protection systems were "broken" within weeks of being publicly released.

Digital rights management (DRM) is a very divisive issue because the rights that the content owner chooses to grant are not necessarily the same as the actual legal rights of the content consumer. Opponents claim that digital restrictions management is a more accurate description of the real functionality.

Attempts to prevent unauthorized access to content that has been delivered via some other electronic channels such as radio and satellite signals, have the same type of technical and management problems. In some jurisdictions it is illegal to receive satellite broadcasts from certain providers, even if the equipment and the content were purchased from the provider.

In lieu of technically feasible solutions that work with the variety of recording/reproducing systems available in the market, publishing industries have resorted to the enforcement of stricter copyright laws such as the **Digital Millennium Copyright Act** (DMCA) which make the attempt of circumventing digital protection systems illegal. Thus the publishers are mainly relying on the justice system to fight unauthorized access and reproduction of copyrighted material.

8.8   Wireless Networks

The introduction of digitally modulated radio-frequency links not only made possible the development of mobile phones, but also opened the door to the concept of mobile computing, i.e., wireless transmission of digital data.

There are differences between **wireless wide area network** (WAN) and **wireless local area network** (LAN) and **wireless personal area network** (PAN) technologies.

However, from the security point of view, the problem of keeping data being transmitted through a wireless link, is a common one for any kind of network.

Cellular Digital Packet Data (CDPD) is a secure, proven, and reliable protocol that has been used for several years by law-enforcement agencies and public-safety officials to securely access critical, private information.

The operation of the CDPD network is as follows: A wireless modem (basically a cell-phone) communicates by radio with the Mobile Data Base Station (MDBS), the cellular tower. The tower transfers this data using a land phone line or microwave to the Mobile Data Intermediate Systems (MD-IS), which processes and sends the information, by Intermediate System gateways (routers), to the appropriate destination.

The manufacturer assigns each wireless modem device a fixed number or Equipment Identifier (EID), which is unique to that modem. The network carrier assigns a Network Entity Identifier (NEI) to each subscriber in the network. This number is uniquely associated with the EID (no two devices in CDPD can have the same EID) which gives the modem an IP address visible to the rest of the Internet. Security is achieved by assigning, at registration, the Authentication Sequence Number (ASN), and the Authentication Random Number (ARN), which together with the EIN form the credentials of that modem. This operation is carried out using the Diffie–Hellman Key Exchange. Although a subscriber can determine her NEI, she cannot obtain the ASN or ARN. In this way there are no chances of "cloning" the modem, because two devices with the same NEI and different ARN and ASN will be detected by the carrier. When a subscriber's modem goes through the authentication procedure during each network registration, the MD-IS checks the subscriber credentials against the current values of the ASN and ARN that are stored in the database. If the stored values do not match those provided by the modem, then the connection is refused. Periodically, the MD-IS generates a new (random) value for the ARN, and increments the ASN by 1. The MD-IS delivers the new ARN to the modem in the final step of the encrypted registration process. The modem stores this ARN internally and increments its local ASN by 1.

Since CDPD is a public wireless data communications service susceptible to eavesdropping, all data transferred between the modem and the MD-IS is encrypted by using the RSA algorithm. The protocol compresses the header information of each IP data packet, segments them for transfer over the CDPD network, and encrypts the segments. Data is encrypted from the modem to the MD-IS. Beyond the MD-IS however, data is generally not encrypted, just as general Internet traffic remains unencrypted unless the end user provides it. If necessary, the carrier or end user may encrypt data traveling over other portions of the network using other mechanisms.

The amount of security that can be achieved is limited by the limited computing power that wireless devices generally have[9], restricting the key-lengths that are practical to use for encryption purposes.

Wireless LAN's have experienced an explosive growth. Almost every laptop computer sold these days in North America includes a wireless network card and the low cost of wireless routers makes them affordable for small networks and even for home use. The introduction of the **hot-spot** technology that provides public wireless access to businesses, academic centers and other repositories of information should multiply the growth rate of wireless networks many-fold.

The technology that provides data security to such networks is known as **WEP** (Wired Equivalent Privacy), covered by the 802.11 standard. The WEP algorithm relies on symmetric key encryption with either 40-bit or 128-bit keys. It has been found that there are a number of flaws in the implementation and key management (at least in the version corresponding to 802.11 and 802.11b standards) of the encryption algorithm, which makes it sensitive to a series of passive and active attacks in which an unauthorized user can retrieve information or gain access to the network.[10]

However, the biggest problem with wireless networks seems to be that unless encrypting traffic is the default setting, many users do not even bother with encryption. Widely published studies indicate that around city centers, up to 60% of corporate wireless LAN's are open to eavesdroppers.

8.9   Communication Protocols

Internet traffic moves over a heterogeneous medium. Physically, signals that represent bits move over copper wires, optic fiber and air. Logically, bits move across a variety of platforms, ranging from cell-phones to supercomputers, which invariably have different ways to encode information. The glue that makes all these systems work together and transport the information without errors is what we call **communication protocols**. These protocols must be sanctioned as standards, so any manufacturer can design and manufacture equipment that works within a common framework. (Digressing, when a new product, e.g. home video cameras, is in the developing stage there is usually a fierce battle between the proponents of different systems, e.g. BETAMAX vs. VHS, to establish their own as the standard. Once the standard is adopted, be it on technical merits or otherwise, the proponents of the winning system usually take a huge share of the market).

One of the protocols that work for the basic layer of data transport is the **ATM** (Asynchronous Transfer Mode) that parses the information into a fixed-size cell for transport

---

[9]Mobility implies a limited energy source, most of which is needed to put enough radio-frequency power on the antenna.

[10]Security of the WEP algorithm, by N. Borisov, I. Goldberg and D. Wagner, http://www.isaac.cs.berkeley.edu/isaac/wep-faq.html

across the network. This protocol supports voice, video and data simultaneously and it is used for applications that need fast transmission.

The **Internet Protocol** (IP) is another data transport protocol that routes packets by including network, source and destination information together with the data. The IP packets are larger and can transport more data than ATM; however, they are not optimized for time-sensitive transport. Despite this limitation IP is used for **Voice Over IP** (VOIP), which is the protocol that makes possible cheap long-distance calls by means of phone-cards.

Another widely used protocol is the one known as **Synchronous Optical NETwork** (SONET) for the transmission of large amounts of data over fiber-optic networks.

WAN's and LAN's historically have used a protocol called Ethernet for the transport of data between nodes in the network. For residential and small business subscribers, Internet providers use a protocol called **Digital Subscriber Loop** (DSL) for high-speed communication over existing copper telephone lines. DSL can achieve speeds of 128 Kbits per second and is based on complex digital signal processing techniques.

The protocols used for exchanging communication are in themselves codes. In a sense, an encrypted message sent over a **TCP/IP** connection can be considered as a code riding on top of another code. Nothing prevents us from designing a protocol (code) that can provide transport, error correction encryption and compression of the message. It is also desirable to have standard protocols that can be adhered to by all the players in the communication industry. Hybrid protocols have been developed, for example, **IPSEC**, to address the inherent security weaknesses of the IP protocol, which was not designed with security concerns in mind.

Parts II and III of this book address information theory and error correction, both of which are integral to modern secure communications.

Part II

Mainly Information Theory

Chapter 9

Information Theory and Its Applications

Goals, Discussion We present a fairly complete but accessible survey of the subject involving ideas from mathematics, physics, and engineering. Not much mathematical background is required.

New, Noteworthy By discussing two of the main results we bridge the gap in Section 9.9 between abstract units of information, known as Shannon bits, and regular bits (= binary digits). We also explain a connection between weighing problems and information theory in Section 9.8. We show in Section 9.4 how information theory can be used in a fundamental way in cryptography. Connections with physics are explored.

9.1 Axioms, Physics, Computation

Information theory can be approached from many different points of view. It has many strands. In some treatments, it is an arcane mathematical subject concerned with axioms, measure theory, and abstract probability theories. In other treatments, it is described in terms of formulae that are the underpinnings of practical questions in modern communication theory in science and engineering. To others it is a subject that is inextricably linked with physics, notably statistical physics, heat, energy and the theory of computation as in Feynmann [Fey99]. Another approach is to tie the subject to complexity and randomness as developed by Solomonoff, Kolmogorov, Chaitin, and others. In this chapter, we will try to touch on these ideas.

As pointed out in Chapter 3, the subject has been dominated by Claude Shannon for over fifty years, but there have been several new applications and developments. For example, information theory is being used in an important way in molecular biology and genetics. In nanotechnology, the parallel theory of quantum information is coming to the fore (Nielsen and Chung [NC]). In this book, we present, in Part III, a new application of information theory using classic physics, to cryptography. This application has recently made its way to the market place and to industrial applications.

Basically, information theory has to do with converting knowledge about probabilities to hard "information," suitably defined, involving units called **Shannon bits**. Ideas of randomness and redundancy follow closely. Rather than launching into definitions we first skirmish, informally, with some ideas.

9.2   Entropy

We start off with numbers. Any positive integer such as 43 can be expressed uniquely as $3 \times 10^0 + 4 \times 10^1 = 43$ using the powers of ten but we can use any other base, or number system, such as binary. For example, in binary, 43 can be written as the binary string of length 6 given by 101011. To explain this, we have $43 = 3 \times 10^0 + 4 \times 10^1$ if we read from the right. Also, reading from the right, $43 = 1 \times 2^0 + 1 \times 2^1 + 0 \times 2^2 + 1 \times 2^3 + 0 \times 2^4 + 1 \times 2^5$.

Now, suppose we are told that a given number is written as a binary string of length 6, and we are asked to guess that number. What are our chances of success? The total number of possibilities is 64, ranging from the number 0 to the number 63. To see this, the rightmost slot can be filled in two ways, and similarly for all the other slots. Thus we have $2 \times 2 \times 2 \times 2 \times 2 \times 2 = 64$ possibilities, ranging from 0 to 63, since $2^6 = 64$. So our probability of success in guessing is $\frac{1}{64}$. In general, if the binary string is of length n, i.e., has n bits, each of which is chosen independently, randomly deciding between 0 and 1 at each stage, there would be 2^n possible numbers from which to choose. Thus n binary digits can indicate any one of 2^n possible numbers. In this case, the **uncertainty** or **entropy** of the string is defined to be n because $n = \log(2^n)$. In this book, log invariably means log to the base 2. In general, we can say that, for a binary string,

$$entropy = \log(number\ of\ possible\ strings) \tag{9.1}$$

Getting back to the case $n = 6$, the number of possible numbers or outcomes would be reduced if we had some extra algebraic information. For example, if we knew that the six bits added up to zero (addition in binary) then any one of the six bits, for example, the sixth bit, is **redundant**. It can be calculated from the other five bits and thus causes no extra uncertainty. In this case, the string of length 6 is **less random** than before. The entropy now is just 5. So we have three principles:

$Entropy$ = $Randomness$

$Entropy$ = $Uncertainty$

$Redundancy$ = $Lack\ of\ Randomness$

But now suppose we just have some **probabilistic** knowledge of the binary sequence. For example, suppose we know that the probability of getting 1 in any bit is, say, equal to 0.85, independent of the other bits. In other words, for each bit a "free choice" is made as to the content, be it 1 or 0, apart from the condition that, overall, a one is chosen with probability 0.85. What are the possibilities now? What is the entropy? This is a very special case of one of the problems solved by Shannon.

Of course, all 2^n binary strings are still possible. For a string of length n, the average number of ones appearing will be $n(0.85)$. However, in all probability, only a **typical sequence** will materialize and the probability of a **nontypical sequence** will tend to zero as n gets large. For example, suppose $n = 100$. In this case, the number of typical sequences is about 2^{61} (see below for details). Because $\log_2(2^{61}) = 61$, the entropy of a typical binary sequence of length 100 (which would be 100 if the sequence were random) is now just 61.

Let us explain. The **Shannon function** $H(p)$ is defined as follows.

$$H(p) = p \log(\frac{1}{p}) + q \log(\frac{1}{q}) = -\Big[p \log p + q \log q\Big], \qquad (9.2)$$

where $q = 1 - p$ and logs are to the base 2. If $p = 0.85$ then $H(0.85)$ is about 0.61 and $100H(0.85)$ is about 61.

A rough sketch of $H(p)$ is as follows.

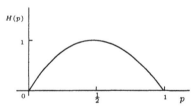

From the diagram we see that $H(p) \le H(\frac{1}{2}) = 1$. Since $p = 0.85$, the average number of ones in a binary sequence of length 100 is $100 \times 0.85 = 85$. *We define a typical sequence here to be one in which the total number of ones is close to 85. The details are in Chapter 12. Shannon's work on entropy, as indicated below, implies that the number of typical sequences will be, in this case, about $2^{nH(p)} = 2^{100 \times 0.61} = 2^{61}$.*

In this particular case, we can show this without explicitly using Shannon's work, as follows. As indicated above, the number of 1s on average is np, where p is the probability of a one appearing. To make life easy, we assume np is an integer like 85 in the above. Then the number of sequences of length n that have exactly np 1s is the binomial coefficient $\binom{n}{np}$. We just choose the np out of n positions where we want the ones to appear. If we take the log of this number to the base 2, then, when n is large, the number $\binom{n}{np}$ is approximately $nH(p)$.

The proof uses an approximation for binomial coefficients known as Stirling's expansion. The details are in Feynmann [Fey99].

In summary then, the entropy has been reduced from n to $nH(p)$.

9.3   Information Gained, Cryptography

Let's look at this from another point of view. Suppose that cryptographic station **A** is sending a secret message M to cryptographic station **B** in the form of a binary string of length n. Let's assume that an eavesdropper E (for Eve) by listening has acquired sufficient information that E knows each bit that **A** has transmitted, independently, with a probability of 0.85. How much information about M does E really have?

Compare the string that E has with that of **A**. The probability that E has the right bit is $p = 0.85$. Thus E knows that with high probability, as above, the total number of strings from which the message M has been chosen is about $2^{nH(p)}$. The calculation is as above. The probability of "getting 1" is analogous, for Eve, to getting "Yes" for a bit, where "Yes" means that E and A agree in that bit. Thus, from the point of view of E, the number of possible strings from which the message M has been chosen has been reduced from 2^n to about $2^{nH(p)}$. Therefore, the entropy has been reduced from n to $nH(p)$ $(= \log_2(2^{nH(p)}))$. Thus the loss of uncertainty is $n - nH(p) = n(1 - H(p))$. Therefore the **gain in information** that E has is $n(1 - H(p))$. When $n = 100, p = 0.85$, Eve gets about $100(1 - 0.61) = 39$ units of information.

We are coming very close to the idea of channel capacity. This was mentioned in the Preface and comes in again in the last chapter.

We can look at it this way. On average, Eve will agree with **A** in 85 positions. However, because she does not know where the corresponding positions of agreement are, Eve only knows "with certainty" about 39 units of information. We record the following principle: *Entropy Removed = Uncertainty Removed = Information Gained.*

Definitions.   It is high time to give some definitions. Suppose that we have a probability space with probabilities $p_1, p_2, \ldots, p_m$. For example, we might be rolling a fair die so $m = 6, p_1 = p_2 = \cdots = p_6 = \frac{1}{6}$. The individual probabilities always lie between 0 and 1 and add up to 1. We define the **entropy**, following Shannon, as follows.

$$H(p_1, p_2, \ldots, p_m) = p_1 \log(\frac{1}{p_1}) + p_2 \log(\frac{1}{p_2}) + \cdots + p_m \log(\frac{1}{p_m}). \qquad (9.3)$$

Examples.   If we roll a fair die so $p_1 = p_2 = \ldots = p_6 = \frac{1}{6}$ then $H = \log 6 = 2.59$. To give another example, suppose we are tossing a coin in which the probability of getting heads is p. Then $H(p, 1 - p) = H(p)$, where $H(p)$ is the Shannon function. If $p = 0.85$ we get

$H(0.85)$ which is equal to 0.61 approximately. Then $H(0.85)$ is the amount of entropy or uncertainty we have about the result of the coin toss *before* the toss. We can also think of $H(0.85)$ as the amount of information received *after* the toss, i.e., the amount of information received when the outcome of the toss is revealed.

Suppose that $p = \frac{1}{2}$. Then

$$H(\frac{1}{2}, \frac{1}{2}) = H(\frac{1}{2}) = \frac{1}{2}\log 2 + \frac{1}{2}\log 2 = 1. \tag{9.4}$$

This is taken as the unit of information: It is one Shannon bit. As Golomb [Gol02] so eloquently puts it: A **Shannon bit** "is the amount of information gained (or entropy removed) upon learning the answer to a question whose two possible answers were equally likely, a priori." With this measure, when the die is rolled as above, a knowledge of the outcome reveals 2.59 Shannon bits of information.

This connection between entropy and "yes or no" questions also suggests the following principle:

Entropy = number of yes or no questions needed, on average, to determine the outcome of an experiment.

Another approach to information and entropy is as follows. Think of information as "surprise." If the probability p of an event is small, then $\frac{1}{p}$ is big so we receive a big surprise if the event occurs. Actually, it is more convenient to work with $\log(\frac{1}{p})$ rather than $\frac{1}{p}$. Another idea is to think of $\log(\frac{1}{p})$ as **the information content** of the event. The formula above for entropy suggests another principle:

Entropy = Average Surprise = Average Information Content.

If p is small, $\log(\frac{1}{p})$ is big, but the product $p\log(\frac{1}{p})$ balances things out in the product when calculating H. In the coin-tossing experiment above, when $p = 0.85$ the entropy $H(0.85)$ is around 0.61. How does one talk about 0.61 questions? The solution is to use "block coding" whereby we toss the coin n times and determine the outcome of the n tosses with about $nH(0.85)$ questions. We can also identify tossing a coin n times with our previous discussion of the entropy of a binary string of length n.

We will always distinguish between a regular bit or binary digit with values 0 or 1 and a Shannon bit, which is a unit of information.

As above, $H(p) = H(p, 1 - p)$ is the information revealed by knowledge of the outcome of one experiment, where p is the probability of getting "heads" or by being informed that a one is in a given position in a binary string. Then, if we perform either experiment n times independently, the information revealed by the outcome is $nH(p)$.

In general, for a binary string of length n, with the probability of getting 1 in any given position being some number $p, 0 \le p \le 1$, the entropy or uncertainty removed or information gained on learning what the string is will be bounded as: $0 \le nH(p) \le nH(frac12) \le n$. In this case, we see an example of the following principle:

The entropy or information content of a binary string of n bits is at most n Shannon bits and is at least 0 bits (because entropy is always nonnegative).

9.4   Practical Applications of Information Theory

Data Compression. Suppose that we want to transmit a file consisting of 1000 characters. Each character is one of five possibilities, say A, B, C, D, or E. Altogether, the symbols A, B, C, D, E appear 400 times, 200 times, 150 times, 150 times, and 100 times, respectively. We want to encode each symbol as a binary string in such a way that the resulting binary "superstring" corresponding to the file is as short as possible. We also impose the condition that the superstring can be unambiguously decoded to give the original file at the receiving end. It turns out that the answer, due to Shannon, is closely related to the entropy $H(0.4, 0.2, 0.15, 0.15, 0.1)$ and that the most economical encoding of the file comes to 2200 bits, where 2200 is close to $1000H(0.4, 0.2, 0.15, 0.15, 0.1)$. This is a big improvement on the uncompressed approach whereby each character is encoded as an 8-bit binary string in ASCII for a total of 8000 bits. One encoding method is **Huffman coding** discussed in Chapter 11.

Channel Capacity. We want to maximize "bandwidth." Suppose that we are sending one of a set of M possible messages across a noisy channel. Our goal is to minimize the probability of error. To take an example, suppose $M = 2$ so that we have just 2 messages, namely, the bits 0 or 1. One way to minimize the probability of error is to encode 0, 1 as the messages $(0, \ldots, 0)$ and $(1, \ldots, 1)$, respectively, both of length n with n large. Then, even if one or two bits get corrupted, the probability of correct decoding (using "majority wins" here) is very high. The downside is that the transmission rate is very small. The rate here is the number of "information bits" M per code word divided by the length n of the code word. So the rate is $\frac{1}{n}$, which goes to 0 as n gets large. Shannon, in his famous noisy channel theorem, showed that the rate does not have to go to 0 for accurate decoding. In fact, so long as the rate is less than a number known as the **channel capacity**, accurate decoding is possible. As an example, suppose that A is transmitting n bits to B and assume that the probability of any given bit being corrupted is p, independent of the other bits. Then the channel capacity is our old friend $n(1 - H(p))$.

In engineering, the channel capacity C of a **continuous channel of bandwidth B hertz**, perturbed by additive white Gaussian noise of power spectral density $\frac{N_0}{2}$ is given by the formula $C = B \log_2(1 + \frac{P}{N_0 B})$ bits per second, where P is the average power. For many people this formula ranks above $E = mc^2$!

Cryptography. Information theory has several applications here. We mention Shannon's "perfect secrecy" criterion in Chapter 15, which quantifies perfect secrecy. Entropy also gives a measure of the "search space" of various cryptographic procedures such as DES, discussed in Chapters 3, 4, and 5. The higher the entropy, the better the security. You don't want to choose a password that is easy to guess. The entropy of the average 8-character password is around 14 bits—bearing in mind that one only has to try about half the key space in a brute-force search (Chapter 3). It is also important to have a good estimate of the entropy of a message, for example, a text message in English. This is discussed in Chapter 14. Another very fundamental application occurs in the new protocols for encryption, described in Part III, where one first has to estimate the information an eavesdropper may have, as we did above in a special case.

9.5 Information Theory and Physics

The word "entropy" was first coined by Clausius in physics around 1865. Entropy is used mainly in statistical mechanics and thermodynamics. Roughly, it measures the amount of energy lost to heat in an irreversible physical process.

It was Boltzmann in 1896 who was the first to show that the physical entropy of a system could be expressed as the average value of the logarithm of the probabilities of the states of a physical system. If the probability of a particular gas configuration is W, we have $S = k \log W$, where k is Boltzmann's constant. As Feynmann [Fey99] puts it *"Generally speaking, the less information we have about a state the higher the entropy."*

If F is the free energy, U the total energy (which remains unchanged), T the temperature, and S the entropy, we have

$$F = U - TS \text{ (see Feynmann [Fey99]).} \tag{9.5}$$

Denoting an infinitesimal change at constant temperature by δ, we have

$$\delta S = -\frac{\delta F}{T} \text{ since } \delta T = 0. \tag{9.6}$$

As in [Fey99] this is "a variant of the standard formula $\delta S = -\frac{\delta Q}{T}$ for the infinitesimal change in entropy resulting from a thermodynamically reversible change in state where at each stage, an amount of heat δQ enters or leaves the system at absolute temperature T. For an irreversible process the equality is replaced by an inequality, ensuring that the entropy of an isolated system can only stay constant or increase—this is the Second Law of Thermodynamics."

In fact, there is an analogy to the idea that "high randomness corresponds to high entropy." In thermodynamics, as Feynmann puts it "the less information we have about a state the higher the entropy."

To conclude this brief discussion, we quote from Shannon and Weaver [SW49], who quote from Eddington's work "The Nature of the Physical World" as follows.

> Suppose that we were asked to arrange the following in two categories—*distance, mass, electric force, entropy, beauty, melody.*
>
> I think there are the strongest grounds for placing entropy alongside beauty and melody, and not with the first three. Entropy is only found when the parts are viewed in association, and it is by viewing or hearing the parts in association that beauty and melody are discerned. All three are features of arrangement. It is a pregnant thought that one of these three associates should be able to figure as a commonplace quantity of science. The reason why this stranger can pass itself off among the aborigines of the physical world is that it is able to speak their language, viz., the language of arithmetic.

9.6   Axiomatics of Information Theory

In our discussions above, we encountered the situation where an eavesdropper E was trying to determine a message M that was being transmitted by cryptographic station **A**. Assume that M is 1000 bits long and that E knows, with probability 0.85, what each bit is in M. Then, on the average, E "knows" 850 bits, i.e., Eve is aware that 850 of her bits are identical to those in M. The problem for Eve is not knowing where the places of agreement and disagreement with M lie in Eve's binary string. In fact, in terms of "hard information," Eve's knowledge only comes out to

$$1000(1 - H(0.85)) = 1000(1 - 0.6098) = 1000(0.3902) \tag{9.7}$$

which is approximately 390 Shannon bits. This seems a bit small compared to the 850 bits above. Perhaps we have the wrong measure for information?

To justify our definition of entropy axiomatically, let us suppose that H is some information measure that assigns a number $H(E)$ to each event E in a sample space. We assume that $H(E)$ only depends on the probability of E and not on the nature of E. Thus we can think of H as being defined on a probability space, so $H = H(p_1, \ldots, p_n)$, with $0 \leq p_i \leq 1, 1 \leq i \leq n$ and $p_1 + \ldots + p_n = 1$.

We make the following 8 assumptions or axioms.

1. $H(p_1, \ldots, p_n)$ achieves a maximum when $p_1 = \ldots = \frac{1}{n}$,

2. $H(p_1, \ldots, p_n)$ is unaffected by the ordering of $p_1, \ldots, p_n$ (so $H(p_1, p_2) = H(p_2, p_1)$ if $n = 2$, etc),

3. $H(p_1, \ldots, p_n) \geq 0$. $H = 0$ if and only if one of $p_1, \ldots, p_n$ is 1 and the rest zero,

4. $H(p_1, \ldots, p_n, 0) = H(p_1, \ldots, p_n)$,

5. $H(\frac{1}{n}, \ldots, \frac{1}{n}) \leq H(\frac{1}{n+1}, \ldots, \frac{1}{n+1})$,

6. $H(p_1, \ldots, p_n)$ is a continuous function in each of $p_1, \ldots, p_n$,

7. $H(\frac{1}{mn}, \ldots, \frac{1}{mn}) = H(\frac{1}{m}, \ldots, \frac{1}{m}) + H(\frac{1}{n}, \ldots, \frac{1}{n})$,

8. Let $p = p_1 + \cdots + p_m$ and $q = q_1 + \cdots + q_n$ where each p_i and q_j are nonnegative. Then if p, q are positive with $p + q = 1$,

$$
\begin{aligned}
H(p_1, \ldots, p_m, q_1, \ldots, q_n) &= H(p, q) + pH(\frac{p_1}{p}, \ldots, \frac{p_m}{p}) \\
&\quad + qH(\frac{q_1}{q}, \ldots, \frac{q_n}{q}).
\end{aligned}
$$

These axioms all have convincing intuitive explanations as pointed out in Welsh [Wel88]. For example, Axiom 1 says that the uncertainty is maximized when all outcomes are equally likely. Axiom 3 says that entropy is always nonnegative. Axiom 5 says there is more uncertainty, in the equiprobable case, when we increase the number of possible outcomes. Axiom 6 says that if we only change the probabilities slightly then H changes only slightly. Axiom 7 has to do with independence. Axiom 8 says that we can break down the total uncertainty involving the uncertainty as to whether one of the p_i or the q_i is chosen (or occurs). Then we have uncertainty involving weighted averages.

We then have the following result (see [Wel88]).

Theorem 9.1 *If H satisfies axioms (1) to (8), then*

$$
H(p_1, \ldots, p_n) = -\lambda \sum_{i=1}^{n} p_i \log p_i \tag{9.8}
$$

where λ is some positive constant and the sum is over all the nonzero numbers $p_j, 1 \leq j \leq n$.

These axioms, originally proposed by Shannon, are not minimal. Many other systems have been published. The axioms are chosen in [Wel88].

9.7 Number Bases, Erdös, and the Hand of God

In this book, we generally work in base 2. However, many of the results hold for *any* number base. We mention in particular the results of Kraft and MacMillian in the next chapter. Also, we mention the important set of examples for source coding given by DNA coding where the alphabet has size 4.

Here we want to talk briefly about number bases and their properties. Let us fix on an arbitrary positive integer $d > 1$. Then every positive integer n can be written as nonnegative combinations of the nonnegative powers of d. Thus $n = c_0 + c_1 d + c_2 d^2 + \ldots + c_t d^t$, where $0 \le c_i < d$. Also, this expression is **unique**. In particular then, the set S of positive integers given by

$$S = \{d^0 = 1, d, d^2, d^3, \ldots\} \tag{9.9}$$

satisfies the following interesting property that we will call property E (for Erdös).

Property E. A set S of positive integers is said to satisfy **Property E** if two different subsets of S must have different sums.

It is clear that the nonnegative powers of d satisfy this property. Otherwise, we would have

$$n = \alpha_0 + \alpha_1 d + \alpha_2 d^2 + \cdots = \beta_0 + \beta_1 d + \beta_2 d^2 + \cdots \tag{9.10}$$

where $\alpha_i, \beta_i \in \{0, 1\}$. Thus n would have two different expansions in powers of d, which is impossible.

If we were trying to construct a set S satisfying Property E one way to do it would be to pick each new number x of the set to be a suitably large number. This will work, but then when x is big, $\frac{1}{x}$ is small. Note that the powers of d "barely work" because each new power of d is the sum of all the previous powers of d plus 1. There are no gaps. This leads naturally to the Erdös problem (or we should say *an* Erdös problem. The legendary Erdös posed many fascinating problems).

Erdös Problem. Given a set S of positive integers satisfying Property E, find a least upper bound for $\sum_{x \in S} \frac{1}{x}$.

Let's try the number bases.

base $= 2$	$1 + \frac{1}{2} + \frac{1}{4} + \frac{1}{8} + \ldots$	which tends to 2
base $= 3$	$1 + \frac{1}{3} + \frac{1}{3^2} + \frac{1}{3^3} + \ldots$	which tends to $\frac{3}{2}$
$\vdots$	$\vdots$	$\vdots$
base $= d$	$1 + \frac{1}{d} + \frac{1}{d^2} + \frac{1}{d^3} + \ldots$	which tends to $\frac{d}{d-1}$

So just looking at the case when S is the set of nonnegative powers of d leads us to conjecture that the least upper bound might be 2. But what about the general case where we only know that S satisfies Property E?

One of the authors, in joint work with David Borwein [BB75], was fortunate enough to solve the problem. The least upper bound is indeed 2. In his book *Mathematical Gems III*, Professor Honsberger devotes an entire chapter, Chapter 17, entitled "A Problem of Paul Erdös," to this problem. In the chapter, he discusses the Borwein–Bruen solution and offers the following comment concerning the solution. "Professor Erdös has the theory that God has a book containing all the theorems of mathematics with their absolutely most beautiful proofs, and when he wants to express the highest application of a proof he explains: This is one from the book! Erdös does not do this very often, but across the top of the note he sent me is the inscription 'I think this proof comes straight out of the book!' "

The proof that the least upper bound is 2 uses nothing but high school algebra. We also point out that the problem was first solved by Erdös and Benkoski. For details, we refer the reader to Honsberger [Hon85].

In fact, the Borwein–Bruen proof goes through for all number bases, not just powers of 2. This result was also rediscovered by Eric Lenza. Here is the result.

Theorem 9.2 (Number Bases Theorem) *Let* $0 < a_1 < \cdots < a_k$ *be a set of positive integers such that all sums* $\sum_i \varepsilon_i a_i$ *are distinct, where* $\varepsilon_i = 0, 1, \ldots, n - 1$. *Then*

$$\frac{1}{a_1} + \cdots + \frac{1}{a_k} \le \left(\frac{n}{n-1}\right)\left(1 - (\frac{1}{n})^k\right) \tag{9.11}$$

with equality exactly when $a_i = n^{i-1}$.

We also mention here a related paper in the binary case by our colleague Richard Guy [Guy82].

9.8 Weighing Problems and Your MBA

Several years ago a colleague of one of the authors, Desmond Ffolliott, was associated with the Ivey business school at the University of Western Ontario. Another colleague, Dave Johnson, an oil and gas expert also lived in London, Ontario and had Ivey connections. In conversation Desmond and Dave mentioned that "the 12 weights problem" was often used to instill the right attitude into prospective MBA students. If they were good they might be able to solve it. If not, the attempt would build character!

Here is the problem. You are provided with a certain number of weights, say 12, and a balance. You are told that exactly one of the weights is defective, being too light or too heavy, you are not told whether the culprit is too light or too heavy.

Problem. Determine which of the 12 weights is defective and determine whether it is too light or too heavy in 3 weighings or less.

Why 12? In fact, the case of 9 weights and three weighings is stated as a problem in Welsh [Wel88]. But we solve the problem here for 12 weights. Why 3 weighings? Can we generalize this problem? What does it have to do with information theory? We will try to outline the answers to some of these questions.

First of all we mention some abstract theory. We have 12 weights numbered 1 to 12. One of these is defective, say the ith weight. Moreover, it is either too heavy, so we have

(i, H), or too light, so we have (i, L). So altogether we have a list of 24 possible outcomes namely, $(1, H), (1, L), \ldots, (12, H), (12, L)$, which are all equally likely. Moreover, exactly one outcome materializes. Think of a horse race with 24 horses in the race and exactly one winner, and all of them equally likely to win. Our uncertainty about the winner is then $H(\frac{1}{24}, \ldots, \frac{1}{24})$, which is $\log 24$. Each weighing gives 3 possible results: heavy to the left or right, or balanced. The maximum amount of information we can get from a weighing (see Axiom 1) is $H(\frac{1}{3}, \frac{1}{3}, \frac{1}{3}) = \log 3$.

This is why it is not a good idea to weight 6 against 6 at the beginning. If we do this there are only 2 possible outcomes instead of 3, and so we won't acquire enough information.

This occurs when the 3 outcomes are equally likely. So with x weighings we get at most $x \log 3$ Shannon bits of information. But from the above we need $\log 24$ Shannon bits of information. Thus $x \log 3 \geq \log 24$. This gives

$$x \geq \frac{\log 24}{\log 3}. \tag{9.12}$$

Logs are taken to the base 2, but we get the same answer no matter which base we choose, because of the ratio. Thus $x \geq 2.89$. So we are going to need at least 3 weighings! In general, with n weights and x weighings we get the following inequality.

General Inequality. $x \log 3 \geq \log(2n)$.

So, for n weights we need at least x weighings, where x satisfies the above inequality.

How can we discover the culprit in at most 3 weighings? First, we choose 4 at random on one side of the balance and another 4 on the other side.

Case 1. They balance. Then the culprit is among the 4 left. Next, we balance 3 of these potential culprits with 3 good weights.

　　Subcase a. They balance. Then we balance the remaining weight, which we know to be the culprit, with one of the good weights, to see if the culprit is too heavy or too light. We have found the answer in 3 weighings.

　　Subcase b. They don't balance. So we have 3 potential culprits, one of which is the real culprit. We can also tell if the culprit is heavy or light. Next, balance one of these 3 against another one of the 3, and all will be revealed whether or not they balance.

Case 2. They do not balance, and the balance tilts to the left say. Label the 4 weights on the left as H_1, H_2, H_3, H_4 and the 4 weights on the right as L_1, L_2, L_3, L_4. Thus either one of the H weights is too heavy or one of the L weights is too light.

Next, we perform the following move. We put H_1, H_2, L_1 on the left against H_3, H_4, L_2 on the right. We have $H_1\, H_2\, L_1 \mid H_3\, H_4\, L_2$.

Subcase a. They balance. The culprit must be either L_3 or L_4. Then put these two opposing each other on the balance to identify the culprit as the lighter of the two.

Subcase b. They do not balance. If the right side rises, the culprit is $H_1, H_2,$ or L_2. If the left side rises the culprit is $H_3, H_4,$ or L_1.

In any case, we have as possible culprits two potentially heavy weights denoted by H_5, H_6 and one potentially light weight, say L.

Now weight H_5 against H_6. If they balance, the culprit is L. If they do not, then the culprit is the heavier of the two.

9.9 Shannon Bits, the Big Picture

It should be emphasized that the abstract unit of information is the Shannon bit. A priori, this is not connected to a regular physical bit or binary digit. However, a wonderful feature of information theory, reaffirming Shannon's genius and the verification of the correctness of his theories, is that Shannon bits and physical bits are, in some ways, the same.

Let us explain. Shannon's first theorem shows that the entropy of a given source, measured in Shannon bits, is arbitrarily close to the number of physical bits on average needed for efficient encoding of the source.

Shannon's second theorem shows that the maximum value of a conditional entropy, i.e. the **channel capacity** measured in Shannon bits, comes arbitrarily close to the maximum transmission rate, measured in physical bits, for accurate communication across the channel.

So, in these two theorems, the motto, roughly speaking, is that:

$$\begin{aligned} Theory &= Practice \\ Shannon\ Bit &= Physical\ Bit \end{aligned}$$

Finally, we should mention the famous Shannon Sampling Theorem for sampling continuous information. This is covered in detail in Chapter 13.

Chapter 10

Random Variables and Entropy

Goals, Discussion In subsequent chapters one of the main topics is various kinds of channels. In preparation for this we need to develop background on random variables, conditional probability, entropy, conditional entropy and mutual information.

We discuss these topics here in an elementary way. Our coverage is quite complete and we offer several worked examples and problems to illustrate the theory. Related topics come in again, notably in Chapter 12

New, Noteworthy We cover the Bayes formula and mutual information in a very transparent way using probability tree diagrams.

Simple inequalities can be extremely useful in entropy arguments. One such inequality is the subadditivity of the function $-x \log x$. This is usually stated (if stated at all) using the assumption that x, y and $x + y$ are probabilities. This assumption tends to obfuscate the inequality which, as we explain, needs no such assumption.

10.1 Random Variables

A **Random Variable** X associates with each possible outcome of an experiment or observation a value of X. For example a source may be emitting symbols chosen from an alphabet with certain probabilities. The value of the random variable X is defined to be the symbol emitted. Or, the experiment might consist of choosing an individual from a set of n people and measuring his or her weight in pounds. In this case the value X is a real number corresponding to an individual's weight in pounds.

Throughout this chapter we assume that X has only finitely many possible values. Generally, X is discrete although the continuous case is discussed via Shannon sampling in Chapter 13.

175

Consider the following examples:

Example 10.1 *Suppose we roll a six-sided die. The number on a side or face of the die is one of 1, 2, 3, 4, 5, 6. If each face is equally likely to show up (with probability $\frac{1}{6}$) the die is said to be* **fair**.

This is really a bit like the source example above. The source, which is the die, selects a number from the set $\{1, 2, 3, 4, 5, 6\}$.

Example 10.2 *We have a set of five individuals. An experiment consists of choosing one of the five at random and measuring that individual's height in inches. Here the sample space is the set $\Omega = \{A, B, C, D, E\}$ consisting of the five individuals. The random variable X is the height of the individual in inches.*

Note that in Examples 10.1 and 10.2 the values of X are real numbers. But this is not always the case as is shown by the following kind of example which is examined in detail in Chapter 11.

Example 10.3 *A source emits a finite number of symbols of an "alphabet" with pre-assigned probabilities. For instance, the symbols might be the binary symbols 0 or 1. Here the sample space can be identified with the symbols emitted. If the possible symbols are $\{x_1, x_2, \ldots, x_n\}$, we say that the set $\{x_1, x_2, \ldots, x_n\}$ is the target space of the random variable X.*

For example, in the string (x_3, x_2, x_4), the source assigns the first, second, and third positions of the string to the symbols x_3, x_2, x_4 respectively.

Abstractly, let Ω be a sample space and X a random variable defined on Ω. Denote the values of X by $x_1, x_2, \ldots, x_n$. Given any value x_i, $1 \leq i \leq n$ we can define "the probability that $X = x_i$" which is denoted $\Pr(X = x_i)$. This is simply the sum of the probabilities of all outcomes e in the sample space which map to x_i. In symbols:

$$\sum_{X(e)=x_i} \Pr(e) = \Pr(X = x_i)$$

Example 10.4 *Suppose each of the 5 individuals in Example 10.2 has an equal probability of being chosen. Since the probabilities add up to 1 this means that each of A, B, C, D, and E has a probability equal to $\frac{1}{5}$ of being chosen. Now assume that the heights in inches of A, B, C, D, and E are respectively 67, 67, 69, 72, and 70. Then*

$$\Pr(X = 67) = \frac{2}{5}, \ \ \Pr(X = 69) = \frac{1}{5}, \ \ \Pr(X = 72) = \frac{1}{5}, \ \ \Pr(X = 70) = \frac{1}{5}$$

In the future we will occasionally be informal and write "$\Pr(x_i)$" instead of "$\Pr(X = x_i)$."

The **average** or **expected value** or **mean** of X is denoted by $E(X)$. It is defined as follows:

$$E(X) = \sum_{i=1}^{n} x_i \Pr(X = x_i)$$

In other words, to calculate the average value of X we multiply each possible value of X by the corresponding probability In Example 10.4, the average value of X is $\frac{1}{5}(67) + \frac{1}{5}(67) + \frac{1}{5}(69) + \frac{1}{5}(70) + \frac{1}{5}(72) = 69$ (inches). Thus the expected value of the height is 69 inches.

Sometimes the symbol μ is used for expected value.

The **Variance** of X is defined as follows:

$$V(X) = \sum_{i=1}^{n} (x_i - E(X))^2 \Pr(X = x_i) = \sum_{i=1}^{n} (x_i - \mu)^2 \Pr(x_i).$$

In other words, the variance of X is the average value of the variable $(x_i - E(X))^2$.

The **Standard Deviation** $\sigma(X)$ is defined to be the positive square root of the variance. Thus,

$$\sigma(X) = \sqrt{V(X)}$$

Example 10.5 *Now suppose we assume that A, B, C, D, E, are chosen in accordance with a different probability distribution from that in Example 10.4 where we used the equiprobable distribution. Let* $\Pr(A) = 0.1$, $\Pr(B) = \Pr(C) = 0.2$, $\Pr(D) = 0.4$, $\Pr(E) = 0.1$

To find $E(X)$, *the average value, we construct the following table. Recall that the heights of A, B, C, D,and E in inches are respectively,* 67,67,69,72,70.

Possible Values of X	67	69	70	72
Corresponding Probabilities	0.3	0.2	0.1	0.4

Then $E(X) = (67)(0.3) + (69)(0.2) + (70)(0.1) + (72)(0.4) = 69.7$ *(inches) We can now calculate the variance of* X. *This involves the average value of* $(x_i - E(X))^2 = (E(X) - x_i)^2$

Possible Values of $(E(X) - x_i)^2$	$(2.7)^2$	$(0.7)^2$	$(0.3)^2$	$(2.3)^2$
Corresponding Probabilities	0.3	0.2	0.1	0.4

$$V(X) = (0.3)(2.7)^2 + (0.2)(0.7)^2 + (0.4)(2.3)^2 + (0.1)(0.3)^2$$

$$= 2.187 + 0.098 + 2.116 + 0.009$$

$$= 4.41$$

Then, $\sigma(X) = \sqrt{4.41} = 2.1$

A formula for calculating the variance which is sometimes easier to use than the one above is as follows:

$$V(X) = E(X^2) - (E(X))^2$$

To calculate $E(X^2)$, the expected value of X^2, we simply multiply the possible values of X^2 by their corresponding probabilities and then sum up. In example 10.5, we know that $E(X) = 69.7$, and we calculate that

$$
\begin{aligned}
E(X^2) &= (0.3)(67^2) + (0.2)(69^2) + (0.1)(70^2) + (0.4)(72^2) \\
&= 4862.5.
\end{aligned}
$$

We finish by finding the variance.

$$V(X) = E(X^2) - (E(X))^2 = 4862.5 - 69.7^2 = 4.41.$$

10.2   Mathematics of Entropy

Given a random variable X let the possible values of X be $x_1, x_2, \ldots, x_n$ and set $p_i = \Pr(X = x_i)$, $1 \le i \le n$. We will only be concerned with the probability space $\{p_1, p_2, \ldots, p_n\}$ and not with the nature of X.

As such we define the **Entropy of X** as follows:

$$H(X) = \sum_{i=1}^{n} p_i \log\left(\frac{1}{p_i}\right) = -\sum_{i=1}^{n} p_i \log(p_i)$$

Here the second equality is obtained by using the fact that $\log\left(\frac{1}{t}\right) = -\log(t)$ for any number $t > 0$. **The logarithms are taken to the base 2**, and this is usually the case throughout this book.

The preceding sum is taken over all *non-zero* p_i. Additionally we take $0 log\left(\frac{1}{0}\right)$ to be zero with respect to our definition of $H(X)$ as above. One justification for this is that the entropy should be continuous and we know from calculus that $\lim_{x \to 0} x \log(x) = 0$.

An especially interesting case occurs when $n = 2$. In this case we have 2 possible outcomes for the random variable, with probabilities p and q where $q = 1 - p$. So we have a probability distribution $[p, 1 - p]$. Then:

$$H(p) = H(p, 1 - p) = p \log\left(\frac{1}{p}\right) + q \log\left(\frac{1}{q}\right) = -(p \log(p) + q \log(q))$$

This is the famous **Shannon Function** whose diagram is given below and whose values are tabulated in the Appendix. We refer also to Chapter 9.

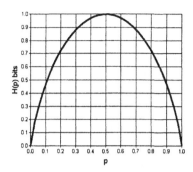

We can think of $H(p)$ as the amount of information received upon learning the outcome of an experiment or trial that has just 2 possible outcomes occurring with probabilities p, q where $q = 1 - p$ and $0 \leq p \leq 1$. $H(p)$ is 0 at $p = 0$ and $p = 1$. In either of these cases we know the outcome already so no information is revealed and $H(p) = 0$. When $p = \frac{1}{2}$, $H(p)$ achieves its maximum value, which is one. **So we receive maximum information, upon learning the outcome, when the outcome is completely random to begin with.** The amount of information received in this case is 1 (**The Shannon Bit**), which is the unit of information and entropy. Thus, information and entropy is measured in Shannon bits. As mentioned in Chapter 9, a *Shannon bit is the amount of information received (or uncertainty removed) upon learning the outcome of an experiment with two possible outcomes, both of which were, a priori, equally likely* (Golomb [Gol02]).

We note that the Shannon function is **concave down** (as can be seen by calculating the second derivative). This means that the graph lies above the line segment joining any 2 points on the graph, and below the tangent line to the graph.

Caution about notation The symbol p is a dangerous one! When we discussed cryptography, p is a prime. Now, p is a probability relating to a source. When we discuss channels, such as the Binary Symmetric Channel (BSC) p will denote the probability of a transmission error.

10.3   Calculating Entropy

The procedure for calculating the entropy of a random variable X is similar to that for calculating the average value or variance of X. We simply make up a table as follows.

Possible Values of X	x_1	x_2	$\cdots$	x_n
Corresponding Probabilities	p_1	p_2	$\cdots$	p_n
Entropy $H(X)$	$p_1 \log(\frac{1}{p_1})$ $+$	$p_2 \log(\frac{1}{p_2})$ $+$	$\cdots$ $+$	$p_n \log(\frac{1}{p_n})$

Note that when X is a source (Chapter 11) the units for $H(X)$ are *Shannon bits per symbol.* Additionally, notice that in the definition of entropy, we only need the probabilities

$\Pr(X = x_i)$ rather than the actual values of X. Thus, for entropy, we only need a probability distribution for its calculation. Then, for a probability distribution $(p_1, p_2, \ldots, p_n)$ we have:

$$H(p_1, p_2, \ldots, p_n) = \sum_{i=1}^{n} p_i \log\left(\frac{1}{p_i}\right) = -\sum_{i=1}^{n} p_i \log(p_i)$$

Here, as usual, logs are taken to the base 2. Another way of looking at the matter is to say that *entropy is the average value of the* log *of one over the probability to the base 2.*

If $p_i \neq 0$, we know that p_i is positive. Now since $p_i \leq 1, \frac{1}{p_i} \geq 1$, and we have that $\log\left(\frac{1}{p_i}\right) \geq 0$. Thus, each term in the sum defining entropy is nonnegative. As will be seen in the next chapter, the maximum value of entropy (for a fixed n) occurs when $p_1 = p_2 = \cdots = p_n = \frac{1}{n}$ in which case we find that $H(X) = \log(n)$.

In summary, given n, we have $0 \leq H(X) \leq \log(n)$.

Example 10.6 *For the six-sided die in Example 10.1, the entropy is:*

$$\begin{aligned} H(X) =& \frac{1}{6}\log(6) + \frac{1}{6}\log(6) + \frac{1}{6}\log(6) + \frac{1}{6}\log(6) + \frac{1}{6}\log(6) + \frac{1}{6}\log(6) \\ =& \log(6) \end{aligned}$$

Since we are using logs to the base 2 we have $H(X) = \log_2(6) = 2.59$ and, recalling that the unit of entropy is the Shannon bit, we have $H(X) = 2.59$ Shannon bits.

10.4   Conditional Probability

In a sample space Ω (= a list of possible outcomes) an **Event** is any set of possible outcomes. Let U, V be the events in a sample space Ω. We want to define $\Pr(U|V)$, the **probability of U given V** or the **Conditional Probability** (of U given V). We define this as follows:

$$\Pr(U|V) = \frac{\Pr(U \text{ and } V)}{\Pr(V)} = \frac{\Pr(U \cap V)}{\Pr(V)} \tag{10.1}$$

where we assume that $\Pr(V) \neq 0$.

We think of $\Pr(U|V)$ roughly as the proportion of outcomes of U lying in V.

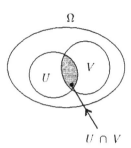

$$U \cap V$$

Example 10.7 *Urn 1 contains 2 black balls and 1 white ball. Urn 2 contains 3 white balls and 1 black ball. A ball is first drawn from Urn 1 and then placed in Urn 2. A ball is then drawn from Urn 2. Let U be the event that a white ball is drawn from Urn 1. Let V be the event that a black ball is drawn from Urn 2. Calculate:*

(a) $\Pr(U)$

(b) $\Pr(V)$

(c) $\Pr(U \cap V)$

(d) $\Pr(U|V)$

(e) $\Pr(V|U)$

Here are the solutions.

(a) Since we have 3 balls in Urn 1 and one of them is white, $\Pr(U) = \frac{1}{3}$.

(b) Suppose we draw a white ball from Urn 1 and put it in Urn 2. Urn 2 now has 5 balls, 4 of which are white. When we now draw, the probability of drawing black is $\frac{1}{5}$. Thus $\Pr$(white ball is from 1 and black ball from 2) is $\frac{1}{3} \cdot \frac{1}{5} = \frac{1}{15}$. Similarly, $\Pr$(black from 1 and black from 2) is $\frac{2}{3} \cdot \frac{2}{5} = \frac{4}{15}$.

Thus the total probability of drawing a black ball from Urn 2 is $\frac{1}{15} + \frac{4}{15} = \frac{1}{3} = \Pr(V)$

(c) $\Pr(U \cap V) = \Pr$(white from Urn 1 and black from Urn 2)$= \frac{1}{15}$

(d) $\Pr(U|V) = \frac{\Pr(U \cap V)}{\Pr(V)} = \frac{\frac{1}{15}}{\frac{5}{15}} = \frac{1}{5}$.

(e) $\Pr(V|U) = \frac{1}{5}$ (see the probability tree below). This is just the probability of moving down to the left having moved down 1 step to the right from the top.

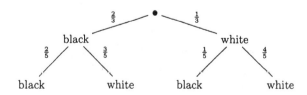

Note that $\Pr(U|V)$ involves the **Bayes formula** whereby we have the "forward probabilities" and we want to find the "backward probabilities". In this example the Bayes formula gives the following result:

$$\Pr(U|V) = \frac{\Pr(U)\Pr(V|U)}{\Pr(U)\Pr(V|U) + \Pr(U^c)\Pr(V|U^c)} \tag{10.2}$$

where U^c denotes the complement of U, so $\Pr(U^c) = 1 - \Pr(U)$. Note also that $\Pr(V|U)$ can be read from the tree diagram.

However, $\Pr(U|V)$ has to be calculated from the formula for conditional probability (10.1). In general the Bayes formula computes "backward probabilities" $\Pr(x|y)$ given the "forward probabilities" $\Pr(y|x)$ as follows.

$$\Pr(x|y) = \frac{\Pr(x)\Pr(y|x)}{\Pr(y)}$$

We say that events U and V are **Independent** if $\Pr(U|V) = \Pr(U)$. This means, from 10.1, that $\Pr(U|V) = \frac{\Pr(U \cap V)}{\Pr(V)} = \Pr(U)$. Thus, for independence, we have:

$$\Pr(U \cap V) = \Pr(U)\Pr(V)$$

In Example 10.7, $\Pr(U|V) = \frac{1}{5} \neq \frac{1}{3} = \Pr(U)$. Thus U and V are **Not Independent**.

Events U, V are independent if the value of V does not affect the value of U, so that V gives no information about U.

It may be the case that we need to combine two (or several) random variables. For example, with a channel, we can measure both the input X and the output Y. In this way we may **combine two random variables** X **and** Y **to obtain a single joint variable** (X, Y) whose values are ordered pairs (x, y) where x is the value of X and y is the value of Y. The probability of (x, y) is the probability that $X = x$ and, simultaneously, $Y = y$. That is, $\Pr(x, y) = \Pr[(X = x) \text{ and } (Y = y)]$.

Theorem 10.8 *From the definition of conditional probability in equation 10.1 we have* $\Pr(x, y) = \Pr(x)\Pr(y|x)$.

This will be important when we discuss channels where we will be given in advance the "forward probabilities" $\Pr(y|x)$.

Example 10.9 *Suppose we roll two fair dice and note the numbers appearing on each, corresponding to X, Y respectively. So our sample space consists of all possible ordered pairs (x, y) with $1 \leq x \leq 6$ and $1 \leq y \leq 6$. There are 36 such pairs. Let Z denote the sum of X and Y. Let U denote the event that X is even, and let V denote the event that Y is odd.*

Then

$$\Pr(U|V) = \frac{\Pr(U \cap V)}{\Pr(V)}.$$

Now to calculate $\Pr(V)$ *we know that the total number of ordered pairs of values is 36. These pairs are* $(1,1), \ldots, (1,6), (2,1), \ldots, (2,6), \ldots, (6,1), (6,2), \ldots, (6,6)$.

Since the event V *corresponds to the pairs where the second component is 1, 3 or 5 we have that* $\Pr(V) = \frac{18}{36} = \frac{1}{2}$. *Similarly,* $\Pr(U) = \frac{1}{2}$. *For* $\Pr(U \cap V)$, *the total number of ordered pairs corresponding to this event is 9, so* $\Pr(U \cap V) = \frac{9}{36} = \frac{1}{4}$. *The event* $U \cap V$ *corresponds to the 9 pairs:* $(2,1), (2,3), (2,5), (4,1), (4,3), (4,5), (6,1), (6,3), (6,5)$.

Thus

$$\Pr(U|V) = \frac{\Pr(U \cap V)}{\Pr(V)} = \frac{\frac{9}{36}}{\frac{18}{36}} = \frac{9}{18} = \frac{1}{2}$$

In Example 10.9 U, V *are independent, since both of* $\Pr(U \cap V)$ *and* $\Pr(U) \times \Pr(V)$ *are equal to* $\frac{1}{4}$.

Example 10.10 *Let* U *be the event that* X *is even and let* V *be the event that* $Z = 4$. *As above, we have* $\Pr(U) = \frac{1}{2}$. *The number of ordered pairs* (x, y) *for which* $x + y$ *is 4 is 3. Thus* $\Pr(U) = \frac{1}{2}$, $\Pr(V) = \frac{3}{36} = \frac{1}{12}$. *The event* $U \cap V$ *is represented by all pairs* (x, y) *with* x *being even and* $x + y = 4$: *this gives the pair* $(2, 2)$. *Therefore* $\Pr(U \cap V) = \frac{1}{36}$. *But* $\Pr(U) \Pr(V) = \frac{1}{24} \neq \frac{1}{36}$. *We conclude that the events* U, V *are not independent.*

Now we move from sample spaces to random variables. If X, Y are random variables they are said to be **independent** or **Statistically Independent** if $\Pr(X = x$ and $Y = y) = \Pr(X = x) \Pr(Y = y)$ for all possible values x of X and y of Y. In the previous example of the two dice X and Y are independent but X and Z are not independent.

In recent years questions regarding independence of events have become a major legal issue. In our discussions of sources and channels in Chapters 11 and 12 we will be assuming independence in various situations. Whether or not X and Y are independent we have that

$$E(X + Y) = E(X) + E(Y)$$

and in general :

$$E(X_1 + X_2 + \cdots + X_n) = E(X_1) + E(X_2) + \cdots + E(X_n)$$

If X, Y are independent then also:

$$E(XY) = E(X)E(Y)$$

$$\text{and } V(X + Y) = V(X) + V(Y)$$

Similarly if $X_1, X_2, \ldots, X_n$ are pair-wise independent then:

$$V(X_1 + X_2 + \cdots + X_n) = V(X_1) + V(X_2) + \cdots + V(X_n)$$

10.5   Bernoulli Trials

Suppose Ω is a sample space with 2 possible outcomes. We can think of tossing a coin with the 2 outcomes being "Heads" and "Tails". Let the probability of "Heads" be p and let "Tails" have probability q with $p+q=1$. If we toss just once and let X_1 denote the number of heads obtained then $E(X_1) = p$ and $V(X_1) = pq$ where $V(X_1)$ denotes the variance of X_1. Recall that the variance of a random variable X means the expected value of the quantity $(X - E(X))^2$.

This can be seen by constructing the following table (to get the average value of any random variable X multiply the possible values of X by the corresponding probabilities)

Possible Values of X_1	1	0
Corresponding Probabilities	p	q

Then $E(X_1) = (1)(p) + (0)(q) = p$. Also, since $V(x_1) = E(X - \mu)^2$ where μ is the average, as in 10.1, we get

$$V(X_1) = (1 - p)^2 p + (0 - p)^2 q = q^2 p + p^2 q = pq(p + q) = pq.$$

If we toss the coin n times corresponding to the n **independent and identically distributed (i.i.d.)** random variables $X_1, X_2, \ldots, X_n$ we are looking at the outcome of n independent trials with 2 possible outcomes, ie. n **Bernoulli trials**. Thus if $Z = X_1 + X_2 + \cdots + X_n$ we have

$$E(Z) = E(X_1 + X_2 + \cdots + X_n) = E(X_1) + E(X_2) + \cdots + E(X_n) = np.$$

$$\text{Also } V(Z) = V(X_1 + X_2 + \cdots + X_n) = nV(X_1) = npq.$$

Then the **standard deviation** of Z, denoted by $\sigma(Z)$ is defined by

$$\sigma(Z) = \sqrt{V(Z)} = \sqrt{npq}.$$

In summary if we have n Bernoulli trials and p is the probability of "success" then the **average number of successes is np and the standard deviation is $\sqrt{npq}$.**

It can be seen that the probability of having exactly t successes in n trials is:

$$\binom{n}{t} p^t q^{n-t}, 0 \le t \le n.$$

The **normal** or **Gaussian curve** is characterized by its two parameters μ and σ. The high point occurs at $x = \mu$. A small σ means that the curve is highly peaked. A large σ means a large dispersion.

From the Law of Large Numbers (section 10.7) it follows that when n is large the outcome of n Bernoulli trials can be estimated by a **Normal** or **Gaussian Curve** ("bell curve") with mean equal to np and standard deviation equal to $\sigma = \sqrt{npq}$.

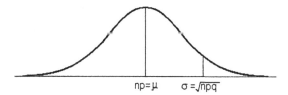

10.6 Typical Sequences

We can also relate Bernoulli trials to "typical sequences". Suppose we obtain a binary sequence of length n from a series of n Bernoulli trials, in which the probability of a one appearing in any given position is p and the probability of a zero appearing in that position is q, with $p+q = 1$. So the average number of ones appearing is np. (To make things convenient assume that np is an integer). The average number of zeros is then $n - np = n(1 - p) = nq$.

Pick any $\epsilon > 0$. Let k be any integer such that $k^2 > \frac{2}{\epsilon}$. Let N denote the number of ones in a sequence. We say that a sequence is **typical** if either $\frac{N-np}{\sigma}$ is less than k or greater than $-k$: in other words the *absolute value* of $\frac{N-np}{\sigma}$ is less than k. From this we have the following result, as shown in Ash [Ash90].

Theorem 10.11 :

(a) *The set of non-typical sequences has total probability less than ϵ.*

(b) *The total number of typical sequences is approximately $2^{nH(p)}$ where $H(p) = p \log \left(\frac{1}{p} \right) + q \log \left(\frac{1}{q} \right)$ is the entropy of the Bernoulli variable X_i with probability of success p, $1 \le i \le n$.*

(c) *Each typical sequence has a probability of occurring that is close to $2^{-nH(p)}$ so that all typical sequences are essentially equiprobable.*

The proof uses the well-known Chebyshev Inequality ("the probability of being very far from the average - for any distribution - is fairly small") together with the Normal approximation for n Bernoulli trials mentioned above.

For reference we include the statement of **the Chebyshev Inequality**, valid for any probability distribution X with mean μ and standard deviation σ as follows.

$$\Pr(|X - \mu| \geq k\sigma) \leq \frac{1}{k^2}.$$

Here, k is any positive real number.

Explanation of typical sequences The concept of a "typical sequence" can be explained, informally, as follows. In a typical sequence, the number of ones and zeros is close to the expected numbers, namely np and $n(1 - p)$ respectively. The difference between the actual and expected numbers is of the order of $\sqrt{n}$ and therefore, *when divided by n*, tends to zero. (Remember that in information theory we are usually talking about *bits per symbol*.)

We should point out that the theory of typical sequences works not just for two possible outcomes (heads and tails) but for an arbitrary number of outcomes.

Lets discuss typical sequences in a slightly different way. Suppose we have a binary memoryless source emitting 1 with probability p and 0 with probability $q = 1 - p$.

Each sequence then has a probability associated with it. For example, the sequence of length 3 given by 101 has probability given by $pqp = p^2q$. This follows from the independence of the source.

For a sequence $\mathbf{x}$ of length n we expect that $\mathbf{x}$ should have roughly np ones and nq zeroes. (Lets assume that np is an integer for convenience.) If $\mathbf{x}$ has exactly np ones and nq zeroes then the probability of $\mathbf{x}$, $\Pr(\mathbf{x})$, is given by the equation $\Pr(\mathbf{x}) = p^{np}q^{nq}$. Then $\log \Pr(\mathbf{x}) = np \log p + nq \log q$. Dividing by n we get $\frac{1}{n} \log \Pr(\mathbf{x}) = p \log p + q \log q$. Thus, $-\frac{1}{n} \log \Pr(\mathbf{x}) = -p \log p - q \log q$, i.e. $\frac{1}{n} \log \left(\frac{1}{\Pr(\mathbf{x})} \right) = p \log \frac{1}{p} + q \log \frac{1}{q} = H(p)$. From this, we can also get that $\log \left(\Pr(\mathbf{x}) \right) = -nH(p)$. Therefore, we have $\Pr(\mathbf{x}) = 2^{-nH(p)}$. In other words, for the "ultra-typical" sequence $\mathbf{x}$ above we have $\Pr(\mathbf{x})$ exactly equal to $2^{-nH(p)}$.

This motivates an **alternative definition** of a typical sequence. A typical sequence $\mathbf{x}$ of length n and error γ is defined to be one for which

$$\left| \frac{1}{n} \log \left(\frac{1}{\Pr(\mathbf{x})} \right) - H(p) \right| < \gamma,$$

where γ is an arbitrarily small given positive number. Thus a typical sequence will have probability of occurrence which is close to $2^{-nH(p)}$, so that typical sequences are close to being equiprobable, the total number of typical sequences being, roughly, $2^{nH(p)}$.

10.7 Law of Large Numbers

First, we want to talk informally about (a special case of) the law of large numbers. Let's suppose we toss a fair coin n times. Then $p = \frac{1}{2}$ is the probability of getting heads on a single toss. Now the probability of getting **exactly** x **heads** in n tosses is $\binom{n}{x}p^x q^{n-x}, q = 1 - p$, which is $\binom{n}{x}(\frac{1}{2})^x(\frac{1}{2})^{n-x} = \binom{n}{x}\frac{1}{2^n}$. To see this, just pick out the x slots where you want heads to appear and multiply by the probabilities. For example, the probability of getting exactly 2 heads in 10 tosses is $\binom{10}{5}\frac{1}{2^{10}} = 0.0439$. The most likely number of heads occuring is approximately $(10)(\frac{1}{2}) = 5$. The probability of getting exactly 5 heads is $\binom{10}{5}\frac{1}{2^{10}} = 0.246$.

As n gets larger, the normal curve estimating these probabilities gets much more peaked. The number of heads may not be exactly equal to $\mu = np$. However, the probability that the total number of heads obtained (*when divided by* n) is close to μ is very high.

In fact, the **law of large numbers** says just this. It has to do with a probability limit. It states that

$$\lim_{n \to 0} \Pr \left(\left| \frac{\text{(total number of heads)}}{n} - \mu \right| > \epsilon \right) - 0$$

where ϵ is any positive number (no matter how small) and $|\ |$ denotes the absolute value. In other words, when n is large, we can say with high probability of being correct that $\frac{\text{total heads}}{n} = \mu$.

Concerning the "Law of Large Numbers" we should mention that Richard Guy makes some amusing and interesting observations on numbers that are not large, but small [Guy88]!

In his paper entitled " The Strong Law of Small Numbers" Guy discusses the difficulty of guaging a mathematical pattern. He points out that "Capricious coincidences cause careless conjectures" and also warns that "initial irregularities inhibit incisive intuition". But we digress.

10.8 Joint and Conditional Entropy

The following lemma (The Entropy Lemma) is proved in Chapter 11.

Lemma 10.12 *Let* $(p_i, 1 \le i \le n)$ *be a given probability distribution with* $0 < p_i \le 1$ *and* $p_1 + p_2 + \cdots + p_n = 1$. *Let* $(q_i, 1 \le i \le n)$ *be another probability distribution with* $0 < q_i \le 1$. *Then*

$$\sum_{i=1}^{n} p_i \log q_i \le \sum_{i=1}^{n} p_i \log p_i$$

with equality if and only if $p_i = q_i$, $1 \le i \le n$.

(In Chapter 11, among other things, this result will be used to show that if X is a random variable with probabilities $p_1, p_2, \ldots, p_n$ then $H(X) \le \log(n)$ with equality if and only if $p_i = \frac{1}{n}, 1 \le i \le n$).

Next, let X, Y be random variables (each having only finitely many values). Then we can construct a random variable (X, Y) from the joint probability distribution (see Theorem 10.8), as follows.

If X has values $x_1, x_2, \ldots, x_s$ (with probabilities $p_1, p_2, \ldots, p_s$), and Y has values $y_1, y_2, \ldots, y_t$ (with probabilities $q_1, q_2, \ldots, q_t$), then (X, Y) has values $x_i y_j, 1 \le i \le s$ and $1 \le j \le t$. We denote $\Pr(X = x_i \text{ and } Y = y_j)$ by p_{ij}. We sometimes put $p_i = \Pr(X = x_i)$ and $q_j = \Pr(Y = y_j)$. From this joint distribution we have

$$\sum_j p_{ij} = p_i, \text{ and } \sum_i p_{ij} = q_j$$

Using the definition of the joint probability distribution (X, Y), we have

$$H(X, Y) = -\sum_i \sum_j p_{ij} \log(p_{ij}).$$

From this definition it follows that $H(X, Y) = H(Y, X)$.

Theorem 10.13 $H(X, Y) \le H(X) + H(Y)$ with equality if and only if X, Y are independent.

Proof.

$$H(X) + H(Y) = -\left(\sum_{i=1}^s p_i \log(p_i) + \sum_{j=1}^t q_j \log(q_j) \right)$$

$$= -\left(\sum_i \sum_j p_{ij} \log(p_i) + \sum_j \sum_i p_{ij} \log(q_j) \right)$$

$$= -\sum_i \sum_j p_{ij} \log(p_i q_j) \qquad \text{since } \log(uv) = \log(u) + \log(v)$$

We now denote the expression $p_i q_j$ by q_{ij}.

At this point, we can invoke Lemma 10.12. The fact that we have a double summation here is not important since we can reduce this to a single summation by re-indexing. So we have

$$H(X) + H(Y) = -\left(\sum_i \sum_j p_{ij} \log(q_{ij}) \right)$$

$$\ge -\sum_i \sum_j p_{ij} \log(p_{ij}) = H(X, Y)$$

From Lemma 10.12, equality occurs if and only if $p_{ij} = q_{ij}$, i.e., if and only if $p_i q_j = p_{ij}$. In other words, we have equality if and only if

$$\Pr(X = x_i \text{ and } Y = y_j) = \Pr(X = x_i)\Pr(Y = y_j)$$

which is equivalent to saying that the random variables X, Y are independent. ∎

Example 10.14 *Going back to our previous example of the two dice (Example 10.9), it is easy to calculate $H(X, Y)$ since X, Y are independent. We have $H(X, Y) = H(X) + H(Y) = \log(6) + \log(6) = 5.18$. We can see that $H(X) = \log(6)$ because there are 6 equiprobable outcomes. $H(Y)$ is calculated in the same way, and is equal to $\log 6$ also.*

Informally, Theorem 10.13 says that the uncertainty of (X, Y) is at most the uncertainty of X plus the uncertainty of Y. But if the value of X is known then it may actually give information about the outcome of Y so that the remaining uncertainty of (X, Y), once X is known, may be less than $H(Y)$. It will only equal $H(Y)$ if X, Y are independent.

Since $H(X, Y) = H(Y, X)$ an analysis like the above can be carried out with the roles of X, Y being interchanged.

Theorem 10.13 may be easily generalized as follows:

Theorem 10.15 *Given n random variables $(X_1, X_2, \ldots, X_n)$ then*

$$H(X_1, X_2, \ldots, X_n) \leq H(X_1) + H(X_2) + \cdots + H(X_n)$$

with equality if and only if the variables $X_1, X_2, \ldots, X_n$ are pairwise independent.

We now are in a position to discuss **Conditional Entropy**.

If X, Y are random variables then, for a given y, the variable $(X|y)$ denotes a random variable with probabilities

$$\Pr(X = x_i|y), \ i = 1, 2, \ldots, n$$

where X has the n values $x_1, x_2, \ldots, x_n$. This is a probability distribution since

$$\sum_x \Pr(x|y) = 1$$

Example 10.16 *In Example 10.9, suppose y has the value 4. Now $\Pr(Y = 4)$ is $\frac{1}{6}$. There are 6 ordered pairs (x, y) with $y = 4$ and $\frac{6}{36} = \frac{1}{6}$. We have, for example,*

$$\Pr(X = 1|Y = 4) = \frac{\Pr(X = 1 \text{ and } Y = 4)}{\Pr(Y = 4)} = \frac{\frac{1}{36}}{\frac{1}{6}} = \frac{1}{6}$$

In fact, $\Pr(X = x_i|Y = 4) = \frac{1}{6}$ for $x_i = 1, 2, 3, 4, 5, 6$.

We can calculate $H(X|y)$ and then obtain $H(X|Y)$ by averaging over y in Y. In this case we get $H(X|Y) = H(X) = \log 6 = 2.59$ Shannon bits since X, Y are independent. This follows from Theorem 10.18.

We can now relate $H(X|Y)$ to $H(X,Y)$. The basic idea for joint probabilities is as follows:

If x_i, y_i are values for X and Y then we have that $\Pr(X = x_i$ and $Y = y_j) = \Pr(X = x_i) \Pr(Y = y_j|X = x_i)$.

To abbreviate we say that $\Pr(x_i y_j) = \Pr(x_i) \Pr(y_j|x_i)$.

Theorem 10.17

$$H(X,Y) = H(X) + H(Y|X) = H(Y) + H(X|Y)$$

Proof.

$$H(X,Y) = -\sum_{i=1}^{m}\sum_{j=1}^{n} \Pr(x_i y_j) \log(\Pr(x_i y_j))$$

$$= -\sum_{i=1}^{m}\sum_{j=1}^{n} \Pr(x_i y_j) \log[\Pr(x_i) \Pr(y_j|x_i)]$$

Using the fact that $\log(uv) = \log(u) + \log(v)$ we get:

$$H(X,Y) = -\sum_{i=1}^{m}\sum_{j=1}^{n} \Pr(x_i y_j) \log(\Pr(x_i)) - \sum_{i=1}^{m}\sum_{j=1}^{n} \Pr(x_i y_j) \log(\Pr(y_j|x_i))$$

$$= -\sum_{i=1}^{m}\sum_{j=1}^{n} \Pr(x_i y_j) \log(\Pr(x_i)) - \sum_{i=1}^{m}\sum_{j=1}^{n} \Pr(y_j|x_i) \log\left(\Pr(y_j|x_i)\right) \Pr(x_i)$$

$$= -\sum_{i=1}^{m}\sum_{j=1}^{n} \Pr(x_i y_j) \log(\Pr(x_i)) + H(Y|X)$$

$$= -\sum_{i=1}^{m} \Pr(x_i) \log \Pr(x_i) + H(Y|X)$$

$$= H(X) + H(Y|X)$$

But $H(X,Y) = H(Y,X)$ so reversing the roles of X, Y we get

$$H(Y,X) = H(Y) + H(X|Y)$$

and this completes the proof. ∎

Theorem 10.17 says that the joint uncertainty that (X,Y) has a particular value equals the uncertainty of Y plus the uncertainty of X given that we know that value of Y.

Theorem 10.17 is called the **chain rule** for entropies. It generalizes to any number of variables. For example, we have:

$$H(X, Y, Z) = H(X) + H(Y|X) + H(Z|X, Y)$$
$$= H(Y) + H(Z|Y) + H(X|Y, Z)$$

From Theorem 10.17 we have $H(X|Y) = H(X, Y) - H(Y)$.

Thus $H(X) - H(X|Y) = H(X) + H(Y) - H(X, Y)$.

We define the **Information conveyed about X by Y** to be

$$I(X : Y) = H(X) - H(X|Y) = H(X) + H(Y) - H(X, Y)$$

Since $H(X) - H(X|Y) = H(X) + H(Y) - H(X, Y)$ and $H(X, Y) = H(Y, X)$ we have the following:

$$I(X : Y) = I(Y : X)$$

and we can call $I(X : Y)$ the **Mutual Information** of X, Y. Thus, the information conveyed about X by Y equals the information conveyed about Y by X. For a homely example we could say that the information conveyed about the political truths by a politician's speech equals the information conveyed about the politician's speech by the political truths. The mutual information can thus be an elusive quantity!

Combining Theorem 10.13 and Theorem 10.17 we have:

Theorem 10.18 $H(X|Y) \leq H(X)$ *with equality if and only if X and Y are independent.*

Proof. From Theorem 10.17, we have

$$H(X|Y) = H(X, Y) - H(Y)$$

From Theorem 10.13, we have

$$H(X, Y) \leq H(X) + H(Y)$$

(with equality if and only if X and Y are independent). Thus

$$H(X|Y) \leq H(X)$$

with equality if and only if X and Y are independent. ∎

Corollary 10.19 *$I(X:Y)=0$ if and only if X, Y are independent.*

Comments The formula for mutual information can be explained as follows. The uncertainty about X has been reduced from its original value of $H(X)$. By observing Y the new uncertainty is now just $H(X|Y)$. Thus the *reduction in uncertainty* is $H(X) - H(X|Y)$.

Another way of looking at this from the Shannon point of view (See Shannon and Weaver [SW49]) is that $H(X) - H(X|Y)$ is the amount of information that we have to add on to the information obtained about X by observing Y to get the full information about X.

Or we can say, thinking of a channel with X transmitting and Y receiving, that $H(X) - H(X|Y)$ is the amount of information transmitted that has been lost by noise in the channel. Since $H(X) = H(X|Y) + (H(X) - H(X|Y))$. The quantity $H(X) - H(X|Y)$ is the amount of information that the "correcting channel" must restore so that full information is received. Once again we can interchange the roles of X, Y in the above.

We should point out that Theorem 10.18 implies that *uncertainty can never be increased by side conditions, i.e. knowing Y cannot make X less certain.*

When could $H(Y|X)$ be zero? This would seem to say that once we know X we know Y so that Y is a function of X. This is indeed the case: we discuss this in the problems.

We can draw a kind of Venn diagram for entropies, where products of variables correspond to unions. (For more than 3 variables the corresponding diagram can be a bit misleading because some of the Venn regions may correspond to negative numbers.)

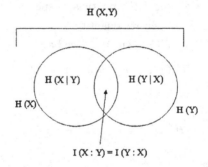

10.9   Applications of Entropy

Entropy is a measure of randomness or uncertainty. Entropy also describes information as the reduction in uncertainty. In cryptography and information security for example we want to choose a password with high entropy to make it harder to guess.

In information theory, as we will see in the next chapter, entropy tells us how efficiently a source coding can be achieved. Also, in a subsequent chapter (Chapter 12) we will see how conditional entropy is the crucial concept for determining the channel capacity, i.e. the largest possible rate for the secure transmission of information.

10.10 Calculation of Mutual Information

The following example may be found in Ash [Ash90] (page 23 - using a different solution than the one here) and provides a nice illustration of $I(X : Y)$.

Example 10.20 *Two coins are available, one being a fair coin and one being a 2-headed coin. A coin is selected at random and tossed twice. The total number of heads is recorded. How much information is conveyed, about which coin has been chosen, by the total number of heads that are obtained?*

Recall that

$$I(X : Y) = H(X) + H(Y) - H(X, Y)$$

Let X be a random variable that has value 0 or 1 according as to whether or not the fair coin or the two-headed coin is chosen. Let Y be the total number of heads obtained when the coin is tossed twice.

Now $H(X) = \log_2(2) = 1$ since the choice of coin is random.

To calculate $H(X, Y)$ we calculate the probabilities corresponding to X=0, Y=0, i.e. the probability of the pair $(0, 0)$, and similarly the probabilities of $(0,1)$, $(0,2)$ and $(1,2)$. Note that if $X = 1$, Y must be 2. Thus

$$H(X, Y) = H\left(\frac{1}{2} \cdot \frac{1}{4}, \frac{1}{2} \cdot \frac{2}{4}, \frac{1}{2} \cdot \frac{1}{4}, \frac{1}{2} \cdot 1\right)$$

$$= H\left(\frac{1}{8}, \frac{1}{4}, \frac{1}{8}, \frac{1}{2}\right) = \frac{1}{8}\log(8) + \frac{1}{4}\log(4) + \frac{1}{8}\log(8) + \frac{1}{2}\log(2)$$

$$= \frac{3}{8} + \frac{2}{4} + \frac{3}{8} + \frac{1}{2} = \frac{14}{8}$$

Now the possible values for Y are 0, 1, 2 with probabilities $\frac{1}{8}, \frac{1}{4}, \frac{5}{8}$, so that

$$H(Y) = \frac{1}{8}\log(8) + \frac{1}{4}\log(4) + \frac{5}{8}\log\left(\frac{8}{5}\right) = \frac{3}{8} + \frac{2}{4} + \frac{5}{8}(\log(8) - \log(5))$$

$$= \frac{3}{8} + \frac{2}{4} + \frac{15}{8} - \frac{5}{8}(\log(5)) = \frac{22}{8} - \frac{5}{8}(\log(5))$$

$$I(X : Y) = H(X) + H(Y) - H(X, Y) = 1 + \frac{22}{8} - \frac{5}{8}\log 5 - \frac{14}{8} = 2 - \frac{5}{8}\log 5 = 0.55$$

For questions on conditional entropy and conditional probability such as the question above it is convenient to use a **probability tree**. For the question above our tree would look like the following:

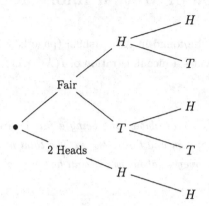

10.11   Mutual Information and Channels

Suppose A is transmitting to B in binary. Assume that the source A is **memoryless** so A chooses a zero with probability α and a one with probability $\beta = 1 - \alpha$, independently of what has already been transmitted.

We also assume that the channel is a **Binary Symmetric Channel** (or **BSC**). Thus the probability of an error in transmission is always a fixed number p, regardless of whether 1 or 0 is transmitted. These channels will be discussed in great detail in subsequent chapters.

Let X denote a binary variable which has the value 1 or 0 depending on whether A chooses 1 or 0. Let Y be a binary variable which has the value 1 or 0 depending on whether or not B receives 1 or 0.

We want to find $H(X) - H(X|Y)$ i.e. the information revealed about X by Y.

One probability tree is as follows:

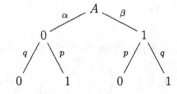

Then

$$H(X) = H(\alpha, \beta) = \alpha \log\left(\frac{1}{\alpha}\right) + \beta \log\left(\frac{1}{\beta}\right)$$

$$H(Y) = H(\alpha p + \beta q, \alpha q + \beta p)$$

since Y receives a one with probability $\alpha p + \beta q$ and a zero with probability $\alpha q + \beta p$.

We want to calculate the information about X that is revealed by knowledge of Y. This information is given by

$$I(X:Y) = H(X) + H(Y) - H(X,Y).$$

The outcome $X = 0, Y = 0$ i.e. the pair $(0,0)$, has probability αq. Similarly the pairs $(0,1), (1,0), (1,1)$ have probabilities $\alpha p, \beta p, \beta q$. Thus (X,Y) gives rise to the probability space $(\alpha q, \alpha p, \beta p, \beta q)$. Then

$$
\begin{aligned}
H(X,Y) &= H(\alpha q, \alpha p, \beta p, \beta q) \\
&= \alpha q \log\left(\frac{1}{\alpha q}\right) + \alpha p \log\left(\frac{1}{\alpha p}\right) + \beta p \log\left(\frac{1}{\beta p}\right) + \beta q \log\left(\frac{1}{\beta q}\right)
\end{aligned}
$$

Using the facts that $\log(uv) = \log(u) + \log(v)$, $\alpha + \beta = 1$ and $p + q = 1$, we finally obtain the mutual information $I(X:Y)$ given by the following formula.

$$I(X:Y) = p\log(p) + q\log(q) - (\alpha p + \beta q)\log(\alpha p + \beta q) - (\alpha q + \beta p)\log(\alpha q + \beta p)$$

10.12   The Entropy of $X + Y$

Earlier on in the chapter we discussed rolling two fair dice, independently. We have that $H(X) = \log(6), H(Y) = \log(6)$.

Thus $H(X) + H(Y) = \log(6) + \log(6) = 2.59 + 2.59 = 5.18$. If $Z = X + Y$ it can be calculated that $H(Z) = 3.2 < H(X) + H(Y)$. This is true in general.

We now show that whether or not the real variables X and Y are independent we have the following result.

Theorem 10.21

$$H(X + Y) \leq H(X,Y).$$

Equality occurs if and only if given any value z in $Z = X + Y$, there is exactly one ordered pair (x, y) of values x in X and y in Y with $x + y = z$.

Sketch of Proof: Suppose that for a given value of z we had say two ordered pairs (x_1, y_1) and (x_2, y_2) with $(x_1 + y_1) = z = (x_2 + y_2)$. Denote the probabilities $\Pr(X = x_i$ and $Y = y_j)$ by p_{ij} with $i = 1, 2$. Then $H(X,Y)$ will contain the term $-p_{11}\log p_{11} - p_{22}\log p_{22}$. The corresponding term when calculating $H(X + Y)$ will be the term $-(p_{11} + p_{22})\log(p_{11} + p_{22})$ which is smaller, as can be seen by using properties of the function $x\log(x)$ (see 10.13). If the pair (x_1, y_1) is the only ordered pair with $x_1 + y_1 = z$ then $H(X,Y)$ and $H(X + Y)$ will both contain just the term $-p_{11}\log(p_{11})$.

Since $H(X,Y) \leq H(X) + H(Y)$ we get a corollary.

Corollary 10.22

$$H(X + Y) \leq H(X) + H(Y)$$

Theorem 10.21 and the corollary can be extended to any number of variables. The reason why Theorem 10.21 holds is that when we "merge two terms" the entropy goes down. An interesting example of this occurred in Example 10.20 and in Theorem 10.21: see also the problems.

10.13   Subadditivity of the Function $-x \log x$

If x, y are positive numbers then

$$x \log x + y \log y \leq x \log(x + y) + y \log(x + y) = (x + y) \log(x + y)$$

since $\log x$ is an increasing function.

The fact that $x \log x + y \log y \leq (x+y) \log(x+y)$ is shown in [Tur76] under the assumption that x, y and $x + y$ are probabilities. However, as is clear from the above, this assumption is not needed. We thank Paul Tarjan for this useful observation.

The usual definition of a **subadditive function** f is that $f(x + y) \leq f(x) + f(y)$. Thus the above implies that $-x \log x$ is subadditive.

10.14   Entropy and Cryptography

In problem 13, we show that $H(f(X)|X) = 0$. **This formula proves that the entropy of, for example, RSA is 0.** To see this put $X = C$, the cipher text, and let f be the decryption function. We are given the cipher text C and $f(C) = C^d = M$, the message. In other words, all the uncertainty of the message is stored in the cipher text, which is known.

What this means is that public key systems, such as RSA, provide only **computational security**. That is, given sufficient time and resources, anyone can break RSA and various other public key systems. However, in practice, the amount of time required seems to be infeasibly large, so that RSA and public key cryptography in general still serve to protect important information.

10.15   Problems

1. A pair of fair die are rolled. If X, Y denote the numbers showing up on the faces find the average value of Z, the sum of the face-numbers.

2. Find the variance and the standard deviation of Z.

3. Find (a) $H(X)$ (b) $H(Y)$ (c) $H(X,Y)$.

4. Find (a) $H(Z)$ (b) $H(Z|X,Y)$.

5. Find $H(X|Z)$, $I(X:Z)$

6. A fair die is tossed 720 times and the random variable X measures the number of sixes obtained in the 720 tosses. Find the mean and the standard deviation of X.

7. Use the normal approximation to find the approximate probability of getting exactly 121 sixes.

8. An urn contains 4 red balls and 3 white balls. A ball is selected at random from the urn and replaced by 5 balls of the other colour. Then a second ball is selected at random from the urn.

   Draw a probability tree diagram for this experiment, labeling all the branch probabilities.

9. Find the probability that both selected balls are of the same colour/

10. Given that both selected balls are of the same colour, what is the probability that they are both white?

11. A source emits a binary string of length 3 subject only to the condition that the binary sum of the 3 digits is zero. What is the entropy of this string?

12. Prove that for any random variable X, $H(X|X) = 0$.

13. Show that $H(Y|X) = 0$ if and only if Y is a function of X.

14. If $\Pr(U|V) = 0$ must $\Pr(V|U) = 0$?

15. A is transmitting binary digits to B. A chooses a one with probability 0.8 and a zero with probability 0.2. Because of transmission error there is a 10 percent chance of error no matter what is transmitted to B. How much information about what A has transmitted is revealed by what B has received?

16. A coin is tossed n times in a row. The probability of getting heads is p at each toss of the coin. Show that

$$\sum_{t=0}^{n} t \binom{n}{t} p^t q^{n-t} = np$$

17. With the same notation as in problem 16, let $n = 2$. Show that $H(p^2, pq, qp, q^2) = 2H(p)$ where H denotes the Shannon function.

18. Can the result in problem 17 be generalized for arbitrary n?

19. Using the notation in problem 17, show that $H(p^2, 2pq, q^2) \leq 2H(p)$.

20. Can the result in problem 19 be generalized for larger values of n?

10.16   Solutions

1. $E(Z) = E(X + Y) = E(X) + E(Y) = 3.5 + 3.5 = 7$

2. It is convenient to use the formula that for the variance given by $V(X) = E(X^2) - (E(X))^2 = \frac{91}{6} - \left(\frac{7}{2}\right)^2 = \frac{35}{12}$. Now $V(X+Y) = V(X)+V(Y)$ since X, Y are independent $V(Z) = V(X + Y) = \frac{35}{6}$ and $\sigma = \sqrt{\frac{35}{6}}$.

3. $H(X) = H(Y) = \log_2 6 = 2.59$. Since X, Y are independent $H(XY) = H(X) + H(Y) = 5.18$.

4. By straight forward calculation we get $H(Z) = H(X + Y) = 3.27$. Once we know X and Y we know Z so $H(Z|XY) = 0$.

5. We need $H(X|Z)$:

$$H(X|Z) = \sum_{z=2}^{12} H(X|Z = z)\Pr(Z = z)$$

$$= \frac{1}{36}\log_2 1 + \frac{2}{36}\log_2 2 + \frac{3}{36}\log_2 3 + \frac{4}{36}\log_2 4 + \frac{5}{36}\log_2 5 + \frac{6}{36}\log_2 6$$
$$+ \frac{5}{36}\log_2 5 + \frac{4}{36}\log_2 4 + \frac{3}{36}\log_2 3 + \frac{2}{36}\log_2 2 + \frac{1}{36}\log_2 1$$
$$\approx 1.9$$

Thus $H(X|Z) = 1.9$. It follows that $I(X : Z) = H(X) - H(X|Z) = 2.59 - 1.9 = 0.69$

6. $E(X) = \frac{1}{6}(720) = 120$, $\sigma(X) = \sqrt{720 \cdot \frac{1}{6} \cdot \frac{5}{6}} = 10$

7. We estimate $\Pr(X = 121)$ by $\Pr(120.5 < X < 121.5)$ i.e. we are estimating a histogram by a piece of an area under the normal curve.

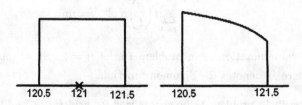

The approximating normal variable Y has mean 120 and standard deviation 10. To convert Y to a standard normal variable Z which we can look up in tables, we proceed as follows.

$$\Pr(X = 120) \approx \Pr(120.5 < Y < 121.5)$$
$$= \Pr\left(\frac{120.5 - 120}{10} < Z < \frac{121.5 - 120}{10}\right)$$
$$= \Pr(0.05 < Z < 0.15)$$
$$= 0.5596 - 0.5199 = 0.0397$$

So there is about a 4% chance of getting exactly 121 heads.

8. Here is the probability tree.

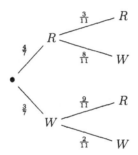

9. $\frac{4}{7} \cdot \frac{3}{11} + \frac{3}{7} \cdot \frac{2}{11} = \frac{18}{77}$

10. Let's calculate $\Pr(\text{both } W | \text{both same color})$:

$$\Pr(\text{both } W | \text{both same color}) = \frac{\Pr(\text{both } W \text{ and both same color})}{\Pr(\text{both same color})}$$
$$= \frac{\Pr(\text{both } W)}{\Pr(\text{both same color})} = \frac{\frac{3}{7} \cdot \frac{2}{11}}{\frac{4}{7} \cdot \frac{3}{11} + \frac{3}{7} \cdot \frac{2}{11}} = \frac{1}{3}$$

11. The entropy is 2 Shannon bits.

12. We have
$$H(X|X) = \sum_{x_0 \in X} H(X|x_0) \Pr(X = x_0)$$

In the calculation of $H(X|x_0)$, the only non-zero probability is 1, corresponding to $x = x_0$, so $H(X|x_0) = 0$ since $\log 1 = 0$. That is,

$$H(X|x_0) = -\sum_{x \in X} \Pr(X = x | X = x_0) \log \Pr(X = x | X = x_0)$$

and since $\Pr(X = x | X = x_0) = 0$ or 1, we have that $H(X | x_0) = 0$. Thus, $H(X | X) = 0$.

13. Remember that

$$H(Y|X) = \sum_{x \in X} H(Y|x) \Pr(X = x)$$

The right hand side is a sum of nonnegative terms. If $H(Y|X) = 0$ then each term in the sum is zero. Since $H(Y|x) = -\sum_y \Pr(y|x) \log \Pr(y|x)$, we know that $\Pr(y|x)$ is defined. From equation 10.1 we have that $\Pr(x)$ must then be nonzero, and so we may restrict to the case where $\Pr(X = x)$ is nonzero. Then, in calculating $H(Y|x)$ all the probabilities must be 0 or 1 since $H(Y|x) = 0$.

To see this, note that a term such as $p \log p$, with p a probability, can only be zero if $p = 0$ or $p = 1$. Now let x be any value of X such that $\Pr(X = x) \neq 0$. It will follow that there is exactly one value y_0 of Y such that $\Pr\left[(Y = y_0)|(X = x)\right] = 1$. For all other $y \neq y_0$, $\Pr(y|x) = 0$. Thus there is determined a function mapping x to y_0 and Y is a function of X. The converse follows easily.

14. For these two conditional probabilities to be defined we must have that $\Pr(V) \neq 0$ and $\Pr(U) \neq 0$. Now

$$\Pr(U|V) = \frac{\Pr(U \text{ and } V)}{\Pr(V)}$$

$$\Pr(V|U) = \frac{\Pr(V \text{ and } U)}{\Pr(U)}$$

Thus if $\Pr(U|V) = 0$ then $\Pr(U \text{ and } V) = 0$ so it follows that $\Pr(V \text{ and } U) = 0$ i.e. $\Pr(V|U) = 0$.

15. Let X be the random variable corresponding to the transmission by A and let Y denote the random variable corresponding to what is received by B.

We have $I(X : Y) = H(X) + H(Y) - H(X, Y)$. Now $H(X) = H(0.2) = 0.7219$.

The possible values for Y are 0, 1 with probabilities given by $\alpha = (0.2)(0.9)+(0.8)(0.1) = 0.26$ and $\beta = 1 - \alpha = 1 - 0.26 = 0.74$. Thus $H(Y) = H(0.26) = 0.8267$.

The possible values for (X, Y) are $(0,0)$, $(0,1)$, $(1,0)$, $(1,1)$ with probabilities $(0.2)(0.9)$, $(0.2)(0.1)$, $(0.8)(0.1)$, $(0.8)(0.9)$. Thus

$$\begin{aligned}
H(X,Y) &= H(0.18, 0.02, 0.08, 0.72) \\
&= -(0.18 \log(0.18) + 0.02 \log(0.02) + 0.08 \log(0.08) + 0.72 \log(0.72)) \\
&= 0.18 \cdot 2.473 + 0.02 \cdot 5.644 + 0.08 \cdot 3.644 + 0.72 \cdot 0.4739 \\
&= 1.1907
\end{aligned}$$

Thus $I(X : Y) = 0.7219 + 0.8267 - 1.1907 = 0.3579$ Shannon bits.

16. Let X be the event corresponding to tossing a coin n times in a row, and let X_i correspond to individual i^{th} coin toss, $1 \leq i \leq n$. To calculate $E(X)$ we can make a table, where X is the number of heads, as follows.

Possible Values of X	1	$\cdots$	t	$\cdots$	n
Corresponding Probability	$\binom{n}{1}p^1q^{n-1}$		$\binom{n}{t}p^tq^{n-t}$		$\binom{n}{n}p^nq^{n-n}$

Then $E(X) = \sum_{t=0}^{n} t\binom{n}{t}p^tq^{n-t} = np$. But $E(X) = E(X_1 + X_2 + \cdots + X_n) = E(X_1) + E(X_2) + \cdots + E(X_n) = p + p + \cdots + p = np$. The result follows.

17. If X_1, X_2 correspond to Bernoulli trials with probability of success p, then $H(X_1, X_2) = H(p^2, pq, qp, q^2)$. Since X_1, X_2 are independent, $H(X_1, X_2) = H(X_1) + H(X_2) = 2H(p)$. The result can also be verified by direct calculation.

18. Yes. Since $X_1, X_2, \ldots, X_n$ are independent we have

$$H(X_1, X_2, \ldots, X_n) = H(X_1) + H(X_2) + \cdots + H(X_n) = nH(p)$$

The expression on the left will be the sum of 2^n terms.

19. We use the subadditivity of the function $-x \log x$ as discussed in the text at section 10.13.

$$
\begin{aligned}
H(p^2, 2pq, q^2) &= -(p^2 \log p^2 + 2pq \log(2pq) + q^2 \log q^2) \\
&= -(p^2 \log p^2 + (pq + qp) \log(pq + qp) + q^2 \log q^2) \\
&\leq -(p^2 \log p^2 + pq \log(pq) + qp \log(qp) + q^2 \log q^2) \\
&= H(p^2, qp, pq, p^2) = 2H(p)
\end{aligned}
$$

20. Yes, using subadditivity from section 10.13, we can say for $n = 3$ that

$$H(p^3, 3p^2q, 3pq^2, q^3) \leq 3H(p)$$

For arbitrary n, we get an analagous result by using the random variable X which denotes the number of successes in n trials. We then have $H(X) \leq nH(p)$. In other words, we have

$$H(x_1, \cdots, x_n) \leq H(x_1) + \cdots + H(x_n) \ (= nH(p)).$$

In the case $n = 2$, this is just Corollary 10.22.

Chapter 11

Source Coding, Data Compression, Redundancy

Goals, Discussion In this chapter we discuss source coding. At this point we are not yet transmitting data over a possibly noisy channel. Instead, we are mainly involved with formatting source words from a source into binary strings that can then be transmitted. The source can have many different forms such as an analog, biological, digital, or other kind of source.

We cover source coding in detail including the basic results of Kraft and McMillan leading to Shannon's First Theorem. This result gives the amazing connection between source coding and entropy, the subject of Chapter 10. The fundamental ideas of compression, which are crucial for such applications as data transfer or downloading from a source such as the Internet, are presented. In particular, arithmetic coding and the fundamental compression algorithms of Huffman and Lempel-Ziv are described. Redundancy is also discussed. The mathematical requirements in this chapter are not unreasonable even though we cover everything in full mathematical detail. One reason for doing this is that the remarkable connection between entropy and encoding must be shown to be believed!

New, Noteworthy The well known Entropy Lemma in Section 11.2 is at the heart of all source coding.

The proof that the Huffman algorithm is optimal is tricky. Various authors make an unwarranted assumption as follows. Given an optimal encoding and a source word of smallest probability, the corresponding code word will have maximal length. This code word will then have a "sibling" on the tree. The assumption, often made, is that the corresponding source word will have the smallest probability or the next smallest probability. In our proof here, we show how to avoid this assumption.

Harking back to Chapter 9 the connection between small probability = big "information content" = long code word is reinforced.

11.1   Introduction, Source Extensions

Let us start off with a **source** which is a mechanism for emitting a continuous stream of **source words** chosen from the set $X = \{x_1, x_2, \ldots, x_m\}$ with corresponding probabilities $p_1, p_2, \ldots, p_m$ where $p_1 + p_2 + \cdots + p_m = 1$ and $0 \leq p_i \leq 1$, $i = 1, \ldots, m$. The set X is known as the **alphabet** of the source. These source words might be actual text words taken from a book, for example, or letters from the English alphabet where $m = 26$ or 27 (depending on whether or not we allow spaces as well as letters). In this case we have $p_1 = 0.064 =$ probability of A, $p_5 =$ probability of $E = \Pr(E)$, etc. The source words might also be signals from a radio antenna. Pierce [Pie79] discusses at length the case where the source emits musical notes. Another example would be a DNA sequence in molecular biology.

Frequently we assume that the source is **memoryless**. This means that the probability p_i that the source emits the symbol x_i is independent of what has been emitted previously. In technical language, the variables x_i are independent and identically distributed random variables (iid). This assumption is not really valid when we write in a language such as English, where the probability that a given letter appears depends very much on the letter or sequence of letters or spaces that precede it. Later on in the book we discuss **ergodic sources**, generalizing from memoryless sources. Ergodic sources yield a better approximation to a written language when regarded as a source.

Given a source Γ with source words chosen from X we can construct a new source, called the sth **order extension of** Γ, denoted by Γ^s. The alphabet of Γ^s consists of all possible strings of length s chosen from the alphabet X. If Z is a word in Γ^s then $Z = y_1, y_2, \ldots, y_s$ with $y_1, y_2, \ldots, y_s$ in X. The probability of Z is defined to be the product of the probabilities of $y_1, y_2, \ldots, y_s$. In symbols, $\Pr(Z) = \Pr(y_1) \cdots \Pr(y_s)$. The sum of the probabilities of elements such as Z add up to 1.

Example 11.1 *Let $X = \{x_1, x_2\}$ with $p_1 = \Pr(x_1) = 0.4$ and $p_2 = \Pr(x_2) = 0.6$. Then the second extension X^2 of X has source words (or alphabet) $X^2 = \{x_1x_1, x_1x_2, x_2x_1, x_2x_2\}$ with corresponding probabilities 0.16, 0.24, 0.24, 0.36. When encoding (see below) it can be more efficient to encode blocks of consecutive source words rather than individual source words, which is why we sometimes use extensions of sources. This process is known as block coding.*

By independence, we have the following important result.

Theorem 11.2 *If Γ is a source with alphabet X, and Γ^s denotes the sth order extension of*

Γ, *then*

$$H(\Gamma^s) = sH(\Gamma).$$

Intuitively, we can think that since we are making s choices from a source with uncertainty $H(\Gamma)$, the uncertainty of the resulting string will be s times the uncertainty of $H(\Gamma)$.

11.2 Encodings, Kraft, McMillan

An **encoding** f maps each source word chosen from X to a string with symbols in the alphabet Y. For example, Y might be the binary alphabet so $Y = \{0, 1\}$, X might be the upper-case English alphabet and f the ASCII encoding which encodes each letter as a binary string of length 8 (see also Chapter 3).

Another possibility is that f might be the encoding given by the **Morse code** so that Y consists of dots, dashes and spaces.

We always assume that for source words x_i, x_j with $i \neq j$, $f(x_i) \neq f(x_j)$. A **message** is defined to be any string of source words from $X = \{x_1, x_2, \ldots, x_m\}$. For example, the message M might be given by $M = x_3 x_1 x_3$. Then M gets encoded by stringing together or concatenating the strings $f(x_3)f(x_1)f(x_3)$. So we have $f(M) = f(x_3)f(x_1)f(x_3)$. The particular strings over Y of the form $f(x_i)$, $1 \leq i \leq m$ are called **code words**. The set of code words $f(x_i)$ is a **code** C.

Example 11.3 *Let X consist of the three source words u, v, w with probabilities $0.3, 0.5, 0.2$ respectively. Then, from Chapter 9,*

$$H(X) = (0.3)\log\left(\frac{1}{0.3}\right) + (0.5)\log\left(\frac{1}{0.5}\right) + (0.2)\log\left(\frac{1}{0.2}\right)$$

$$= 0.5211 + 0.5 + 0.4644 = 1.4855.$$

An encoding f from X to Y with $Y = \{0, 1\}$ is given as follows:

$$f(u) = 01, \quad f(v) = 1, \quad f(w) = 101.$$

Then if $m = vu$, $f(m) = f(v)f(u) = 101$. The **average length** of an encoded source word is $(0.3)(2) + (0.5)(1) + (0.2)(3) = 1.7$.

An encoding f is said to be **uniquely decipherable** (u.d.) if there do not exist two different messages M_1, M_2 with $f(M_1) = f(M_2)$. Note that, in Example 11.3, f is not u.d, because there are 2 messages that get encoded to the same code word. For example, $f(vu) = f(w) = 101$. The encoding f is an **instantaneous code** (or **prefix code**) if there do not exist two distinct source words x_i, x_j such that $f(x_i)$ is a prefix of $f(x_j)$. This means that we cannot have $f(x_j) = f(x_i)y$ where y is some string over Y. Thus, a prefix code can be uniquely decoded from left to right without "look ahead".

Remark 11.4 *It could be argued that the term "prefix" should be replaced by "prefix free".*

Lemma 11.5 *If f is instantaneous, then f is u.d.*

Proof. Suppose M_1, M_2 are two messages with $f(M_1) = f(M_2)$, so the two encodings are equal. Let $M_1 = x_a z_1$ and $M_2 = x_b z_2$ where z_1, z_2 are strings over X, and x_a, x_b are source words. Then, by definition of the encoding function f, we have $f(x_a)f(z_1) = f(x_b)f(z_2)$, because $f(M_1) = f(M_2)$. Suppose $x_a \neq x_b$. Then $f(x_a) \neq f(x_b)$. Then, depending on the length, either $f(x_a)$ is a prefix of $f(x_b)$ or $f(x_b)$ is a prefix of $f(x_a)$. We conclude from this that $x_a = x_b$. Continuing, if $z_1 = x_c u$ and $z_2 = x_d v$ we get $x_c = x_d$. Proceeding we get that $M_1 = x_a x_c u$ and $M_2 = x_b x_d u$. Continuing in this way we conclude that if $f(M_1) = f(M_2)$ then $M_1 = M_2$. ∎

Lemma 11.6 *There exist u.d. codes which are not instantaneous.*

Proof. Let $X = \{a, b\}$ with $f(a) = 1$, $f(b) = 10$. Then $f(a)$ is a prefix of $f(b)$ so f is not instantaneous. However, f is u.d. as we can see by working to the left each time a zero appears in the output. For example, the string 1101 is the encoding of the string aba. ∎

Example 11.7 *Let $X = \{a, b\}$, $f(a) = 1$, $f(b) = 1110$. Here again f is u.d. but not instantaneous because $f(a)$ is a prefix of $f(b)$.*

This example brings up essential differences between u.d. and prefix codes. In a prefix code we can decode *"on line"* moving from left to right as indicated in the proof of Lemma 11.5. Thus, given a message M, and an encoding $f(M)$, proceed from left to right until a code word $y = f(x)$ is formed. Then x must be the first source word in M: then we just iterate this procedure.

A fundamental result which we only prove now for the binary alphabet but which can be immediately generalized to any alphabet is as follows.

Theorem 11.8 (Kraft's inequality) *A necessary and sufficient condition for the existence of an instantaneous encoding $f : X \rightarrow Y^*$ with word-lengths $l_1, l_2, \ldots, l_m$ is that $\sum_{i=1}^{m} 2^{-l_i} \leq 1$. Here Y^* denotes the set of all possible strings over the alphabet Y.*

Proof. For convenience of notation we assume that $l_1 \leq l_2 \leq \cdots \leq l_m$, and construct an associated **binary tree** of depth l_m. We can also think of this tree as a **decision tree** with, say, 0 corresponding to "No" and 1 corresponding to "Yes". For example, let $m = 2$, $X = \{x_1, x_2\}$ and let $f(x_1) = 1$, $f(x_2) = 01$. Then $l_1 = 1$, $l_2 = 2$ and the corresponding tree is in Figure 11.1.

This tree has depth 2. There are $2^2 = 4$ **terminal points** labeled $11, 10, 01$ and 00. The **initial vertex** or **root** is V. Any prefix code gives rise to a binary tree such that, if $u \neq v$,

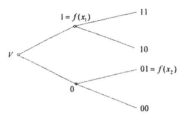

Figure 11.1: Decision tree

the vertex $f(u)$ (or $f(v)$) is not a predecessor of $f(v)$ (or $f(u)$ on the tree i.e. there is no path on the tree between $f(u)$ and $f(v)$.

To prove Theorem 11.8, we first assume the existence of a binary instantaneous code with word-lengths $l_1 \leq l_2 \leq \cdots \leq l_m$. We associate with this code a binary tree of depth l_m.

A code word of length l_1 excludes $\frac{2^{l_m}}{2^{l_1}} = 2^{l_m - l_1}$ terminal points, because this is the number of terminal points to the right of the code word, i.e., the number of terminal points for which the code word is a predecessor. Note that if u, v are distinct code words of length l_i, l_j then, by the prefix condition, any terminal point excluded by u is different from a terminal point excluded by v.

Thus, given a prefix code with word-lengths $l_1, l_2, \ldots, l_m$, the total number of excluded terminal points is $\sum_{i=1}^{m} 2^{l_m - l_i}$. Since the total number of terminal points being 2^{l_m} and since $l_1 \leq l_2 \leq \cdots \leq l_m$, we have that

$$\sum_{i=1}^{m} 2^{l_m - l_i} \leq 2^{l_m}.$$

Dividing by 2^{l_m} gives $\sum_{i=1}^{m} 2^{-l_i} \leq 1$.

For the converse, suppose we are given positive integers $l_1, l_2, \ldots, l_m$ with $1 \leq l_1 \leq l_2 \leq \cdots \leq l_m$ and satisfying the condition that $\sum_{i=1}^{m} 2^{-l_i} \leq 1$.

To construct a prefix code with word lengths $l_1, l_2, \ldots, l_m$, choose any point P on a binary tree of depth l_1 where P is l_1 steps from the root or initial vertex V.

As above, P excludes $2^{l_m - l_1}$ terminal points. In fact, no vertex of the tree on a path from P to one of these terminal points can correspond to another code word.

Since $\sum_{i=1}^{m} 2^{-l_i} \leq 1$ we get, multiplying by 2^{l_m}, that $\sum_{i=1}^{m} 2^{l_m - l_i} \leq 2^{l_m}$. It follows that $2^{l_m - l_1} < 2^{l_m}$. Thus, at least one terminal point is not excluded by P. It follows that any point on a path from V to one of the non-excluded terminal points can be chosen as another point on the tree. We choose such a point Q which has distance l_2 from the root (or left end-point) V of the tree. Then, proceeding, we get the required prefix tree with code-lengths $l_1, l_2, \ldots, l_m$ with $l_1 \leq l_2 \leq \cdots \leq l_m$. ∎

Note For a code alphabet of size D the Kraft inequality becomes $\sum_{i=1}^{m} D^{-l_i} \leq 1$.

Remark 11.9 *If the number of code words of length i is denoted by λ_i the condition in Theorem 11.8 can be written as $\sum_{j=1}^{t} \lambda_j 2^{-j} \leq 1$ where $t = l_m$ is the maximum length of a code word.*

In the sequel, the following lemma will be fundamental.

Lemma 11.10 (The Entropy Lemma) *Let p_i, $1 \leq i \leq n$ be a given probability distribution so that $0 < p_i \leq 1$ with $p_1 + p_2 \cdots + p_n = 1$ and let q_i, $1 \leq i \leq m$ be another probability distribution with $q_i > 0$. Then, taking logs to the base 2, we have $\sum_{i=1}^{n} p_i \log q_i \leq \sum_{i=1}^{n} p_i \log p_i$ with equality if and only if $p_i = q_i$, $1 \leq i \leq n$.*

Proof. Recall from calculus the function $y = \ln x = \log_e x$. When $x = 1$, $y = \ln x = 0$. The equation of the tangent line is $y - 0 = 1(x - 1)$, i.e., $y = x - 1$. Since $y'' < 0$, the function y is concave down. From calculus, we know that the graph of a function that is concave down lies below the tangent line at P (no matter where P is chosen on the curve). Thus, by taking the tangent at $x = 0$ we see that, for all x, $\ln x \leq x - 1$, with equality if and only if $x = 1$. So, $\ln\left(\frac{q_i}{p_i}\right) \leq \frac{q_i}{p_i} - 1$, with equality if and only if $q_i = p_i$, $1 \leq i \leq n$. Therefore

$$\sum_{i=1}^{n} p_i \ln\left(\frac{q_i}{p_i}\right) \leq \sum_{i=1}^{n} p_i \left(\frac{q_i}{p_i} - 1\right) \leq \sum_{i=1}^{n} (q_i - p_i) = \sum_{i=1}^{n} q_i - \sum_{i=1}^{n} p_i = 0.$$

Thus $\sum_{i=1}^{n} p_i \ln(q_i) - \sum_{i=1}^{n} p_i \ln(p_i) \leq 0$ giving

$$\sum_{i=1}^{n} p_i \ln(q_i) \leq \sum_{i=1}^{n} p_i \ln(p_i)$$

with equality if and only if $p_i = q_i$, $1 \leq i \leq n$. Now, since for any number v, $\log_e v \log_2 e = \log_2 v$ (and $\log_2 e$ is positive) the result follows. ∎

We can also use the above argument to prove our next result. Here $X = \{x_1, x_2, \ldots, x_n\}$ is a source with probabilities $p_1, p_2, \ldots, p_n$, and $H(X)$ is the entropy of X (see Chapter 9).

Theorem 11.11 $H(X) \leq \log_2 n$ *with equality if and only if $p_1 = p_2 = \cdots = p_n = \frac{1}{n}$ so that X is equiprobable. In other words, to maximize the entropy make the probabilities equal.*

Proof. In Lemma 11.10 put $q_1 = q_2 = \cdots = q_n = \frac{1}{n}$. Then, from the above

$$\sum_{i=1}^{n} p_i \log\left(\frac{1}{n}\right) \leq \sum_{i=1}^{n} p_i \log(p_i).$$

Since, for any number v, $\log\left(\frac{1}{v}\right) = -\log v$ we get that

$$\sum_{i=1}^{n} p_i \log\left(\frac{1}{p_i}\right) \leq \sum_{i=1}^{n} p_i \log n = \log n,$$

because $\sum_{i=1}^{n} p_i = 1$. Thus, $H(X) \leq \log n$ with equality if and only if $p_1 = p_2 = \cdots = p_n = \frac{1}{n}$. (Intuitively, to increase randomness, equalize the probabilities.) ∎

The next result gives a fortuitous reduction from u.d. to prefix codes.

Theorem 11.12 (McMillan's inequality) *A uniquely decipherable code with word-lengths $l_1, l_2, \ldots, l_n$ exists if and only if an instantaneous (or prefix) code exists with these word-lengths. This is in turn equivalent to the statement*

$$2^{-l_1} + 2^{-l_2} + \cdots + 2^{-l_n} \leq 1.$$

Proof. The condition can be written as $\sum_{j=1}^{t} \lambda_j 2^{-j} \leq 1$, as described in the remark above, where the longest code word has length t.

The "if" part of the theorem is clear, since every prefix code is u.d. We proceed to the converse. Thus we assume that we have a u.d. code with word-lengths $l_1, \cdots, l_n$. We will show that there exists a prefix code with these word-lengths by proving that the Kraft inequality above holds.

Write

$$\left(\sum_{j=1}^{t} \lambda_j 2^{-j}\right)^s = \sum_{k=s}^{ts} N_k 2^{-k} \tag{11.1}$$

where s is an arbitrary positive integer that in the sequel goes to infinity. Equation (11.1) is obtained by multiplying out the s terms on the left. From this we see that N_k counts the total number of source messages M obtained by stringing together, or concatenating, exactly s (not necessarily distinct) source words $u_1 u_2 \cdots u_s = M$ such that when we encode M we get a binary string of length exactly equal to k, so that $|f(M)| = k$.

Now the encoding f is u.d. Thus, the same binary string $f(M)$ never appears again when we multiply the s terms on the left. For if it did, two different source messages would encode to the same binary string.

Thus the biggest possible value that N_k can have corresponds to the case in which every possible binary string of length exactly k shows up as $f(M)$. But the total number of binary strings of length k is 2^k. To summarize we have

$$N_k \leq 2^k.$$

Substituting in equation (11.1) we have

$$\left(\sum_{j=1}^{t} \lambda_j 2^{-j} \right)^s \leq \sum_{k=s}^{ts} 2^k 2^{-k} = ts - s + 1 \leq ts.$$

Raising both sides to the power $\frac{1}{s}$ we get

$$\sum_{j=1}^{t} \lambda_j 2^{-j} \leq t^{1/s} s^{1/s}.$$

Letting s go to infinity (and using some calculus) gives us the result that $\sum_{j=1}^{t} \lambda_j 2^{-j} \leq 1$, as required. ■

We now proceed to one of the main results for source coding called the

Noiseless Coding Theorem . The setup is this. A memoryless source emits source words chosen from $X = \{x_1, x_2, \ldots, x_m\}$ with corresponding probabilities $p_1, p_2, \ldots, p_m$. We have a binary u.d. encoding f which minimizes the average length t of $\{|f(x_i)|\} = \{n_i\}$, where $|f(x_i)| = n_i$ is the length of $f(x_i)$, $1 \leq i \leq m$. Our goal is to obtain an estimate for t. The entropy of a binary string of length t is at most t (see Chapter 9). So we suspect that $t \geq H(X)$ since $H(X)$ is the average uncertainty of a source word. This turns out to be the case.

Theorem 11.13 (Noiseless Coding Theorem) *If a memoryless source has entropy H then the average length of a binary, uniquely decipherable, encoding of that source is at least H. Moreover, there exists such a code having average word-length less than $1 + H$, on the assumption that the emission probability p_i of each source word is positive, $1 \leq i \leq m$.*

Proof. We have $H = -\sum_{i=1}^{m} p_i \log p_i$. The average length of a code word is $t = \sum_{i=1}^{m} p_i n_i$. From Theorem 11.8, we have $A = \sum_{i=1}^{m} 2^{-n_i} \leq 1$. Now define $q_i = \frac{2^{-n_i}}{A}$. Then $q_i \geq 0$ and $\{q_i\}$ gives a probability distribution. From Lemma 11.10 we have $H \leq -\sum_{i=1}^{m} p_i \log q_i = -\sum_{i=1}^{m} p_i(-n_i - \log A)$. Thus:

$$H \leq \sum_{i=1}^{m} p_i n_i + \left(\sum_{i=1}^{m} p_i \right) \log A.$$

Since $A \leq 1$ and $\sum_{i=1}^{m} p_i = 1$ we have $H \leq \sum_{i=1}^{m} p_i n_i$, as required since $\log A \leq 0$. To prove the last sentence of the theorem we choose $n_1, n_2, \ldots, n_m$ such that, for each i, n_i is the smallest positive integer satisfying $\frac{1}{p_i} \leq 2^{n_i}$. Since $p_1 + p_2 + \cdots + p_m = 1$ this gives $\sum_{i=1}^{m} 2^{-n_i} \leq 1$. Thus there exists a u.d. code (in fact, a prefix code) with these word lengths. From the definition of n_i we have $n_i \geq -\log p_i$, and by minimality of n_i we get

$n_i < -\log p_i + 1$. Then

$$t = \sum_{i=1}^{m} p_i n_i < \sum_{i=1}^{m} p_i \left(-\log p_i\right) + \sum_{i=1}^{m} p_i = H + 1$$

so $t < H + 1$. ∎

Note Suppose we use, instead of the binary alphabet, a general alphabet of size D. Then the lower bound for the average length becomes $\frac{H}{\log D}$. The upper bound becomes $1 + \frac{H}{\log D}$.

11.3 Block Coding, The Oracle, Yes-No Questions

Now suppose that instead of assigning a code word $f(x_i)$ to each source word x_i we use "**block coding**" to construct the s-fold extension or s^{th} extension of X denoted by X^s and assign a code word to each source word Z in X^s. In other words, we take a sequence of s independent measurements of X and assign a code word, using f, to a source word Z in X^s as follows. For $Z = (y_1, y_2, \ldots, y_s)$, put $f(Z) = f(y_1)f(y_2)\cdots f(y_s)$. We have that $\Pr(Z) = \Pr(y_1)\Pr(y_2)\cdots\Pr(y_s)$ where Pr denotes "probability". If $|X| = m$ then $|X^s| = m^s$. Let t_s denote the average length of an encoded word of Z. From Theorem 11.13 we have $H(Z) \leq t_s < H(Z) + 1$. Since Z is obtained from s **independent** measurements of X, $H(Z) = sH(X)$. Then $sH(X) \leq t_s < sH(X) + 1$.

Dividing by s (which is > positive 0) we get

$$H(X) \leq \frac{t_s}{s} < H(X) + \frac{1}{s}.$$

Now $\frac{t_s}{s}$ represents the *average code word length per value of* X and can be made as close as we wish to $H(X)$ by increasing s. Thus $H(X)$ can be regarded as the minimum number of binary digits needed on average to encode a source word from X.

Harking back to Chapter 10, if we use a binary tree for encoding a source X, $H(X)$ can be thought of as the minimum number of "yes-no" questions needed, on average, to find out the value of X. Given the tree, we determine the outcome by asking "Yes or No" questions, for example, "Is the first entry of the code word 0?", "Is the second 0?", etc. from an "oracle". This is also closely related to the **binary search algorithm** in computer science.

Example 11.14 *We have a biased coin in which* $\Pr(Head) = 0.7$ *and* $\Pr(Tail) = 0.3$. *We toss it twice. To find out the final outcome, we could just ask for the outcome of each toss separately. This would take 2 questions.*

However, if we asked, "Is it heads then heads?", we would get a yes answer with probability $0.7 \times 0.7 = 0.49$. *If we got a no answer, we can ask "Is it heads then tails?", which* *would be correct with probability $0.7 \times 0.3 = 0.21$. Finally, if the answer to that is no, we could ask "Is it tails then heads?", which is correct with probability 0.21 and incorrect with probability equal to 0.09. On average this yields $1 \times 0.49 + 2 \times 0.21 + 3 \times 0.21 + 3 \times 0.09 = 1.6$ questions to determine the outcome of the two tosses which is fewer questions than the naïve approach.*

Note: This setup is the encoding function f where $f(HH) = 1$, $f(HT) = 01$, $f(TH) = 001$, $f(TT) = 000$.

11.4   Optimal Codes

We now look to the problem of actually constructing encodings f and codes C that minimize the average length t of the code words. If C is the code that minimizes $t = L(C)$ in the class of prefix codes we claim that C minimizes $L(C)$ in the class of all u.d. codes. For, if C_1 minimizes t in this larger class then there exists a *prefix code* C_2 with the same code word lengths as C_1. The claim now follows from the optimality of C.

(Technically, one needs to prove that an optimal code C actually exists, i.e., that we don't have to get involved in some kind of limiting process involving the greatest lower bound. In fact, the existence can be shown using the Kraft inequalities above).

So we restrict attention to prefix codes C (= instantaneous codes). We assume that C is an encoding of a source, with source words drawn from $X = \{x_1, x_2, \ldots, x_m\}$ with the corresponding probabilities

$$p_1 \geq p_2 \geq \cdots \geq p_{m-1} \geq p_m.$$

The source word x_i is encoded by an encoding function f, and we denote the length of the binary string $f(x_i)$ by N_i, $1 \leq i \leq m$. Our notation is also chosen so that source words with the same probability are listed in increasing order of code word length. We claim that if C is optimal (i.e., minimizes the average code word length) then C satisfies the following two properties.

Property 1 *Let $\mathbf{c}$ denote any longest code word in an optimal encoding by a code C. Then there must also exist another code word $\mathbf{d}$ agreeing with $\mathbf{c}$ in all digits except the last, so $\mathbf{c}$ and $\mathbf{d}$ are siblings. (Think of a family tree.)*

Proof. Suppose this were not true. Then we could cancel the last digit of $\mathbf{c}$ and still have a prefix code which would then have a shorter average length than C, contradicting the optimality of C. $\blacksquare$

Property 2 *Higher probability symbols have shorter code words. Thus, if $p_i > p_j$ then $N_i \leq N_j$.*

Proof. If this were not the case we could construct an instantaneous code C_1 by interchanging the code words $f(x_i)$ and $f(x_j)$. The average length of C_1 minus the average length of C is $p_j N_i + p_i N_j - (p_j N_j + p_i N_i) = (p_j - p_i)(N_i - N_j)$, which is negative (given that $p_i > p_j$) or if we assume that $N_i > N_j$. That is, C_1 has shorter average word length than C, but now we have a contradiction to the optimality of C, i.e., to the minimality of $L(C)$. ∎

Concerning the construction of codes C, with $L(C)$ small, where $L(C)$ denotes the average length we see one way to do this, which is embedded in the proof of Theorem 11.13. Namely, assuming that all p_i are positive, we define N_i to be the smallest positive integer satisfying $\frac{1}{p_i} \leq 2^{N_i}$. Then, from the Kraft inequality, we can construct a prefix code C so that $L(C) < 1 + H$ because $\sum_{i=1}^{m} 2^{-N_i} < \sum_{i=1}^{m} p_i = 1$. The actual construction of C can be carried out as in the proof of Theorem 11.8.

This method of constructing codes with small average code word length is called **Shannon-Fano** encoding. For the Shannon-Fano encoding, $L(C)$ is within 1 of the optimal length.

In fact, it is possible to construct an optimal encoding by a remarkably simple procedure called **Huffman coding**.

11.5 Huffman Coding

This procedure will lead to an instantaneous (or prefix) code C such that $L(C)$ is less than or equal to $L(C_1)$ for any u.d. code C_1 associated with the given source.

We start off with a given source $S = S_0$ that emits source words which are drawn from $X_0 = X = \{x_1, x_2, \ldots, x_m\}$. Our notation is as in the previous section, so $p_1 \geq p_2 \geq p_3 \geq \cdots \geq p_{m-1} \geq p_m$, where $p_i = \Pr(x_i)$, $1 \leq i \leq m$.

First we "merge" the two source words with smallest probability into a single new source word. Thus in the first iteration we merge x_{m-1} and x_m to give a new "heavier" vertex W_1 with probability $y = x_m + x_{m-1}$.

Simultaneously we are constructing, inductively, a graph. When we start off we have m distinct points in the graph and no edges. Then, when we merge, we have an edge from each of the vertices representing x_m and x_{m-1} to W_1.

Thus initially our graph G_0 looked like Figure 11.2

Figure 11.2: Initial Graph

After the first merge we have a new "source" S_1 with $m-1$ source words $x_1, x_2, \ldots, x_{m-2}, W_1$. Our graph G_1 now has vertices $x_1, x_2, \ldots, x_{m-2}, x_{m-1}, x_m, W$ and looks like Figure 11.3

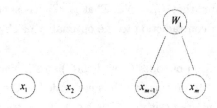

Figure 11.3: After One Merge

The net effect is that we have *increased* the number of vertices in the graph by one, and the number of edges by two. Simultaneously the number of source words in our source has *decreased* by one.

At the next stage we iterate the process. Thus we adjoin a new vertex W_2 to the existing graph G_1 together with two new edges emanating from the two source words in S_1 with smallest probability, where S_1 draws its source words from $X_1 = \{x_1, x_2, \ldots, x_{m-2}, W_1\}$. In what follows, given any graph G we denote the vertices and edges of G by $V(G)$ and $E(G)$.

Associated with our Huffman procedure we have the following table arising from the iterative procedure above, which terminates when the number of source words is 1.

Source	Number of Source Words	Graph G	$|V(G)|$	$|E(G)|$
S_0	m	G_0	m	0
S_1	$m-1$	G_1	$m+1$	2
S_2	$m-2$	G_2	$m+2$	4
$\vdots$				
S_i	$m-i$	G_i	$m+i$	$2i$
$\vdots$				
S_{m-1}	1	G_{m-1}	$2m-1$	$2(m-1)$

Note that at the end of our procedure we have a connected graph G_{m-1}, which is a **tree** because the number of edges is 1 less than the number of vertices. The set of new vertices is $\{W_1, W_2, \ldots, W_{m-1}\}$. Using the tree we can encode the original source S and obtain a prefix encoding f of S–which turns out to be optimal. Each source word of S will correspond to a **leaf** on the tree, i.e.: a vertex with just one edge on it. All other vertices (or nodes) on the tree are called **internal vertices** (or nodes).

Note that the tree and the encoding will not be unique. For one thing we may have ties among the probabilities. There is also a left-right ambiguity. A small example can now be presented.

Let $X_0 = X = \{a, b, c, d, e\}$ where the corresponding probabilities are $0.45, 0.2, 0.15, 0.1, 0.1$. Our initial graph, and the corresponding probabilities, look as follows

After the first merge we have the following picture

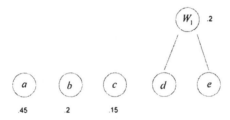

Our probability set now is $\{0.45, 0.2, 0.2, 0.15\}$. So we have a choice. We can merge b and c or we can merge W_1 and c. Either procedure works to give an optimum code, although the coding will be different.

Our diagram after the second merge will be more manageable if we merge b and c. We then have the following.

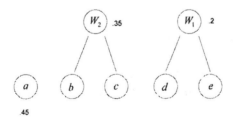

After this merger our probability set is $\{0.45, 0.35, 0.2\}$. So we merge W_1 and W_2. When we do this, we have the following diagram.

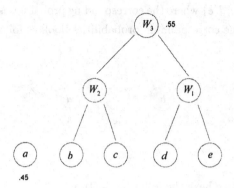

Our probability set is now $\{0.55, 0.45\}$. So we merge a and W_3 to get our encoding tree.

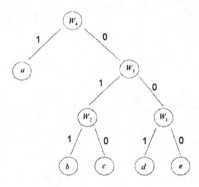

Now starting from the top vertex W_4 we construct our encoding f. Imagine the encoder starting out from W_4 and coming down. If the encoder goes right she puts in a 0; if she goes left she puts in a 1. So with this convention we have $f(a) = 1$, $f(b) = 011$, $f(c) = 010$, $f(d) = 001$, $f(e) = 000$. Note that $H(X) = (0.45)\log\left(\frac{1}{0.45}\right) + (0.2)\log\left(\frac{1}{0.2}\right) + 0.15\log\left(\frac{1}{0.15}\right) + 2(0.1)\log\left(\frac{1}{0.1}\right) = 1.7$. The average length is $(0.45)(1) + 3[0.2 + 0.15 + 0.1 + 0.1] = 2.1$.

Calculating the average length   There is a shortcut here. One way of getting the average length is by multiplying the length of $f(x_i)$ by p_i, and adding, as we have just done. But there is a shorter approach, as follows. Each time we merge we get a new vertex W_i, $i = 1, 2, \ldots, m-1$ with associated probability, say q_i. We have the following result.

Theorem 11.15 *The average length is $q_1 + q_2 + \cdots + q_{m-1}$.*

Proof. Our final graph which, being connected and having no circuits is a tree G_{m-1}, also gives a prefix encoding f and a sequence of codes $C_1, C_2, \ldots, C_{m-1}$ for the sources $S_1, S_2, \ldots, S_{m-1}$.

The source C_{m-1}, containing just one source word with probability 1, is encoded by the null string. (In our example W_4 is encoded by the null string, a is encoded as 1, W_3 is encoded by 0, W_2 by 01, etc.). Denoting the average length of C_i by $L(C_i)$ we have the following.

$$L(C) = L(C) - L(C_1) + [L(C_1) - L(C_2)] + \cdots + [L(C_{m-2}) - L(C_{m-1})] + L(C_{m-1})$$

Since $L(C_{m-1}) = 0$ we need only calculate $[L(C_{i-1}) - L(C_i)]$. To get from the graph G_{i-1} to the graph G_i we merge two words with probability u, v say into a single vertex W_i with probability $u + v$. Thus the average length is changed, as follows. For some integer ℓ, which is the length of the code word corresponding to the node W_i, we have

$$L(C_{i-1}) - L(C_i) = (\ell + 1)u + (\ell + 1)v - \ell(u + v) = u + v.$$

So we get that $L(C_{i-1}) - L(C_i) = u + v$, which is the probability associated with the vertex W_i, $1 \le i \le m - 1$. ∎

Remark 11.16 *In the previous example, associated with W_1, W_2, W_3, W_4 we have the probabilities $0.2, 0.35, 0.55, 1$ for a total of 2.1, which agrees with our previous calculation for the average of length of $L(C)$.*

Remark 11.17 *Note, too, that when merging we can put either one of the merged nodes on the left and the other on the right, giving different encodings.*

Example 11.18 *Let X be a memoryless source which emits heads with probability 0.7 and tails with probability 0.3. Let X^2 denote the second extension of X. We want to find an optimal encoding of X^2.*

The possibilities after two coin tosses are HH, HT, TH and TT with respective probabilities 0.49, 0.21, 0.21 and 0.09. When we run the Huffman algorithm, we come up with a tree as follows.

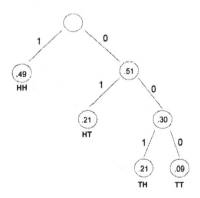

Which offers the encoding f where $f(HH) = 1$, $f(HT) = 01$, $f(TH) = 001$ and $f(TT) = 000$, which is exactly the encoding we had in example 11.14 of section 11.3.

11.6 Optimality of Huffman Coding

Proof of the Optimality of Huffman Coding . As mentioned in the proof of Theorem 11.15, the tree constructed from the Huffman encoding C_0 of a source $S = S_0$ yields a sequence of Huffman encodings $C_1, \cdots, C_{m-1}$ for the sources $S_1, \cdots, S_{m-1}$.

To prove the result we induct on m where the source $S_0 = S$ chooses source words from $X = \{x_1, \cdots, x_m\}$.

If $X = \{x_1, x_2\}$ consists of just two source words, then the Huffman encoding gives an encoding function f with $f(x_1) = 1$, $f(x_2) = 0$ which is optimal.

Let us construct the Huffman encoding for S_0. Moving down from the top of the corresponding tree, assume that the source S_i is optimally encoded by C_i (so that the average length $L(C_i)$ is minimal) and that the code C_{i-1} is not an optimal encoding of S_{i-1}. Then there exists an optimal prefix encoding f of S_{i-1} and a code D obtained from f such that $L(C_{i-1}) > L(D)$.

Let l be the maximum length of $f(\mathbf{x})$ where $\mathbf{x}$ is any source word. Then there must exist a code word $\mathbf{u}$ in S_{i-1} of minimal probability such that $|f(\mathbf{u})| = l$ where $|\ |$ denotes length. For if not, we could construct a new prefix code by interchanging $f(\mathbf{x})$ with $f(\mathbf{u})$, where $|f(\mathbf{x})| = l$ and $\mathbf{u}$ has minimal probability. Then using the same argument as in Property 2 of Section 11.4 we could show that D is not optimal.

From Property 1 in Section 11.4 there exists a sibling code word $f(\mathbf{v})$ to $f(\mathbf{u})$ in D where $\mathbf{v}$ is another source word in S_{i-1}.

We want to show that we may assume that either $\Pr(\mathbf{v}) = \Pr(\mathbf{u})$ or else that $\Pr(\mathbf{v})$ has the next smallest probability of any source word in S_{i-1}. Suppose there exist $\mathbf{w}$ in S_{i-1} with $\Pr(\mathbf{v}) > \Pr(\mathbf{w})$. Then by Property 2 in Section 11.4, $|f(\mathbf{v})| = |f(\mathbf{w})|$. By interchanging $f(\mathbf{v})$, $f(\mathbf{w})$ we then have a prefix code E with $L(D) = L(E)$ with E being optimal and having the desired property.

So we can assume that there exists an optimal prefix encoding f of S_{i-1} with code D such that the following holds. There are two source words $\mathbf{u}, \mathbf{v}$ having the 2 smallest probabilities q_1, q_2 in S_{i-1} such that the code words $f(\mathbf{u})$ and $f(\mathbf{v})$ are siblings.

By merging the 2 source words $\mathbf{u}$ and $\mathbf{v}$ we get a new source T and an encoding D_1 of T. Now $\mathbf{u}, \mathbf{v}$ may not be the same two source words of smallest probability in S_{i-1} that were merged to form the source S_i. However, as sources, T and S_i are the same, and D_1 gives an encoding of S_i.

So we have, with L denoting the average code word length, that:

$$L(D) = L(D_1) + q_1 + q_2$$
$$L(C_{i-1}) = L(C_i) + q_1 + q_2$$

Thus $L(D) - L(C_{i-1}) = L(D_1) - L(C_i)$.

If $L(D) < L(C_{i-1})$ then $L(D_1) < L(C_i)$. But this contradicts the optimality of the code C_i. We conclude that $L(D) = L(C_{i-1})$ so that C_{i-1} is an optimal encoding of S_{i-1}.

This completes the induction and the proof of the optimality of Huffman encoding.

11.7 Data Compression, Lempel-Ziv Coding, Redundancy

Data compression is a major area because of the many applications when transmitting or downloading text, audio, video etc. Like shorthand, it is a method for encoding data in an economical way, that is, a noiseless economical encoding of a source that can easily be decoded to give the original message.

We have seen the fundamental theoretical connection between the average length of a code word when the source is encoded into a code C and the entropy of the source. If the source has alphabet X and none of the probabilities are 0 and $L(C)$ is the average code word length, we have

$$H(X) \leq L(C) < H(X) + 1.$$

If we allow zero probabilities, the above statement must be modified and we get

$$H(X) \leq L(C) \leq H(X) + 1.$$

(To go to the zero probability case we can use a limiting process, but then we have to allow equality in the upper bound.)

Arithmetic Coding . We have already met two important compression algorithms, namely Shannon-Fano encoding and Huffman encoding. In section 11.2, we talked about "block coding", where we use the s-fold extension of a source to show that the average word length is close to $H(X)$. Another way of encoding, suitable for the s-fold extension, is called "**Arithmetic Coding**". Let's give the main idea.

Suppose our source is $X = \{x_1, x_2\}$ with probabilities given by $\frac{3}{5}, \frac{2}{5}$, and we want to encode length 2 messages. We can write down the $2^2 = 4$ possible source words lexicographically giving the words $x_1x_1, x_1x_2, x_2x_1, x_2x_2$ with probabilities $\frac{9}{25}, \frac{6}{25}, \frac{6}{25}, \frac{4}{25}$. We can make the following diagram

x_1x_1	x_1x_2	x_2x_1	x_2x_2

0        $\frac{9}{25}$ $\frac{15}{25}$ $\frac{21}{25}$ 1

The probabilities above in decimal form are $0.36, 0.24, 0.24, 0.16$ and the cumulative probabilities are $0.36, 0.60, 0.84, 1$. To encode a message, we just need to indicate the corresponding segment unambiguously. For example, the point $\frac{1}{4}$ is in the first segment, the point $\frac{1}{2}$ is in the second, the point $\frac{3}{4}$ is in the third segment and the point $\frac{7}{8}$ is in the fourth segment. This is represented in Figure 11.4.

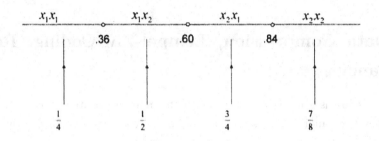

Figure 11.4: Arithmetic Coding

Now all we have to do is to calculate $\frac{1}{4}, \frac{1}{2}, \frac{3}{4}, \frac{7}{8}$ as *binary decimals*. We have $\frac{1}{4} = 0.01$, $\frac{1}{2} = 0.1$, $\frac{3}{4} = 0.11$, $\frac{7}{8} = 0.111$. This gives the following encoding.

$$x_1x_1 \longrightarrow 01$$
$$x_1x_2 \longrightarrow 1$$
$$x_2x_1 \longrightarrow 11$$
$$x_2x_2 \longrightarrow 111$$

The average length (per alphabet symbol) is

$$\frac{1}{2}\{(0.36)2 + (0.24)1 + (0.24)2 + (0.16)3\} = 0.96$$

On the other hand, $H(X) = \frac{3}{5}\log\left(\frac{5}{3}\right) + \frac{2}{5}\log\left(\frac{5}{2}\right) = 0.9710$.

For arithmetic coding the number of bits is determined by the size of the interval. In general, a shorter interval will have smaller probability and will require more bits, giving a nice tie-in with Shannon's measure of $\log(\frac{1}{p})$ for entropy. It can be shown that arithmetical coding is optimal in the limit as the size of the source (= the length of the message) goes to infinity.

Lempel-Ziv Coding . The Lempel-Ziv (LZ) algorithm has, in large part, displaced the Huffman algorithm for data compression. Haykin [Hay01] reports that LZ achieves a compression of around 55 percent when applied to ordinary English text as opposed to a figure of 43 percent for Huffman. One reason is that the Huffman encoding does not seem to exploit statistical dependencies in English as well as LZ. In general, a disadvantage of Huffman (as opposed to LZ) is the fact that one needs to know in advance (or estimate) the probabilities of the source words. The LZ algorithm is simple and easy to implement. **Unix compression** uses general LZ methods, as do the various **zip** and **unzip** algorithms.

The LZ technique is remarkably simple. The method is this. We parse the source stream into segments that are the **shortest** subsequences not yet encountered. These new subsequences are longer by one symbol than previously encountered sequences, giving rise to compression by storing pointers to the data. We can also construct trees (or "tries") for encoding/decoding. Each new subsequence not yet encountered will be equal to an old subsequence with a single letter (innovation symbol) added on at the end.

An example will clarify the method. Suppose our alphabet has just two letters x and y and that our sample input stream is

$$\boxed{x\,y\,y\,y\,x\,x\,y\,x\,x\,x\,x\,y\,x\,y\,x\,x\,y\,x\,x\,x}$$

Our algorithm or procedure is as follows.

Procedure . Proceeding from the left, break up the remaining stream into segments that represent the shortest subsequences not yet encountered. Also, index these segments. This gives us the following.

x	y	$y\,y$	$x\,x$	$y\,x$	$x\,x\,x$	$y\,x\,y$	$x\,x\,y$	$x\,x\,x\,x$
1	2	3	4	5	6	7	8	9

These index numbers from 1 to 9 are used to label the segments and construct the tree.

The empty string corresponding to the start of the text has index 0. The segment numbered 1 is $\{x\}$. Thus this segment gets the label $0x$.

Similarly, the segment y in slot 2 gets labeled $0y$. The string in slot 3 is $\{yy\}$. This can be regarded as the concatenation of the old string in slot 2, namely y, with the new or innovation symbol y. So slot 3 is labeled $2y$. Slot 4 becomes $1x$. Slot 5 is $2x$. Slot 6 is $4x$. Slot 7 is $5y$. Slot 8 is $4y$. Slot 9 is $6x$. So we now have the following

Labels	$0x$	$0y$	$2y$	$1x$	$2x$	$4x$	$5y$	$4y$	$6x$
Slots	1	2	3	4	5	6	7	8	9

Now we see the compression at work. For example, in slot 8 we have (uncompressed) 3 letters xxy which cost $3(8) = 24$ bits when encoded in ASCII. With the compression we only need to encode $4y$, which takes just 13 bits. In effect, we are replacing long strings of text with just a number and a text letter for each slot.

We can also construct a tree associated with the encoding. This tree will have as its root the label 0 corresponding to the empty string. It will have, in addition, 9 vertices corresponding to the 9 indexes 1 to 9.

To save space on the tree we have not included the vertex 9, which is obtained from the vertex 6 by adjoining x.

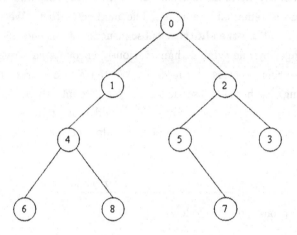

11.8   Problems

1. Carry out the other Huffman encoding for the source with probabilities $0.45, 0.2, 0.15, 0.1, 0.1$.

2. Is the following code C optimal where C is a set of code words given by $C = \{1, 100, 101, 1101, 1110\}$?

3. What is the maximum number of source words in a prefix binary code with maximum word length 5?

4. Find a Huffman code for a source with probabilities $0.1, 0.15, 0.15, 0.2, 0.4$.

5. Let X be the source which emits heads with probability 0.8 and tails with probability 0.2. Find an optimal encoding for X^2, the second extension of X. What is the average word length?

6. Let X be as in problem 5. Find an optimal encoding for X^3, the third extension of X. What is the average word length?

weight 1?

7. If a source with N source words is encoded as an instantaneous code and the code word lengths are $\ell_1, \ell_2, \ldots, \ell_N$, show that $\ell_1 + \ell_2 + \cdots + \ell_N \geq N \log_2 N$.

8. Construct the LZ tree for the sequence

$$y\ x\ x\ y\ y\ x\ y\ y\ x\ x\ x\ y\ y\ x\ y\ x\ y\ y\ x$$

9. What does the Huffman code look like for a source with 2^m source words where each source word is equally likely?

11.9 Solutions

1. First begin by merging 0.1 and 0.1 to give a new source $\{0.45, 0.2, 0.15, 0.2\}$. Combine the 0.15 to the 0.2 to get $\{0.45, 0.2, 0.35\}$. Merge 0.2 and 0.35 to get $\{0.45, 0.55\}$. Finally merge those to get the full tree. Now we get the following encoding.

$$
\begin{aligned}
0.45 &\rightarrow 0 \\
0.2 &\rightarrow 10 \\
0.15 &\rightarrow 110 \\
0.1 &\rightarrow 1110 \\
0.1 &\rightarrow 1111
\end{aligned}
$$

Notice that this is an optimal code as defined in Section 11.4. The average word length is 2.1.

2. No. The longest strings, 1101 and 1110, should be siblings.

3. You can't fill any of the spots above the 5th level without blocking off other code word spots at the 5th level. So the maximum number of source words occurs when the whole 5th level is full, which is $2^5 = 32$ source words.

4. Combine 0.1 and 0.15 to get a new source $\{0.4, 0.2, 0.15, 0.25\}$. Now merge 0.15 and 0.2 to get $\{0.4, 0.35, 0.25\}$. Join 0.25 and 0.35 to get $\{0.4, 0.6\}$, and then join the last two. This yields an encoding of:

$$
\begin{aligned}
0.4 &\rightarrow 0 \\
0.2 &\rightarrow 100 \\
0.15 &\rightarrow 101 \\
0.15 &\rightarrow 110 \\
0.1 &\rightarrow 111
\end{aligned}
$$

The average word length is 2.2.

5. The source X^2 has possibilities HH, HT, TH, TT with associated probabilities 0.64, 0.16, 0.16 and 0.04. An optimal encoding is

$$
\begin{aligned}
HH &\rightarrow 1 \\
HT &\rightarrow 01 \\
TH &\rightarrow 001 \\
TT &\rightarrow 000
\end{aligned}
$$

with average word length 1.56.

6. The source X^3 is tabulated as follows.

source symbol	HHH	HHT	HTH	THH	HTT	THT	TTH	TTT
probability	0.512	0.128	0.128	0.128	0.032	0.032	0.032	0.008
code word	1	011	010	001	0001	00001	000001	000000

The average word length is 2.192.

7. The given encoding will still give an instantaneous encoding for any probability distribution, including the equiprobable case. In that case, we have $H(X) \leq t$ where t is the average word length. Thus,

$$
\log N = \sum_{i=1}^{N} \frac{1}{N} \log N = H(X) \leq t = \frac{\ell_1}{N} + \frac{\ell_2}{N} + \cdots + \frac{\ell_N}{N}
$$

and from this the conclusion follows.

8. Here is the LZ tree:

y	x	$x\,y$	$y\,x$	$y\,y$	$x\,x$	$x\,y\,y$	$x\,y\,x$	$y\,y\,x$
1	2	3	4	5	6	7	8	9
$0y$	$0x$	$2y$	$1x$	$1y$	$2x$	$3y$	$3x$	$5x$

9. All source words will end up in the bottom level, because after merging any 2 source words, you will get a node whose probability is bigger than any other pure source word. So you will get a code C where each code word has length m, and, furthermore, C will contain *all* codewords of length m.

Chapter 12

Channels, Capacity, the Fundamental Theorem

Goals, Discussion The material here is fundamental and intricate. We want to give detailed explanations of these topics, not just statements of theorems. In Chapter11 we have seen how data can be efficiently encoded in binary form. We discussed how to get rid of "bad" redundancy that adds nothing to the information content of the source.

Here is an example. Let a source S emit symbols a, b, c, d, e with probabilities 0.45, 0.2, 0.15, 0.1, 0.1. As we saw in Chapter 11, the Huffman encoding, which is the most efficient coding, is such that, on average, 2.1 bits per emitted symbol are required. If we are constructing a file of 1000 successive source symbols from S we require about 2100 bits. We point out that the individual source letters have been encoded into binary strings of *variable length*.

Using, for example, a Huffman encoding, we can get rid of "bad redundancy". But now we introduce "*good redundancy*"! Let us explain.

We want to transmit our file of approximately 2100 bits over some kind of **noisy channel**, denoted by Γ.

What we do is this. We divide our data stream of 2100 bits into blocks of a *fixed* size k. We then "encode" our block of length k–so we have k **information digits**, or **message digits** (or **bits**)–into a longer block of length n, called a **code word**, by adjoining $n - k$ bits which add "good redundancy". These $n - k$ bits, in the case of linear codes, are "**check bits**" or "**parity bits**" or "**parity checks**". For example, with our 2100 bits we could let $k = 7$ and adjoin just one parity bit. This single parity bit adjoins 0 or 1 as the case may be to ensure that *the total number of ones in the enlarged message of length 8 is an even number*. For example, the block $(0\,1\,1\,1\,1\,0\,1)$ which has 5 ones, gets encoded as $(0\,1\,1\,1\,1\,0\,1\,1)$ which has 6 ones. On the other hand, a block $(0\,1\,1\,1\,0\,0\,1)$ that has 4 ones, gets encoded as $(0\,1\,1\,1\,0\,0\,1\,0)$ which still has 4 ones. Then, if a binary string with an odd number of ones is received we know that there has been a transmission error and can ask for a resend. This is an example of **error detection**.

In the general case, each check bit will be a binary sum (or linear combination, for bigger fields) of message bits. The idea is that if only a few bits of a code word are corrupted by the channel the receiver can decode correctly (perform **error correction**) or **detect** errors and ask for a resend.

Our task then is to determine exactly the maximum rate of transmission whereby the code words can be decoded in such a way that the probability of error is vanishingly small over a given channel. This famous problem was solved by Claude Shannon in the 1940's. His result is the fundamental result in communication theory. A more engineering-type aspect of the same problem is discussed in Chapter 13.

New, Noteworthy We explain the fundamental concepts of information rate and channel capacity by means of several examples. We show how to calculate channel capacity in a very easy manner using elementary probability tree diagrams. We are able to quickly obtain the limiting value of the **cascade** of a channel with itself n times by using some of the mathematical results behind Markov chains. The basic idea is that, mathematically, certain kinds of channel matrices can be regarded as the matrix of a Markov chain.

12.1   Abstract Channels

We start off with a **channel** Γ which is a mechanism for transmitting data messages. We will give several examples shortly.

Abstractly, the input and output of the channel correspond to discrete random variables X, Y. Our channels will be discrete and X, Y **will have only finitely many values**. Y will not be a function of X in the sense that, knowing X, we cannot predict exactly what Y will be. We think of Y as a "noisy version" of X. This noise could correspond to thermal noise in a circuit, for example. X might correspond to a source, or to a source obtained by encoding X. For each input x in X there is an output determined by the **forward probabilities** or **transition probabilities** $\Pr(y|x)$ for y in Y. If these probabilities do not change with time and are independent, i.e., they correspond to independent random variables, the channel is **memoryless** (and discrete). Then, the probability of a given symbol being output depends only on the input symbol.

Thus, given a channel Γ, there corresponds a pair X, Y of statistical variables and a set of forward probabilities, $\Pr(Y = y|X = x)$.

On the other hand, given a join pair (X, Y) of random variables, we can construct a discrete channel with forward probabilities

$$\Pr(y|x) = \Pr((Y = y)|(X = x)).$$

That is, knowing $\Pr(x, y)$ and $\Pr(x)$ we can find $\Pr(y|x)$ using the formula for conditional probability of chapter 10, which says that

$$\Pr(y|x) = \frac{\Pr[(X = x) \text{ and } (Y = y)]}{\Pr(X = x)}, \Pr(X = x) \neq 0.$$

We can summarize as follows.

Theorem 12.1 *Channels are equivalent to discrete joint probability distributions. That is, given a channel, we can construct a discrete joint probability distribution, and vice versa.*

12.2 More Specific Channels

Let us examine the situation in more concrete terms. We suppose that the channel accepts input symbols from an alphabet $A = \{a_1, \ldots, a_m\}$ and outputs symbols from the alphabet $B = \{b_1, \ldots, b_n\}$. Given any input symbol a_i there is a certain probability that b_j is the output, namely $p_{ij} = \Pr(b_j|a_i)$. These probabilities are independent and never change with time. Given these probabilities we can form the **channel matrix** $P = (p_{ij})$, which is a matrix with m rows and n columns in which each entry lies between 0 and 1 and the sum of the entries in each row is 1. Very often, $A = B = \{0, 1\}$.

Example 12.2 *Let $A = B = \{0, 1\}$. Suppose that the channel Γ has the following two properties.*

(a) Whenever 0 is the input, then 0 is the output with probability 0.8.

(b) Whenever 1 is the input, then 1 is the output with probability 0.9.

The channel matrix P is then as follows.

$$P = \begin{matrix} & 0 & 1 \\ 0 & \\ 1 & \end{matrix} \begin{pmatrix} 0.8 & 0.2 \\ 0.1 & 0.9 \end{pmatrix}$$

The general binary channel matrix P for a memoryless channel will look like this:

$$P = \begin{pmatrix} q_1 & p_1 \\ p_2 & q_2 \end{pmatrix}.$$

Here p_1, p_2, q_1, q_2 are probabilities, i.e., they are real numbers lying between 0 and 1. Moreover, $p_i + q_i = 1, i = 1, 2$. We see that p_1 (p_2) is the probability that 0 (respectively 1) is incorrectly received.

As a very important special case, suppose $p_1 = p_2 = p$. Then $q_1 = q_2 = q$. Our matrix P is now as follows.

$$P = \begin{pmatrix} q & p \\ p & q \end{pmatrix} = \begin{pmatrix} 1-p & p \\ p & 1-p \end{pmatrix}.$$

The number p is then called the **parameter** of the channel. **Thus p is the probability of a transmission error** regardless of whether 0 or 1 is sent over the channel Γ. This channel is then called the **binary symmetric channel** (BSC), because P is now a **symmetric matrix**.

We can diagram this channel as follows.

If $p > \frac{1}{2}$ then, by reversing the output, i.e., by interchanging 0,1 in the output, we have a new channel with parameter less than or equal to $\frac{1}{2}$. **Thus we can always assume that $p \leq \frac{1}{2}$.** If $p = \frac{1}{2}$, the channel is **completely random**.

12.3   New Channels from Old, Cascades

Given two channels Γ_1, Γ_2 with the number of outputs of Γ_1 equalling the number of inputs of Γ_2, we can compose them and form the **cascade** of the two channels by putting the channels in series with the input of Γ_2 being the output of Γ_1 in order to obtain a new channel called the cascade of the two original channels.

$$\text{input} \ \rightarrow \Gamma_1 \rightarrow \Gamma_2.$$

Example 12.3 *Suppose Γ_1, Γ_2 are channels with respective channel matrices*

$$P_1 = \begin{pmatrix} 0.6 & 0.4 \\ 0.4 & 0.6 \end{pmatrix}$$

and

$$P_2 = \begin{pmatrix} 0.8 & 0.2 \\ 0.1 & 0.9 \end{pmatrix}.$$

We have the following probability tree diagram for the cascade of the two channels.

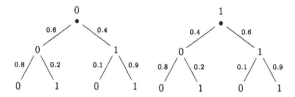

We note the following:

If 0 is the input, the probability of outputting 0 is $(0.6)(0.8) + (0.4)(0.1)$.

If 0 is the input, the probability of outputting 1 is $(0.6)(0.2) + (0.4)(0.9)$.

If 1 is the input, the probability of outputting 0 is $(0.4)(0.8) + (0.0)(0.1)$.

If 1 is the input, the probability of outputting 1 is $(0.4)(0.2) + (0.6)(0.9)$.

Thus the channel matrix for the cascade $\Gamma_1\Gamma_2$ is

$$P = \begin{pmatrix} (0.6)(0.8) + (0.4)(0.1) & (0.6)(0.2) + (0.4)(0.9) \\ (0.4)(0.8) + (0.6)(0.1) & (0.4)(0.2) + (0.6)(0.9) \end{pmatrix} = \begin{pmatrix} 0.52 & 0.48 \\ 0.38 & 0.62 \end{pmatrix}.$$

Note that for any channel matrix the rows must add up to 1 but the columns need not add up to 1. But another key fact here is that the matrix P for the cascade of the two channels is just the **matrix product** of the two channels. In general, we have the **cascade theorem**.

Theorem 12.4 (Cascade Theorem) *The matrix Q for the cascade of s memoryless channels $\Gamma_1, \ldots, \Gamma_s$ with channel matrices $P_1, \ldots, P_s$ of size $n \times n$ is the matrix product $P_1 \cdots P_s$. In particular, if $\Gamma_1 = \cdots = \Gamma_s = \Gamma$ has matrix P of size $n \times n$, then $Q = P^s = \underbrace{P \cdots P}_{s\ times}$ is the cascade matrix.*

This last fact is very useful, for the following reason. The matrix P is what is called a **Markov transition matrix**, it is then easy to use Markov theory and obtain an approximation of P^s. P^s will in fact tend to a matrix such that every row is equal to **the fixed probability vector of P**.

Example 12.5 *Let*

$$P = \begin{pmatrix} 0.6 & 0.4 \\ 0.7 & 0.3 \end{pmatrix}.$$

Find an approximation for P^6.

To do this we calculate the fixed probability vector $\mathbf{w}$ of P. This is a vector $\mathbf{w} = (x, y)$ determined by the following 2 conditions:

(1) $\mathbf{w}P = \mathbf{w}$

(2) $x + y = 1$.

It is a remarkable fact of Markov theory that –assuming P is irreducible (Chapter 14)– these equations can always be solved uniquely so that x, y are **positive**. *The vector* $\mathbf{w}$ *will be a probability vector. We have*

$$(x, y) \begin{pmatrix} 0.6 & 0.4 \\ 0.3 & 0.7 \end{pmatrix} = (x, y).$$

This gives

$$\begin{aligned} 0.6x + 0.3y &= x \\ 0.4x + 0.7y &= y \end{aligned}$$

and

$$x + y = 1.$$

The second equation can be seen to follow from the third and first equations by subtracting them, so we can ignore it. From the first and third equations we get $x = \frac{3}{7}$ and $y = \frac{4}{7}$. Thus P^s tends to the matrix $\begin{pmatrix} \frac{3}{7} & \frac{4}{7} \\ \frac{3}{7} & \frac{4}{7} \end{pmatrix}$ *as s gets large. In particular, P^6 is approximately equal to* $\begin{pmatrix} \frac{3}{7} & \frac{4}{7} \\ \frac{3}{7} & \frac{4}{7} \end{pmatrix}$.

Example 12.6 *Let Γ be the BSC with parameter $p \le \frac{1}{2}$ and channel matrix*

$$P = \begin{pmatrix} 1-p & p \\ p & 1-p \end{pmatrix}.$$

Describe the channel that is obtained by composing Γ with itself, i.e., the channel that corresponds to the cascade of Γ with Γ.

We calculate the matrix

$$PP = \begin{pmatrix} q & p \\ p & q \end{pmatrix} \begin{pmatrix} q & p \\ p & q \end{pmatrix} = \begin{pmatrix} q^2 + p^2 & 2pq \\ 2pq & p^2 + q^2 \end{pmatrix},$$

where $q = 1 - p$. This gives that

$$PP = P^2 = \begin{pmatrix} 1 - 2p(1-p) & 2p(1-p) \\ 2p(1-p) & 1 - 2p(1-p) \end{pmatrix}.$$

This matrix is symmetric so the cascade is again a BSC. It has parameter $2p(1-p)$. This parameter is the probability of a transmission error in the cascade channel $\Gamma\Gamma$. Examining the quadratic $2p(1-p)$ we see that $2p(1-p) \leq \frac{1}{2}$. Since $p \leq \frac{1}{2}$, we have $1-p > \frac{1}{2}$. Multiplying by $2p > 0$, we thus have $p < 2p(1-p) \leq \frac{1}{2}$. This means that the cascade has parameter $\leq \frac{1}{2}$ but is less reliable *because the probability of a transmission error is larger.*

In fact the reliability decreases as we construct the cascades $\Gamma\Gamma, \Gamma\Gamma\Gamma, \ldots\ldots$. In the limit we get the result (discussed in the problems) that the parameter tends to $\frac{1}{2}$ so that the channel becomes completely unreliable, i.e., random, in the limit, with parameter $\frac{1}{2}$.

12.4 Input Probability, Channel Capacity

So far we have only discussed the channel. We also need to pin down some facts about the source and the input probability. Having done so we can define the **channel capacity**.

In general, let X and Y be random variables corresponding to the input and output as in Section 12.1. Recall that from Chapter 10 we can calculate the entropy $H(X)$, which is defined as $-\sum_{x \in X} \Pr(x) \log \Pr(x)$. We will restrict attention in this chapter to **memoryless** sources X as defined in Chapter 11. This just means that, if a is an input symbol, $\Pr(X_i = a)$ does not vary with i and is independent of the probability of emission of other symbols. We can think of starting off with a source S, as in the introduction, and then encoding S in binary or in some other alphabet. We now have a new binary source X from the encoding that provides the channel input. Recall that there is associated with the channel the "forward probabilities" $\Pr(y|x)$. But Y can also be regarded as a "source" emitting symbols with certain probabilities, as follows.

Given any y in Y we have $\Pr(y) = \sum_x Pr(y \text{ and } x)$. From Chapter 10, we have

$$\Pr(U|V) = \frac{\Pr(U \cap V)}{\Pr(V)},$$

where $\Pr(V) \neq 0$. Applying this we get

$$\Pr(y) = \sum_x \Pr(y|x) \Pr(x).$$

Then we can calculate $H(Y), H(X|Y), H(X,Y)$ and

$$I(X:Y) = H(X) + H(Y) - H(X,Y).$$

We define the **capacity** of the channel as follows.

$$\textbf{capacity} = \max_{\Pr(X)} I(X:Y),$$

where $\Pr(X)$ denotes the probability distribution of the input X. In other words, we take the maximum, over all possible input probability distributions, $\Pr(X)$, of the mutual information $I(X : Y)$. (Strictly speaking, in rigorous mathematical terms, we should define capacity to be $\sup_{\Pr(X)} I(X : Y)$, i.e., the supremum of $I = I(X : Y)$ taken over all input probabilities. However, in our case, because I is a continuous function of X there will exist an input distribution X which actually attains the supremum. So we are justified in using the maximum in the definition. We encountered this kind of issue before in Chapter 11 when discussing optimal encodings.)

The capacity will be a number that *will depend only on the channel matrix*. Its **units** are Shannon bits per symbol or Shannon bits per time unit depending on the context.

To understand the concept more fully, let us do some calculations. Our first concern is the binary symmetric channel (BSC).

Channel Capacity for Binary Symmetric Channels.   As usual, let p denote the parameter of a BSC, i.e., the probability of a transmission error.

The beginning of the calculation here is the same – apart from notation – as that in 10.11.

Let x denote the probability that 0 is input so that $1 - x$ is the probability that 1 is input. As usual, all our logs will be to the base 2.

Therefore, we have

$$H(X) = -(x \log x + (1 - x) \log(1 - x)).$$

Setting $q = 1 - p$, we get that the probability that a zero is output is $x(1 - p) + (1 - x)p = \alpha$ and the probability that a one is output is $(1 - x)(1 - p) + xp = \beta$. Therefore,

$$H(Y) = H(\alpha, \beta) = -(\alpha \log \alpha + \beta \log \beta).$$

The possible outcomes (x, y), corresponding to the random variable (X, Y), are easily obtained.

The possible values for the input X are $0, 1$ with probabilities $x, 1 - x$. The possible values for the output Y are $0, 1$ with probabilities α, β. Thus the possible values for (X, Y) are $\{(x, y)\} = \{(0, 0), (0, 1), (1, 0), (1, 1)\}$. Then, $\Pr((x, y) = (0, 0)) = \Pr(0$ is input and 0 is output$) = x(1 - p)$. Probabilities for the other values of (x, y) can be calculated in a similar fashion. It follows that

$$H(X, Y) = H(x(1 - p), xp, (1 - x)p, (1 - x)(1 - p)).$$

We then have

$$I(X : Y) = H(X) + H(Y) - H(X,Y).$$

$I(X : Y)$ is now just a function of a single variable x, denoted by $f(x)$, where x denotes the probability that the source emits a zero. What we have to do is to find the maximum value of $f(x)$, where $0 \le x \le 1$. By a well-loved calculus principle this maximum can be obtained where the derivative of f is zero (having first checked the value of f at the end points 0 and 1). We need first to simplify using the fact that $\log(uv) = \log u + \log v$. Set $q = 1 - p$. We have

$$
\begin{aligned}
H(X) - H(X,Y) &= -x \log x - (1-x) \log(1-x) + xq \log(xq) + xp \log(xp) \\
&\quad + (1-x)p \log((1-x)p) + (1-x)q \log((1-x)q) \\
&= \log x(-x + xq + xp) + \log(1-x)(x - 1 + (1-x)p + (1-x)q) \\
&\quad + \log q(xq + (1-x)q) + \log p(xp + (1-x)p) \\
&= q \log q + p \log p \text{ (since the first two terms are 0).}
\end{aligned}
$$

Thus $I(X : Y) = H(X) + H(Y) - H(X,Y) = p \log p + q \log q - (\alpha \log \alpha + \beta \log \beta)$.

As before, we have $f(x) = I(X : Y)$. We want to find the maximum value of $f(x), 0 \le x \le 1$.

Suppose $x = 0$. Then $\alpha = p, \beta = q$ and

$$I(X : Y) = p \log p + q \log q - (p \log p + q \log q) = 0.$$

Suppose $x = 1$. Then $\alpha = q, \beta = p$. We have

$$I(X : Y) = p \log p + q \log q - (q \log q + p \log p) = 0.$$

Next, differentiating $f(x)$ and setting the derivative to 0 gives $x = \frac{1}{2}$. Then $\alpha = \frac{1}{2}$ and $\beta = \frac{1}{2}$. So

$$I(X : Y) = p \log p + q \log q - \left(\frac{1}{2} \log \left(\frac{1}{2} \right) + \frac{1}{2} \log \left(\frac{1}{2} \right) \right).$$

Since $\log \left(\frac{1}{2} \right) = -\log 2 = -1$, we get

$$I(X : Y) = 1 + p \log p + q \log q = 1 - H(p).$$

We have now our channel capacity.

Theorem 12.7 (Capacity Theorem for the Binary Symmetric Channel) *The channel capacity Λ of the binary symmetric channel with parameter p is equal to $1 - H(p) = 1 + p \log p + q \log q$, where $q = 1 - p$.*

A diagram for this capacity function $\Lambda = \Lambda(p)$ is as follows.

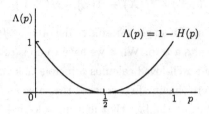

Remarks.

When $p = \frac{1}{2}$, the channel is purely random and $\Lambda = 0$.

When $p = 1$, the channel is completely reliable and $\Lambda = 1$.

When $p = 0$, the channel is completely unreliable but also $\Lambda = 1$.

We refer to the Appendix for tables of values of $H(p)$.

12.5 Capacity for General Binary Channels, Entropy

In general, capacities are difficult to calculate. Let us examine the general binary memoryless channel with channel matrix P as follows.

$$P = \begin{pmatrix} q_1 & p_1 \\ p_2 & q_2 \end{pmatrix}$$

with

$$\begin{cases} p_1 + q_1 &= 1, \\ p_2 + q_2 &= 1. \end{cases}$$

Let the input probability for 0 be denoted by x so that the input probability for 1 is $1 - x$. Using a probability tree as in the previous section, we deduce the following:

The probability that 0 is output is $x q_1 + (1 - x) p_2 = \alpha$.

The probability that 1 is output is $x p_1 + (1 - x) q_2 = \beta$.

Then the capacity Λ is given by the maximum value of $f(x)$ where

$$f(x) = H(x, 1 - x) + H(\alpha, \beta) - H(x q_1, x p_1, (1 - x) p_2, (1 - x) q_2).$$

To actually calculate this maximum is slightly complicated and is best done using Lagrange multipliers. However, in any given case we can calculate the capacities as indicated below.

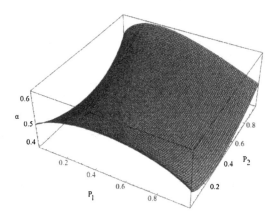

A discussion of the general case is given by Ash ([Ash90], page 304) who uses the inverse of the channel matrix. The conclusion is as follows. Let the capacity be Λ. Then

$$\Lambda = \log(2^u + 2^v),$$

where $u = \frac{q_2 H(q_1) - p_1 H(p_2)}{p_2 - q_1}$, $v = \frac{-p_2 H(q_1) + q_1 H(p_2)}{p_2 - q_1}$.

Let us return to the symmetric case with $p_1 = p_2 = p$, $q_1 = q_2 = q$ so that we have a BSC with parameter p. Note that Λ above then simplifies to $1 - H(p)$ as expected.

Using the notation of Section 12.4, the input entropy is

$$H(X) = -(x \log x + (1 - x) \log(1 - x)) = H(x).$$

The entropy of the output Y is

$$H(Y) = H(x(1 - p) + (1 - x)p, xp + (1 - x)(1 - p)) = H(x(1 - p) + (1 - x)p).$$

(See also below.) We now have the following result for the BSC.

The entropy of the input of a binary symmetric channel is less than or equal to the entropy of the output.

To see this, let us study the graph of the Shannon function $y = H(x)$. As can be seen from the graph (or by taking the second derivative and verifying that it is negative between 0 and 1) this function is (strictly) concave down (=convex). This implies that if we join 2 points A, B on the graph, the y value at a point on the curve whose x-value lies between the x-values of A, B is greater than the corresponding y-value on the line AB.

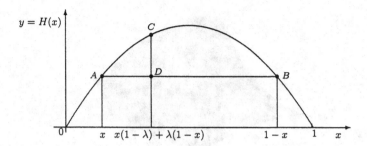

Here, the y-value of C is greater than the y-value of $D = H(x)$. A point on the axis between x and $1-x$ has x-coordinate equal to $x + \lambda(1-x-x) = x(1-\lambda) + \lambda(1-x)$ for some number λ with $0 \leq \lambda \leq 1$. Replacing λ by p, the result above follows.

Algebraic Relation Between Input and Output.   Over a binary channel let $(x, 1-x)$ be the input probability distribution so that the probability of 0 being transmitted is x. Then, if P is the channel matrix, the output is obtained by calculating $(x, 1-x)P$. For a BSC with parameter p we get

$$(x, 1-x) \begin{pmatrix} 1-p & p \\ p & 1-p \end{pmatrix} = (x(1-p) + (1-x)p, \, xp + (1-x)(1-p))$$

for the output, where the first element of the pair is the probability that 0 is output.

12.6   Hamming Distance

Suppose $\mathbf{x}$ and $\mathbf{y}$ are two **vectors** of length n, i.e., two strings of length n over some alphabet. Then the **Hamming distance** between $\mathbf{x}$ and $\mathbf{y}$, denoted by $d(\mathbf{x}, \mathbf{y})$, is defined to be the number of positions in which $\mathbf{x}$ and $\mathbf{y}$ *disagree*.

Example 12.8 *Suppose the alphabet is the binary alphabet, $n = 4$, $\mathbf{x} = (1011)$, $\mathbf{y} = (0101)$. Then $\mathbf{x}$ and $\mathbf{y}$ differ in the first three positions, agreeing on the fourth, so the Hamming distance is 3, i.e., $d(\mathbf{x}, \mathbf{y}) = 3$.*

The important point about this definition is that the Hamming distance is a distance in the mathematical sense, similar to the Euclidean distance. That is, it satisfies the following three properties.

(a) $d(\mathbf{x}, \mathbf{y}) \geq 0$. Also $d(\mathbf{x}, \mathbf{y}) = 0$ if and only if $\mathbf{x} = \mathbf{y}$.

(b) $d(\mathbf{x}, \mathbf{y}) = d(\mathbf{y}, \mathbf{x})$ (symmetry).

(c) $d(\mathbf{x}, \mathbf{z}) \leq d(\mathbf{x}, \mathbf{y}) + d(\mathbf{y}, \mathbf{z})$ for any 3 vectors $\mathbf{x}$, $\mathbf{y}$, $\mathbf{z}$ of the same length over the same alphabet.

Property (c) is called the **triangle inequality**. It says that the distance between two points of a triangle is less than or equal to the sum of the other two distances in the triangle. See the following picture.

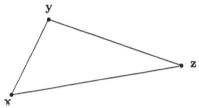

We can easily visualize what is going on in the binary case by drawing an n-cube. For $n = 3$ we get the following picture.

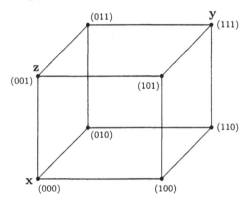

We have $2^3 = 8$ possible binary strings of length 3 represented by the 8 vertices of a cube.

The Hamming distance between any 2 vertices is not the Euclidean distance but rather the number of cube edges that one has to traverse in going from one of the vertices to the other. For example, if $\mathbf{x} = (000)$ and $\mathbf{y} = (111)$ then $d(\mathbf{x}, \mathbf{y}) = 3$ because the shortest path of edges between $\mathbf{x}$ and $\mathbf{y}$ contains 3 edges. Similarly, if $\mathbf{z} = (001)$ we get $d(\mathbf{x}, \mathbf{z}) = 1$, $d(\mathbf{y}, \mathbf{z}) = d(\mathbf{z}, \mathbf{y}) = 2$.

12.7 Improving Reliability of a Binary Symmetric Channel

Suppose A is transmitting binary vectors or strings of length n to B over a binary symmetric channel Γ with parameter p. We want to calculate some error probabilities and to devise means for improving the reliability of the channel.

Example 12.9 *Let $n = 1$, so A has just two possible messages, namely, "yes" and "no" encoded as 1 and 0, respectively. Then the probability of a message getting corrupted is p.*

Example 12.10 *Let $n = 3$ and suppose A has eight possible messages corresponding to the eight possible binary strings of length 3. We calculate the probability that a given message of length 3 will undergo a transmission error, as follows. The probability that the first bit gets transmitted properly is $q = 1 - p$. Independently, the probability that the second and third bits are transmitted correctly is also q. So the probability that all 3 bits are correctly received is q^3. Thus the probability of at least one transmission error is $1 - q^3$. For example, if $p = 0.01$ then the probability of at least one error is $1 - (0.99)^3 = 0.0297$. This is almost 3p.*

In general, these error probabilities can become unacceptably large. We want to devise a strategy to reduce the error.

A related problem concerns the *average* number of transmission errors.

Example 12.11 *A transmits a binary string of length n to B over the channel Γ. On the average, how many transmission errors will there be?*

On average, as follows from Chapter 10, there will be np transmission errors.

12.8   Error Correction, Error Reduction, Good Redundancy

A wants to transmit messages chosen from a set M of possible messages to **B** and to minimize the possibility of error. Each message is presumed to be a binary string of fixed length k. Thus the total number M of possible messages is at most $(2) \cdots (2) = 2^k$.

The strategy is this. A encodes each message **x** of length k in a longer message $f(\mathbf{x})$ of length n, using some encoding rule f. So we have

$$\mathbf{x} = (x_1, \ldots, x_k) \xrightarrow{f} (y_1, \ldots, y_k, y_{k+1}, \ldots, y_n).$$

Thus the encoding function now adds "good redundancy" for correct decoding as opposed to the "bad, wasteful" redundancy of Chapter 11. The string or vector $f(\mathbf{x})$ is called a **code word**.

We can often assume that $x_1 = y_1, \ldots, x_k = y_k$ and that the extra bits $y_{k+1}, \ldots, y_n$ are just parity bits, i.e., they are linear combinations of the **message bits** or **information bits** $x_1, \ldots, x_k$.

Example 12.12 *Suppose W is the message set consisting of all possible binary strings of length 2. So $W = \{(0,0), (0,1), (1,0), (1,1)\}$ and the number of messages M is 4. Our encoding function f maps (x_1, x_2) to $(x_1, x_2, x_1, x_2, x_1 + x_2)$. Then*

$$(00) \xrightarrow{f} (00000)$$
$$(01) \xrightarrow{f} (01011)$$
$$(10) \xrightarrow{f} (10101)$$
$$(11) \xrightarrow{f} (11110)$$

In general, we need to make several assumptions about the encoding function f. Of course, we assume that no errors occur in this encoding. Also, f only makes sense if it is one to one. This means that if $x_1 \neq x_2$ then $f(x_1) \neq f(x_2)$. Then, *given any code word there is exactly one message corresponding to it so if* **B** *decodes the correct code word,* **B** *can then calculate the message.*

The total number of code words is equal to M, which is the total number of messages. *Transmitter* **A** *and receiver* **B** *are both equipped with a list of the set of all possible code words* $f(x)$. This list C is called the **code list** or simply, the **code**. *Since f is one to one, the number of words in C equals the number of messages M. In symbols, $M = |C|$.*

We want the Hamming distance between any 2 code words in C to be large. We have seen that, on the average, only a certain number of transmission errors, namely, np errors, can be made. Now, when **B** receives a string z of length n it is likely that this string came from (corresponds to, is a corruption of) the code word closest to z, i.e., the code word u such that $d(z, u) \leq d(z, v)$ for any v in C.

In other words, we use **nearest neighbour decoding** (sometimes called **minimum distance decoding**). In the event of a tie we make an arbitrary choice between the competing code words. Having found u, we can then reconstruct the original message because f is one to one.

Example 12.13 *Let us assume, using the code C in the previous example, that we transmit the code word* $u = (01011)$ *(over the BSC channel Γ with parameter p) and that the string $z = (11011)$ is received. Then $d(z, u) = 1$. Moreover, $d(z, v) \geq 2$ if v is in C with $v \neq u$. Thus z is decoded as u.*

Probability of Error. Suppose we did not first encode the 4 messages of length 2 in Example 12.12 into longer code words of length 5 but, instead, tried a direct transmission. The probability that no error occurs is $q^2 = (1 - p)^2$. Thus the probability of at least one error when a message of 2 bits is transmitted is $1 - (1 - p)^2$. For example, if $p = 0.05$ this is equal to 0.0975. This represents the probability that an incorrect message is received.

Now suppose we encode into code words of length 5 in the code C of Example 12.13. As can be checked, the Hamming distance between any 2 words in C is at least 3. So the receiver decodes correctly with nearest neighbour decoding if there is at most 1 error. This follows from the triangle inequality. In general, if the Hamming distance between any 2 words in C is at least $2e + 1$, then decoding is correct provided that there are at most e transmission errors. (This and similar issues will be discussed in Part III.)

We want to find the probability that the wrong message is received. The probability of zero errors is q^5, where $q = 1 - p$. The probability of exactly one error is $\binom{5}{1}pq^4$. So the probability of at most one error is $q^5 + 5q^4p = q^4(q+5p)$. For $p = 0.05$, we have $q = 0.95$ and $q^4(q + 5p) = (0.95)^4(0.95 + 0.25) = 0.9774$. Thus the probability that the wrong message will be received has been reduced from 0.0975 to $1 - 0.9774 = 0.0226$.

We obtain an even more dramatic improvement in our next example.

Example 12.14 *Suppose our message set W contains just two messages, namely, 0 and 1. Then the probability that the wrong message is received over Γ is p. If $p = 0.05$ this means that there is a 5 percent chance that the wrong message is received.*

Now, use the encoding function f, which maps $\mathbf{x} = (x_1)$ to $f(\mathbf{x}) = (x_1, x_1, x_1)$. The resulting code, which has just 2 code words, namely $(0, 0, 0)$ and $(1, 1, 1)$, is called the **binary repetition code** *(of length 3). Thus we have*

$$(0) \xrightarrow{f} (000)$$
$$(1) \xrightarrow{f} (111)$$

The Hamming distance between the 2 code words is 3. Using nearest neighbour decoding we ask the following question: What is the probability that an incorrect message is received?

The correct message will be decoded provided that there are no errors or exactly one error. The probability of this happening is

$$q^3 + 3pq^2 = q^2(q + 3p) = (0.95)^2(0.95 + 0.15) = 0.9928,$$

where we assume again that $p = 0.05$. Thus the probability of an error has been reduced from 0.05 to $1 - 0.9928 = 0.0072$.

In Examples 12.13 and 12.14, we have shown how to drastically reduce the probability of receiving the wrong message. But there is a price to be paid. In Example 12.13 we are using 5 bits for encoding a message with 2 bits. In Example 12.14 we use 3 bits to encode a 1-bit message.

In general, let W denote the set of all possible messages to be transmitted, with each message being of length k. Each message in W is first encoded to a code word in a code C. So $|C| = |W| = M$, say. Each code word has length n. Since the maximum number of binary strings of length k is 2^k, we get $M \leq 2^k$. For example, suppose $M = 2^k$. Then we can define the information rate or rate of the code C as $\frac{k}{n} = \frac{\log M}{n}$. We want to handle the more general case when M may be less than 2^k. We have $M = 2^{\log M}$ and we can think of W as containing all possible binary strings of "length" $\log M$ – even though $\log M$ may not be an integer.

This motivates the general definition of the **transmission rate** or simply the **rate** of C. It is defined as $\frac{\log M}{n}$, where M is the total number of possible messages being transmitted, which is also the total number of code words in the code list (or code) C, and n is the length of each code word in C.

Now we are encountering one of the most fundamental issues in coding theory, which is the tension between two competing goals for a code C. These goals are as follows.

(1) A large Hamming distance between any 2 code words in C, but

(2) a large transmission rate for C.

Note that in Example 12.13 the rate is $\frac{2}{5}$. In Example 12.14, the rate is $\frac{1}{3}$. Again, the code in Example 12.14 is a binary repetition code of length 3. We can use the encoding formula given by

$$f(\mathbf{x}) = f(x_1) = (x_1, \ldots, x_1)$$

with 3 replaced by n, where the parity bits are $x_1, x_1, \ldots$. Then the probability of receiving the wrong message is arbitrarily small, since the Hamming distance between the two code words in C is n. This is because the two code words are $(0, \ldots, 0)$ and $(1, \ldots, 1)$. However, the rate is $\frac{1}{n}$, which tends to zero as n gets large.

The question then becomes the following.

To get the probability of incorrect decoding to be arbitrarily small must the transmission rate tend to zero? If not, how large can the rate be?

This question is at the heart of communication theory. We pursue it in the next section.

12.9   The Fundamental Theorem of Information Theory

We are now in a position to reconcile the two competing considerations for a noisy channel, namely,

1. a high transmission rate and

2. a low error probability.

In this section, we assume that we have a memoryless source denoted by an input random variable X. The source emits binary strings of length n over a channel Γ. Any set of these binary strings is called a code, and the strings in the code are code words. Although the result that we discuss will work over any channel, we assume that Γ is a binary symmetric channel. Later on in chapter 14, we show how the requirements on the source can be relaxed. As discussed earlier in this chapter, one way in which these binary strings arise is as follows. We have a data stream arising from a previous source. This stream is blocked off into binary strings of length k and then we adjoin a total of $n - k$ parity bits (as discussed in detail also in Part III) to end up with code words of length n.

The result that we want to discuss, and try to understand, is as follows.

The Fundamental Theorem of Information Theory. *Let Γ be a BSC with parameter $p < \frac{1}{2}$ and resulting capacity $\Lambda = \Lambda(p) = 1 + p \log p + q \log q$. Let R be any information rate with $R < \Lambda$. Let $\varepsilon > 0$ be an arbitrarily small positive quantity. Then, if $N = N(\varepsilon)$ is sufficiently large, there is a code C of length N with the following properties. C has rate R. Moreover, the average probability of error, using nearest neighbour decoding, is less than ε.*

This theorem involves two basic questions.

1. Why does the capacity Λ give the upper bound for accurate communication?

2. having surmised this, how can we prove it?

Rather than launching into a formal proof, which may not give much understanding of what is going on, we prefer to take several different approaches.

Approach 1. We are given an input distribution X, an output distribution Y and a channel defined by the forward probabilities $\Pr(y|x)$. Y can also be regarded as a source using the fact that

$$\Pr(y) = \sum_{x \in X} \Pr(y|x) \Pr(x).$$

Basically we want to find the **maximum number of distinguishable inputs**. Two inputs are **distinguishable** if their "output fans" do not overlap.

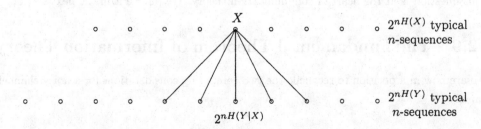

X is transmitting binary strings of length n. The elements in the string are chosen independently in accordance with some probability distribution. The source X emits $2^{nH(X)}$ typical sequences of length n. Similarly, Y will emit $2^{nH(Y)}$ typical sequences of length n. (As a side remark, we have seen that $H(Y) \geq H(X)$ for any BSC.) Also, fixing x, we have that the number of typical sequences received when x is transmitted, is about $2^{nH(Y|X)}$ sequences, which are equiprobable. To see this, recall that $H(Y|X)$ is the average of $H(Y|X)$, averaged over x in X. Equiprobability follows as in our discussion of typical sequences in Chapter 10. So the output fan is about $2^{nH(Y|X)}$. Thus if we have M inputs and we want their output fans not to overlap we must have our "**sphere-packing condition**", namely, that

$$M 2^{nH(X|Y)} \leq 2^{nH(Y)}.$$

Thus,

$$M \leq 2^{n(H(Y)-H(X|Y))}.$$

Then $M \leq 2^{nI(Y:X)}$ because $I(Y:X)$ represents the maximum value of $H(Y)-H(X|Y)$. Moreover it will turn out that this upper bound can be achieved. In the case of the BSC with parameter p we have $I(Y:X) = \Lambda(p)$ the channel capacity. Thus our argument suggests that $M \leq 2^{n\Lambda(p)}$ so that $\log M \leq n\Lambda(p)$. Since the rate R is $\frac{\log M}{n}$ we get $R \leq \Lambda(p)$. Our argument above suggests the following Fundamental Principle.

Fundamental Principle: **The capacity of a channel is the log of the maximum number of distinguishable inputs.**

Let us reconcile this principle with our result above. Above we got that $\log M \leq n\Lambda$. The binary symmetric channel accepts one bit and transmits one bit. When a binary string of length n is successively transmitted (and received) over a BSC the correct channel for this is the **n-th extension of the binary symmetric channel**, which has capacity $n\Lambda$ (rather than Λ). This channel transmits binary n-strings to binary n-strings. We can think of it as n copies of a single BSC operating independently and in parallel.

Approach 2. This is a bit like Approach 1, but we will also sketch a proof as to why any rate less than the channel capacity will work for accurate transmission. It is similar to Shannon's original informal argument, and is described by Ash [Ash90].

As above we have approximately $2^{nH(X)}$ typical input n-sequences; similarly, we have $2^{nH(Y)}$ typical output n-sequences. We are always assuming that n is suitably large. Then there will also be $2^{nH(X,Y)}$ typical pairs of input-output n-sequences. Here a typical pair is obtained by first choosing a typical output n-sequence $\mathbf{y}$ and then selecting a typical input n-sequence $\mathbf{x}$ such that $(\mathbf{x}, \mathbf{y})$ is a typical pair.

There are approximately $2^{nH(Y)}$ typical output n-sequences $\mathbf{y}$. Because the total number of typical pairs $(\mathbf{x}, \mathbf{y})$ is $2^{nH(X,Y)}$, it follows that for each typical output sequence $\mathbf{y}$ there are, on average,

$$2^{nH(X,Y)-H(Y)} = 2^{nH(X|Y)}$$

input sequences $\mathbf{x}$ for which $(\mathbf{x}, \mathbf{y})$ is a typical pair. So the input fan of $\mathbf{y}$ has about $2^{nH(X|Y)}$ sequences when n is large.

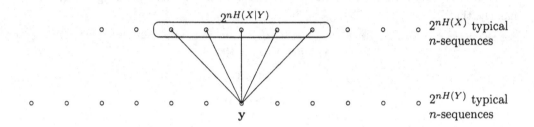

For accurate decoding of a code C we want disjoint fans. So, if W is the set of code words in C, with $|W| = M$ we have

$$M2^{nH(X|Y)} \le 2^{H(X)}.$$

Then

$$M \le 2^{n(H(X)-H(X|Y))} \le 2^{nI(X:Y)}.$$

Thus $\frac{\log M}{n} \le I(X : Y)$. Again, this shows that the rate is bounded above by the capacity.

Remark. In Approach 1, we had that the allowable rate is at most $I(Y : X)$. Here we get the rate to be at most $I(X : Y)$. So we have another way of looking at the fact that $I(X : Y) = I(Y : X)$.

Now we want to give a plausible argument as to why an accurate code *can* be constructed with a rate R, $R < \Lambda$, Λ being the capacity. Construct a code C randomly as follows. Let C be a set of $2^{nR} < 2^{n\Lambda}$ typical n-sequences with the sequences chosen successively, at random, such that the $2^{nH(X)}$ possible typical n-sequences are equally likely to be chosen at each stage. Here X is chosen in accordance with the distribution that achieves channel capacity. Suppose a typical sequence $\mathbf{u}_i$ is transmitted and the typical sequence $\mathbf{y}$ is received. We can only have a decoding error if at least one other typical sequence or code word $\mathbf{u}_j \ne \mathbf{u}_i$ belongs to the set T of possible typical input n-sequences that output to $\mathbf{y}$.

Now, the probability that at least one other sequence $\mathbf{u}_j$ belongs to T is less than or equal to the sum

$$\sum_{j=1,j\neq i}^{2^{nR}} \Pr(\mathbf{u}_j \text{ in } T) = \theta, \text{ say.}$$

(Here, we are invoking the fact that the probability of a union is at most the sum of the probabilities. For example, $\Pr(A \text{ or } B) \leq \Pr(A) + \Pr(B)$). Now

$$\theta \leq (2^{nR} - 1)\frac{2^{nH(X|Y)}}{2^{nH(X)}} < \frac{2^{nR}}{2^{n\Lambda}}$$

since $\frac{H(X|Y)}{H(Y)}$ gives the fraction of the typical input n-sequences that are in T. Thus the probability, on average, that $\mathbf{x}_i$ is decoded incorrectly tends to zero as n gets large because $R < \Lambda$.

Remark. In upper-bounding θ we appeal to our choice of X.

Approach 3. This is a bit like the first part of Approach 2. **A** is transmitting code words chosen from a code C consisting of binary n-sequences to **B** over a binary symmetric channel Γ with parameter p. Suppose that a sequence $\mathbf{y}$ of length n is received by **B**. **B** knows that $\mathbf{y}$ is the corruption of $\mathbf{x}$, where $\mathbf{x}$ is a code word of length n. On average, the number of transmission errors is $np = \lambda$. For convenience, let's assume that np is an integer. Thus the Hamming distance between $\mathbf{x}$ and $\mathbf{y}$ is np, so $\mathbf{x}$ differs from $\mathbf{y}$ in about np positions. For example, if $\mathbf{y}$ is the all-zero vector then $\mathbf{x}$ in all likelihood is an n-string with about np ones in it and $n - np$ zeros. There are $\binom{n}{np}$ ways of choosing the np positions where the ones occur, giving about $\binom{n}{np}$ choices for $\mathbf{x}$, where $\binom{n}{np}$ is the binomial coefficient. Thus the input fan associated with $\mathbf{y}$ is about $\binom{n}{\lambda}$. Now if the input distribution associated with X were random, the total number of typical input n-sequences would be 2^n. (In general, it is $2^{nH(X)}$.) So, if the input fans are disjoint, we get that $|C|\binom{n}{\lambda} \leq 2^n$. Now, using Stirling's expansion as in Chapter 9, we get that $\log\binom{n}{\lambda} = \log\binom{n}{np}$, which is approximately $nH(p)$. Thus $\log|C| \leq n(1 - H(p)) = n(1 + p\log p + q\log q)$. This gives that the largest transmission rate that C can have is the capacity of Γ, namely, $\Lambda(p) = 1 + p\log p + q\log q$.

Stirling's Expansion. For reference, we include here Stirling's expansion. It says that

$$\ln n! \approx \frac{1}{2}\ln n + n\ln n - n$$

where $\ln$ denotes the natural logarithm, and $n!$ means n factorial, or $n(n-1)(n-2)\cdots(2)(1)$. We also refer to [].

Approach 4. Again, this is similar to Approach 2. We choose a code C at random. Suppose code word $\mathbf{x}$ is transmitted over the BSC with parameter p. Let $\mathbf{y}$ be the received string. We have the following facts.

Fact 1. The Hamming distance between $\mathbf{x}$ and $\mathbf{y}$, on average, is np.

Fact 2. The probability that another code word $\mathbf{x}_i$ is within a Hamming distance t of $\mathbf{y}$ is, roughly,

$$\sum_{i=0}^{t} \binom{n}{i}.$$

To see Fact 2, we count the total number of words at distance $0, 1, \ldots, t$ from $\mathbf{y}$.

Fact 3. The upper bound for the probability that there is a decoding error when $\mathbf{y}$ is received is around $\frac{(|C|-1)}{2^n} \sum_{i=0}^{t} \binom{n}{i}$. This is bounded by $\frac{|C|}{2^n} \sum_{i=0}^{t} \binom{n}{i}$, because $|C|-1 < |C|$. To get the upper bound, note that the total number of n-sequences is 2^n. We are choosing the words in C at random. The argument now is the same as when we obtained θ in Approach 2.

Fact 4. Let $0 < p < \frac{1}{2}$. Then $\sum_{i=0}^{np} \binom{n}{i} \leq 2^{nH(p)}$. This is an interesting result on the "tail" of the binomial coefficients. We refer the reader to Welsh [Wel88].

From the above, then, since $|C| = 2^{nR}$, by setting $t = np$, the approximate upper bound on the probability of a decoding error (on average) is

$$\frac{2^{nR}}{2^n} 2^{nH(p)} = \frac{2^{nR}}{2^{n\Lambda}}.$$

This tends to zero when n gets large because $R < \Lambda$.

Remark. If you look at this proof in Approach 4, there is a lot of averaging and estimating going on. To really make the proof rigorous we have to average over all possible random codes, as Shannon did, and show that the decoding error tends to zero. The theorem will then follow. We refer the reader to Ash [Ash90] and van Lint [vL98] for details.

Approach 5. This is discussed by Feynmann [Fey99]. The idea is as follows. We have a message consisting of k bits. We adjoin $n - k$ parity check bits so that our code word now has n bits. The n-k parity check bits need to be able to describe the location of all the possible error patterns as we need to **correct** all errors when decoding. Let $\lambda = np$. Let's assume that it is an integer. So λ is the average number of errors. The total number of error patterns is then around $\binom{n}{\lambda}$. So we get that

$$2^{n-k} \geq \binom{n}{\lambda}.$$

Using Stirling's expansion, we get that

$$n - k \geq n[H(p)].$$

Thus $1 - \frac{k}{n} \geq H(p)$, so $\frac{k}{n} \leq 1 - H(p)$. Since the transmission rate R is $\frac{k}{n}$, we get

$$R \leq 1 + p \log p + q \log q.$$

Thus $R \leq \Lambda(p)$, the capacity of the binary symmetric channel Γ.

Feynmann [] goes on to develop two other arguments in support of the fundamental theorem! One of these is a "sphere-packing" argument relating to error-correction. The other argument relates to physics and free energy.

We want to make some further comments on the fundamental theorem.

1. As mentioned earlier we need not restrict ourselves to a binary symmetric channel. Also, the source need not be memoryless. In practice, ergodic sources are extensively used.

2. The length n of the code words may have to be quite large to get the error-rate close to zero.

3. No specific recipe is given for constructing the codes in the theorem.

4. We can think of the theorem as saying that for the BSC the maximum transmission rate is just under the capacity $\Lambda(p)$, where $0 \leq \Lambda(p) \leq 1$. So we can safely send $\Lambda(p)$ bits per second across the channel if, physically, the channel can transmit 1 bit per second.

5. In Theorem 1, we assume that $R < \Lambda$. This is, in fact, the best rate possible, i.e., we have a kind of Converse Theorem.

Converse. For a memoryless channel of capacity Λ and rate $R > \Lambda$ there cannot exist a sequence of codes C_n, $n = 1, 2, \ldots$ with the property that C_n has 2^{nR} code words of length n and error probability tending to zero as n gets large.

We refer the reader to Welsh [Wel88]. A much stronger result due to Wolfowitz [Wol61] shows that the maximum probability of error tends to 1 as n gets large. It can be shown by using the **Hamming codes** that the error probability p can tend to 1 in the limit.

6. It can be shown that the fundamental theorem holds if we restrict ourselves just to **linear codes**. We refer the reader to Feynmann [Fey99].

12.10   Summary, the Big Picture

We want to summarize again one of the main guiding principles in this chapter. The transmitter (but also the receiver!) has a list C of allowable code words. The receiver decodes by nearest neighbour decoding. In the event of a tie, the receiver makes a random decision among the competing code words. If the code words in the list are very well spread out, their output fans will be disjoint and we will have correct decoding. But, if they are too far apart, the transmission rate, which is related to the number of code words in the list, becomes too small. The biggest transmission rate that you can get away with for accurate decoding is the channel capacity. *Capacity is the upper bound to accurate communication.* The main theorem guarantees the existence of a code C with information rate less than the channel capacity such that no matter what code word from C is transmitted the receiver will decode it accurately.

Again we remind the reader that the Fundamental Theorem holds also even if we restrict to just linear codes.

We should point out that the capacity of other structures such as graphs has also been of considerable interest. A particularly famous problem suggested by Shannon was to find the capacity of the 5-cycle, C_5. This problem was finally solved by Laci Lovasz [Lov79].

12.11   Problems

1. A binary symmetric channel with parameter 0.2 is used for transmitting the code word 011. What is the probability that there is a transmission error?

2. The code $C = \{(1,0),(0,1)\}$ is being transmitted over a binary symmetric channel with parameter p. Find the probability that the code word $(1,0)$ is decoded as $(0,1)$ by using nearest neighbour decoding.

3. A binary source emitting binary strings (x_1, x_2, x_3, x_4) is encoded as the code word $(x_1, x_2, x_3, x_4, x_1 + x_3, x_2 + x_4)$ of length 6 in the code C and transmitted sequentially over a binary symmetric channel with parameter p. Decoding is carried out using nearest neighbour decoding.

   (a) Does the information rate depend on p?

   (b) Does the information rate depend on the decoding rule?

   (c) Does the channel capacity depend on p?

4. What is the rate in Problem 3?

5. A message of N bits is transmitted over a BSC with parameter p. An "error burst" is defined to be a sequence of 3 consecutive bits each of which is incorrectly transmitted. Find the average number of error bursts in the transmission.

6. What is the probability that 2 random binary strings of length N have Hamming distance at most 4?

7. A binary memoryless channel Γ transmits 0 correctly with a probability of 0.8 and transmits 1 correctly with probability of 0.7. What is the capacity of Γ?

8. If we take the cascade of the binary symmetric channel with itself n times and let n go to infinity, what is the capacity of the resulting channel?

9. See [Wel88]. Prove that if $\frac{1}{2} < \lambda < 1$ and λn is an integer, then

$$\sum_{k=\lambda n}^{n} \binom{n}{k} \leq 2^{nH(\lambda)}.$$

12.12 Solutions

1. The probability that there is no error is $(0.8)^3 = 0.512$. Thus the probability that there is at least one error is $1 - 0.512 = 0.488$.

2. We transmit $(1,0)$. The probability that $(0,1)$ is received is p^2. If $(0,1)$ is received we decode it as $(0,1)$. Suppose $(0,0)$ is received: the probability for this is pq. The Hamming distance of $(0,0)$ from $(1,0)$ and $(0,1)$ is 1. So with probability $\frac{1}{2}$ we decode $(0,0)$ as $(0,1)$. Similarly, with probability $\frac{1}{2}$ we decode $(1,1)$ as $(0,1)$. Therefore, the probability of incorrect decoding is $p^2 + \frac{1}{2}pq + \frac{1}{2}pq = p^2 + pq = p(p+q) = p$.

3. (a) No.   (b) No.   (c) Yes, since for a BSC the capacity is $\Lambda(p) = 1 + p \log p + q \log q$.

4. The size M of the code (or code list) is 2^4. The rate is $\frac{\log M}{n} = \frac{4}{6} = \frac{2}{3}$.

5. There are $(N-3) + 1 = N - 2$ sequences of 3 consecutive bits –don't forget the first 3. The probability of a burst error is p^3. So the average number of burst errors is $(N-2)p^3$.

6. Having Hamming distance at most 4 means that there is either 0 disagreements or 1 disagreement or 2 or 3 or 4. In general the probability of exactly x disagreements is $\binom{N}{x}(\frac{1}{2})^x(\frac{1}{2})^{N-x} = \binom{N}{x}(\frac{1}{2})^N$. So we get

$$\frac{1}{2^N}\left[\binom{N}{0} + \binom{N}{1} + \binom{N}{2} + \binom{N}{3} + \binom{N}{4}\right].$$

7. We can calculate directly, using a probability tree or using the formula in the text from Ash (page 304), with $p_1 = 0.2, q_1 = 0.8, p_2 = 0.3, q_2 = 0.7$. Our formula is $\log(2^u + 2^v)$, where

$$u = \frac{q_2 H(q_1) - p_1 H(p_2)}{p_2 - q_1}, v = \frac{-p_2 H(q_1) + q_1 H(p_2)}{p_2 - q_1}.$$

which gives a capacity of 0.1912.

8. We have the transmission matrix P given by the following formula

$$P = \begin{pmatrix} 1-p & p \\ p & 1-p \end{pmatrix}.$$

Let $\mathbf{w} = (x, y)$ be the fixed probability vector. Then, since $\mathbf{w}P = \mathbf{w}$ we get

$$(x, y) \begin{pmatrix} 1-p & p \\ p & 1-p \end{pmatrix} = (x, y).$$

Then

$$\begin{aligned} x(1-p) + yp &= x \\ xp + y(1-p) &= y \end{aligned}$$

and also $x + y = 1$. The second equation is obtained by subtracting the first equation from the third equation. Using the first and third equations we get

$$x(1-p) + (1-x)p = x.$$

This gives $x = \frac{1}{2}$ so $y = \frac{1}{2}$. Then the fixed vector is $\left(\frac{1}{2}, \frac{1}{2}\right)$. It follows that P^n tends to $\begin{pmatrix} \frac{1}{2} & \frac{1}{2} \\ \frac{1}{2} & \frac{1}{2} \end{pmatrix}$ –see Chapter 13. Then the capacity tends to $1 + \frac{1}{2}\log(\frac{1}{2}) + \frac{1}{2}\log(\frac{1}{2}) = 0$, since the parameter of P^n tends to $\frac{1}{2}$, leading to a purely random channel.

9. By the symmetry of the binomial coefficients we have $\binom{n}{k} = \binom{n}{n-k}$. Then the sum on the left is equal to

$$\sum_{k=0}^{(1-\lambda)n} \binom{n}{k}.$$

Since $\frac{1}{2} < \lambda < 1$, we have that $0 < 1 - \lambda < \frac{1}{2}$. Putting $p = 1 - \lambda$ the sum now becomes

$$\sum_{k=0}^{np} \binom{n}{k},$$

$0 < p < \frac{1}{2}$. By the tail inequality in the text this sum is bounded above by

$$2^{nH(p)} = 2^{nH(1-\lambda)} = 2^{nH(\lambda)},$$

since $H(\lambda) = H(1-\lambda) = H(\lambda, 1-\lambda)$. (It is not really necessary to assume that λn is an integer. We can just replace it by the ceiling function in the statement.)

Chapter 13

Signals, Sampling, SNR, Coding Gain

Goals, Discussion We prove the sampling theorem and use it to give an argument in support of the band-limited information capacity theorem. We present a short discussion of "coding gain".

New, Noteworthy The argument for the channel capacity connects up with our discussion of capacity and the fundamental principle in Chapter 12. We also show exactly how to get the spheres that occur in the proof of the information capacity theorem, using the underlying statistical distributions.

13.1   Continuous Signals, Shannon's Sampling Theorem

We will be dealing with a continuous signal of finite energy that is band-limited, having no frequency component higher than W hertz.

The **sampling theorem**, which Feynmann [Fey99] refers to as "another of Claude Shannon's babies" has two equivalent parts that apply to the transmitter and the receiver of a pulse modulation system, respectively.

Theorem 13.1 (The Sampling Theorem) *(a) A band-limited signal of finite energy, which has no frequency components higher than W hertz, is completely described by specifying the values of the signal at instants of time separated by $\frac{1}{2W}$ seconds.*

(b) A band-limited signal of finite energy, which has no frequency components higher than W hertz, may be completely recovered from a knowledge of its samples taken at the rate of $2W$ samples per second.

The sampling rate of $2W$ samples per second for a signal bandwidth of W hertz is called the **Nyquist rate** and the reciprocal $\frac{1}{2W}$ (in seconds) is called the **Nyquist interval**.

In practice, an information signal will not be band-limited, leading to **undersampling**. Two correcting mechanisms can be used.

1. The high-frequency components of the signal can be attenuated by a filter.

2. The filtered signal is sampled at a rate somewhat higher than the Nyquist rate.

In what follows we outline a proof of the sampling theorem. Although Shannon developed applications for the sampling theorem, it seems that the result in mathematical form goes back to Whittaker [Whi15]. The non-specialist is encouraged to skip the proof and just remember the statement, which is of fundamental importance in information theory. The method of proof is based on a combination of the Fourier series and the Fourier integral in their complex form.

Mathematically, a signal is a function $f(t)$ defined for all of t. We have the **Fourier integral** representation of $f(t)$ as follows:

$$f(t) = \int_{-\infty}^{\infty} g(\lambda)e^{i\lambda t}d\lambda \qquad (13.1)$$

This is easier than it looks. We are just saying that $f(t)$ is a "sum" of orthogonal exponential functions e^{iwt}. For example, we can write any vector $\mathbf{f}$ in 3 dimensions as a linear combination of 3 basic vectors. Then

$$\mathbf{f} = g_1\mathbf{u} + g_2\mathbf{v} + g_3\mathbf{w}, \qquad (13.2)$$

where $\mathbf{u}, \mathbf{v}, \mathbf{w}$ are pairwise orthogonal vectors (of length 1, say) in 3 Euclidean dimensions.

To find the number g_1 we take the dot product of both sides with $\mathbf{u}$. We get

$$\mathbf{f} \cdot \mathbf{u} = g_1\mathbf{u} \cdot \mathbf{u} + g_2\mathbf{v} \cdot \mathbf{u} + g_3\mathbf{w} \cdot \mathbf{u}. \qquad (13.3)$$

This gives, by using the orthogonality, that $\mathbf{f} \cdot \mathbf{u} = g_1$. Similarly, $\mathbf{f} \cdot \mathbf{v} = g_2$, $\mathbf{f} \cdot \mathbf{w} = g_3$ and we have then found the coefficients g_1, g_2, g_3. Taking the dot product is analogous to integrating. So we get

$$g(\lambda) = \frac{1}{2\pi} \int_{-\infty}^{\infty} f(t)e^{-i\lambda t}dt, \qquad (13.4)$$

where the factor of $\frac{1}{2\pi}$ is simply a normalizing factor. We assume that the signal is **band-limited**, meaning that $g(\lambda) = 0$ for $|\lambda| > w$ where w is the **cut off frequency**. Then equation (13.1) becomes

$$f(t) = \int_{-w}^{w} g(\lambda)e^{i\lambda t}d\lambda. \qquad (13.5)$$

Now, since $g(\lambda)$ vanishes outside $[-w, w]$ we can extend g periodically, with period $2w$ and then use a Fourier series, as follows.

$$g(\lambda) = \sum_{n=-\infty}^{\infty} c_n e^{\frac{in\pi\lambda}{w}}, \quad -w < \lambda < w. \tag{13.6}$$

The coefficients c_n, which are complex numbers, are given by

$$c_n = \frac{1}{2w} \int_{-w}^{w} g(\lambda) e^{\frac{-in\pi\lambda}{w}} \, d\lambda. \tag{13.7}$$

The integral for c_n is actually the value of $f(t)$ for some t. In fact, from equation (13.5) and the restriction on λ we get

$$c_n = \frac{1}{2w} f\left(\frac{-n\pi}{w}\right). \tag{13.8}$$

Then, from (13.6),

$$g(\lambda) = \frac{1}{2w} \sum_{n=-\infty}^{\infty} f\left(\frac{-n\pi}{w}\right) e^{\frac{in\pi\lambda}{w}} \tag{13.9}$$

$$= \frac{1}{2w} \sum_{n=-\infty}^{\infty} f\left(\frac{n\pi}{w}\right) e^{\frac{-in\pi\lambda}{w}}, \quad -w < \lambda < w. \tag{13.10}$$

From equation (13.5) we have

$$f(t) = \int_{-w}^{w} g(\lambda) e^{i\lambda t} \, d\lambda = \frac{1}{2w} \sum_{n=-\infty}^{\infty} f\left(\frac{n\pi}{w}\right) \int_{-w}^{w} e^{\frac{-in\pi\lambda}{w}} e^{i\lambda t} \, d\lambda. \tag{13.11}$$

We now integrate and use the fact that

$$\sin z = \frac{e^{iz} - e^{-iz}}{2i}. \tag{13.12}$$

This gives

$$f(t) = \sum_{-\infty}^{\infty} f\left(\frac{n\pi}{w}\right) \frac{\sin(wt - n\pi)}{wt - n\pi}. \tag{13.13}$$

Note that, when $t = \frac{n\pi}{w}$ we also get $f(\frac{n\pi}{w})$ on the right-hand side (using the fact from calculus that $\lim_{x \to 0} \frac{\sin x}{x} = 1$) since all other values in the sum over n are zero.

The above formula (13.13) says that f can be reconstructed exactly by sampling f at $t = 0, \pm\frac{\pi}{w}, \pm\frac{2\pi}{w}, \ldots$. Now as in Feynmann [Fey99] the bandwidth W is given as $W = \frac{w}{2\pi}$, so $\frac{1}{2W} = \frac{\pi}{w}$. Thus, by sampling the signal at time instants separated by $\frac{1}{2W}$ seconds, we capture the entire signal. This completes the proof of the sampling theorem.

This result will be the basic building block in the proof of Shannon's information capacity theorem for band-limited, power-limited Gaussian channels.

13.2   The Band-Limited Capacity Theorem, an Example

We consider a continuous stationary random variable $X = X(t)$, where the mean of X is zero and where X is band-limited to W hertz. We let $X_1, \ldots, X_m$, with $m = 2W$ be the random variables corresponding to uniform sampling of $2W$ samples per second (the Nyquist rate). The samples are transmitted over a noisy channel that is also band-limited to W hertz. If the duration of transmission is T seconds then, altogether, $2WT$ samples are transmitted.

We have seen in Section 12.4 that a continuous signal of bandwidth W can be completely characterized by its amplitude at $2W$ sample points per second with the amplitude measured in units such as volts. Using suitable units, we can define the **energy** as the square of the sample amplitude.

Conversely, given the samples we can construct a band-limited signal of band width W passing through $2W$ sample points per second.

The variables $X_1, \ldots, X_m$ corresponding to the $m = 2W$ amplitude samples are independent. (See Chapter 14.) *The signal source is assumed to be ergodic.* Define the **average transmitted power** P as the average of the variable $X_i^2, i = 1, \ldots, 2W$. In symbols,

$$E(X_i^2) = P, i = 1, \ldots, 2W. \tag{13.14}$$

The channel output is perturbed by **additive white Gaussian noise (AWGN)** with mean zero and variance $\sigma^2 = \frac{N_0}{2}$, where $\frac{N_0}{2}$ is the power spectral density. Let $(X_1, \ldots, X_m)$ denote an input signal and $(Y_1, \ldots, Y_m)$ its received signal. We have

$$Y_i = X_i + N_i, i = 1, \ldots, m = 2W \tag{13.15}$$

with N_i being a normal (Gaussian) variable with mean zero and variance equal to $\frac{N_0}{2}$, with the noise also being band-limited to W hertz.

We assume also that the variables X_i, N_i are independent.

From equation (13.14) we also have

$$E(X_1^2 + \ldots + X_m^2) = mP, \quad m = 2W. \tag{13.16}$$

If $(x_1, \ldots, x_m)$ denotes observed amplitude values we have that, on average,

$$x_1^2 + \ldots + x_m^2 = mP. \tag{13.17}$$

Thus the points $(x_1, \ldots, x_m)$ lie close to the surface of a hypersphere in a Euclidean

m-dimensional space, $m = 2W$, with center the origin and radius $\sqrt{mP}$.

Our input signal $(X_1, \ldots, X_m)$ has band-limit W, and the sampling rate is $2W$ samples per second. If the signal duration is T seconds and the average power per sample is denoted by P, it follows that the points representing the samples at different times lie close to the surface of a sphere having radius $\sqrt{2WPT}$.

The received message corresponding to Y lies in a small sphere *centered at the input signal* with radius equal to $\sqrt{2WT\sigma^2}$. To see this we have $Y_i = X_i + N_i$, so $Y_i - X_i = N_i$. Thus, $E(Y_i - X_i) = E(N_i)$, where E denotes the expected value. Since N_i has mean 0 and variance σ^2, we have from Chapter 10 $\sigma^2 = E(N_i - 0)^2 = E(N_i^2)$. Then $E[(Y_1 - X_1)^2 + (Y_2 - X_2)^2 + \ldots + (Y_m - X_m)^2] = m\sigma^2$. Thus, on average the received message corresponding to Y lies on a sphere centered at X with radius equal to $\sqrt{2WT\sigma^2}$.

In a time T, the total received energy is $2WT(P + \sigma^2)$ and the point representing whatever signal was sent plus whatever noise was added to it lies within a hypersphere of radius $\sqrt{2WT(P + \sigma^2)}$ and close to the surface of that hypersphere.

To see this, take $T = 1$. We have $Y_i = X_i + N_i$ so $Y_i^2 = (X_i + N_i)^2$. Thus $E(Y_i^2) = E(X_i + N_i)^2 = E(X_i^2 + N_i^2 + 2X_i N_i)$. Since X_i, N_i are independent, $E(X_i N_i) = E(X_i)E(N_i) = E(X_i)0 = 0$ (see Chapter 10). Therefore we have $E(Y_i^2) = E(X_i^2) + E(N_i^2) = P + \sigma^2$. Thus, $E(Y_1^2 + Y_2^2 + \ldots + Y_m^2) = m(P + \sigma^2)$.

So we have a collection of M small hyperspheres of radius $\sqrt{2WT\sigma^2}$ all lying within a larger hypersphere of radius $\sqrt{2WT(P + \sigma^2)}$ in m dimensions where $m = 2WT$. For accurate decoding these spheres should be disjoint.

How large can M be in order that the hyperspheres be disjoint? The argument is reminiscent of our work in Chapter 11, where we were filling up a finite set (e.g., of size $2^{nH(X)}$) with a family of disjoint subsets, all of the same size (e.g., of size $2^{nH(X|Y)}$).

In the case $n = 2$ we have a collection of small circles, each having the same radius, being packed into a larger circle with no two of the small circles overlapping.

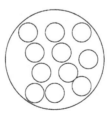

We think of each sample as an input source word. As discussed in Chapter 12, we have the following relation.

Channel capacity $= \log(\textit{number of distinguishable inputs}))$.

The inputs are distinguishable here if their output fans, in this case the small spheres, are disjoint.

Now in m dimensions (with $m = 2WT$ here) the volume of a hypersphere is proportional to R^m, where R is the radius.

In symbols, the volume is $V_m(m(P + \sigma^2))^{\frac{m}{2}}$, where V_m is a constant depending on m. The volume of each smaller sphere is $V_m(m\sigma^2)^{\frac{m}{2}}$.

The ratio is

$$\left(1 + \frac{P}{\sigma^2}\right)^{\frac{m}{2}} = 2^{\frac{m}{2}\log(1 + \frac{P}{\sigma^2})}, \qquad (13.18)$$

where logs are taken to the base 2, as usual. It follows that if M is the maximum number of disjoint small spheres we get

$$M \le 2^{\frac{m}{2}\log(1 + \frac{P}{\sigma^2})}. \qquad (13.19)$$

Since $m = 2WT$, we get

$$M \le 2^{WT\log(1 + \frac{P}{\sigma^2})}. \qquad (13.20)$$

Taking logs we get

$$\log M \le WT\log(1 + \frac{P}{\sigma^2}). \qquad (13.21)$$

As the message is T seconds in duration, then, dividing by T, we get an upper bound on the channel capacity as $W\log(1 + \frac{P}{\sigma^2})$. Shannon was then able to show that the upper bound was achieved. Thus we get the following result.

Information Capacity Theorem. *The information capacity C is given by the formula $C = W\log(1 + \frac{P}{\sigma^2})$ Shannon bits per second, where P is the average power.*

Since $\sigma^2 = N_0 W$, we get the following equivalent formulation for C.

Capacity Formula 2. $C = W\log(1 + \frac{P}{N_0 W}) = W\log(1 + \frac{P}{N})$ *Shannon bits per second, where $N = N_0 W$ and P is the average power.*

Capacity Formula 3. $C = B\log(1 + S/N)$ *This is the traditional formulation where B stands for "Bandwidth" and SNR denotes the "Signal to Noise Ratio".*

This formula is of great practical importance and gives the precise connection between the information capacity, the bandwidth and the average power. Again, the **information capacity** is defined as the maximum rate at which information can be transmitted across the channel without error, which is the log of the number of distinguishable inputs. We can regard the term $\frac{P}{N_0 W}$ as the average **"signal to noise ratio"** (SNR).

From the capacity formula, C varies in a linear fashion with W versus only a logarithmic growth with P. It follows that to increase capacity, it is easier to increase C by widening the band width rather than by increasing the power for a prescribed noise variance. For further details, we refer the reader to Pierce [Pie79] and Haykin [Hay01].

13.3 The Coding Gain

Figure 13.1: Professor J. H. van Lint, one of the world's leading coding theorists. [Photo courtesy Professor H. van Tilborg]

In many practical situations there is a choice between the number U of user bits that need to be accurately transmitted through a noisy channel and the available power P. There is a trade-off. For example, in deep space transmission, U determines the number of pictures that can be sent and P is the power available from solar panels. The transmitter will have an average energy of

$$E_b = \frac{U}{P}. \tag{13.22}$$

When we encode we use up more bits. Given that the available power is fixed, there is then a loss of energy per bit and the "dividing line" between 0 and 1 becomes blurred. The probability of a bit error will therefore increase. In van Lint [vL98], the author gives the example of the 32×32 Hadamard code used in the spacecraft Mariner '69. In the case of no coding, the bit error probability is about $\frac{10^{-4}}{6}$. With coding, the bit error probability increases by a factor of about 2000 to about 0.036. However, the *message error probability* improves from about 10^{-4} (with no coding) to 1.4×10^{-5} with coding!

Another way of looking at the situation is as follows. *We can use coding not to reduce message error probability but to improve the SNR, keeping the error probability the same.* In the Mariner example this allows for the reduction of the solar panels by almost 15%. For further details we refer the reader to van Lint [vL98].

Chapter 14

Ergodic and Markov Sources, Language Entropy

Goals, Discussion   So far, our sources have been discrete. In the real world this need not be the case. Sometimes we need to utilize continuous sources such as continuous electromagnetic signals used in engineering. Also, even in the discrete case, memoryless sources can be too restrictive. Ergodic sources form a bridge. For discrete sources we have a hierarchy: Memoryless sources are the most specialized. Next in generality are stationary sources, then special kinds of Markov sources, then ergodic sources. Languages can be modeled as ergodic sources. From this we can estimate the redundancy of a language such as English. This has obvious ramifications in cryptography. Later on we will discuss other examples connected to biology and the genetic code.

New, Noteworthy   Conceptually these topics are complex. We give several examples in the text and in the problems to illustrate some of the subtleties. On the Markov side we give clear statements about the main results. We discuss the fixed probability vector for an irreducible Markov chain and emphasize that this probability vector must have all components **positive,** other accounts of the theory notwithstanding. We use the fixed probability vector to compute the entropy of a Markov source.

14.1   General and Stationary Sources

Let S be any source emitting symbols from some alphabet A. We denote the random variables corresponding to the values of the alphabet emitted by the source by $X_1, X_2, \ldots$, where X_n corresponds to the nth letter that the source outputs.

Then, for any n, we will show how $H(X_1, X_2, \ldots, X_n)$ can be calculated in various ways.

Example 14.1 *Suppose S produces symbols from $A = \{0,1\}$ such that S independently emits 0 with probability 0.3 and 1 with probability 0.7 in each position. Calculate $H(X_1, X_2)$.*

The possible values of (X_1, X_2) are $(0,0)$, $(0,1)$, $(1,0)$, $(1,1)$ with corresponding probabilities .09, .21, .21, .49, respectively. Then

$$H(X_1, X_2) = -[(.09)\log(.09) + (.21)\log(.21) + (.21)\log(.21) + .49\log(.49)].$$

Alternatively, $H(X_1, X_2) = 2H(X_1) = 2(H(0.7)) = 2(0.8813) = 1.7626$.

In general we want to define $H(S)$, the entropy of an arbitrary source S. This is done as follows. **The entropy of S**, denoted by $H(S)$, is defined to be a limit, namely $\lim_{n\to\infty} \frac{H(X_1, X_2, \ldots, X_n)}{n}$, if such a limit exists.

Note that if S is memoryless, as in example 14.1, then

$$\frac{H(X_1, X_2, \ldots, X_n)}{\cdot\ n} = \frac{H(X_1) + H(X_2) + \cdots + H(X_n)}{n} = \frac{H + H + \cdots + H}{n} = H$$

where $H = H(X_i)$, $1 \le i \le n$.

To see that $H(X_1, X_2, \ldots, X_n) = H(X_1) + H(X_2) + \cdots + H(X_n)$ we use the fact that the X_i are *independent*.

Here is another formula that we might propose for the definition of entropy.

$$H(S) = \lim_{n\to\infty} H(X_n \mid X_1, X_2, \ldots, X_{n-1}) \quad \text{(if this limit exists)}.$$

In fact, we have the following result.

Theorem 14.2 *If $\lim_{n\to\infty} H(X_n \mid X_1, \ldots, X_{n-1})$ exists, then $\lim_{n\to\infty} \frac{H(X_1, X_2, \ldots, X_n)}{n}$ exists and the two limits are equal.*

Proof. We refer to the problems for a proof. ∎

However, as pointed out by Welsh [Wel88] it may be the case that $\lim_{n\to\infty} \frac{H(X_1, X_2, \ldots, X_n)}{n}$ exists even though $\lim_{n\to\infty} H(X_n \mid X_1, X_2, \ldots, X_{n-1})$ does not exist. Here is an example.

Example 14.3 *Let a source S emit a sequence $(X_1, X_2, X_3 \ldots)$ in such a way that $X_2, X_4, X_6, \ldots, X_{2m}, \ldots$ are zero and such that $(X_1, X_3, X_5, \ldots)$ are chosen independently at random from the alphabet $\{0, 1\}$.*

Then (see problems) $H(X_1, X_2, X_3, \ldots, X_n)$ is $\frac{n+1}{2}$ if n is odd and $\frac{n}{2}$ if n is even. It follows that $\lim_{n\to\infty} \frac{H(X_1, X_2, \ldots, X_n)}{n}$ exists and is equal to $\frac{1}{2}$. However, if n is even, $H(X_n \mid X_1, X_2, \ldots, X_{n-1}) = 0$. If n is odd, then $H(X_n \mid X_1, X_2, \ldots, X_{n-1}) = 1$. Thus $\lim_{n\to\infty} H(X_n \mid X_1, X_2, \ldots, X_{n-1})$ does not exist.

We now come to **stationary sources**. The source S is **stationary** if it is "time-invariant". This means that $\Pr(X_{i_1} = a_1, X_{i_2} = a_2, \ldots, X_{i_n} = a_n) = \Pr(X_{i_1+w} = a_1, X_{i_2+w} = a_2, \ldots, X_{i_n+w} = a_n)$ where $w \geq 0$ is any nonnegative integer and, and $i_1, i_2, \ldots, i_n$ is any set of nonnegative indices.

Let's examine this condition. Take the alphabet to be binary. As a very special case we have that $\Pr(X_i = 1) = \Pr(X_{i+w} = 1)$ for any integer $w \geq 0$. Thus $\Pr(X_1 = 1) = \Pr(X_2 = 1) = \cdots = \Pr(X_n = 1) = p$. Doesn't this say that the sequence is in fact memoryless? Not quite. We refer to the problems.

Another example of the definition is this. The probability that both the 8th and the 11th symbol are 1 equals the probability that both the 15th and the 18th symbol are 1. We get this by taking $w = 7$.

Any memoryless source is stationary. This follows because in the memoryless case the variables X_i are, by definition, independent and identically distributed random variables.

Example 14.4 *Let S be a binary memoryless source where the probability of emitting 1 is 0.7, and the probability of emitting 0 is 0.3. Then*

$$\Pr(X_8 = X_{11} = 1) = \Pr(X_8 = 1)\Pr(X_{11} = 1) = (0.7)^2 = 0.49$$
$$\Pr(X_{15} = X_{18} = 1) = \Pr(X_{15} = 1)\Pr(X_{18} = 1) = (0.7)^2 = 0.49.$$

However, not all stationary sources are memoryless

Example 14.5 *A source S emits $\{X_1, X_2, \ldots\}$ as follows. An unbiased coin is tossed. If the result is heads then $X_n = 1$ for all n. If the result is tails then $X_n = 0$ for all n. We pose the following questions*

(a) Is S memoryless?

(b) Is S stationary?

The answer to (a) is "No", but it is worth discussing, and we do so in the problems. The answer to (b) is "Yes".

To see this, observe that $\Pr(X_{i_1} = X_{i_2} = X_{i_3} = \cdots = 0) = \frac{1}{2} = \Pr(X_{i_1+w} = X_{i_2+w} = X_{i_3+w} = \cdots = 0)$. If we replace 0 by 1 above the conclusion is the same. For example, $\Pr(X_1 = X_2 = 1) = \Pr(X_9 = X_{10} = 1) = \frac{1}{2}$. We remark that the only subsequence that has a nonzero probability of occurring is either the all-zero or the all-one subsequence.

Theorem 14.6 *Any stationary source has an entropy. This entropy is the limit given by* $\lim_{n \to \infty} H(X_n \mid X_1, X_2, \ldots, X_{n-1})$.

Proof. $H(X_n \mid X_1, X_2, \ldots, X_{n-1}) \leq H(X_n \mid X_2, \ldots, X_{n-1})$ as can be seen from the fact that side-information, in this case X_1, "never increases entropy" –see Chapter 10. By stationarity the right side equals $H(X_{n-1} \mid X_1, X_2, \ldots, X_{n-2})$. Therefore, $v_n = H(X_n \mid X_1, X_2, \ldots, X_{n-1}) \leq H(X_{n-1} \mid X_1, X_2, \ldots, X_{n-2}) = v_{n-1}$. So $\{v_n\}$ is a *decreasing* sequence, and $v_n \geq 0$. By a basic property of the real numbers the limit of $\{v_n\}$ exists as n tends to infinity, since $\{v_n\}$ is bounded below by zero. The result now follows from Theorem 14.2. ■

14.2   Ergodic Sources

The fact that a source S is stationary pins down S to some extent. But we need more structure in the source. The type of source that we need is called an **ergodic source**.

We assume that S is a stationary source emitting symbols $(X_1, X_2, \ldots, X_n)$ according to some known probability distribution over some alphabet A. Let $\mathbf{a} = (a_1, \ldots, a_t)$ denote some sequence over A. We define $f_n(\mathbf{a})$, the **frequency** of $\mathbf{a}$ for an output $(X_1, X_2, \ldots, X_n, \ldots)$ of the source, to be the number of times that $\mathbf{a}$ occurs in the first n terms of the output.

Example 14.7 *Let* $\mathbf{a} = (0, 1)$ *with* S *binary and* $(X_1, X_2, \ldots, X_n) = (100110100110)$. *Here* $n = 12$, $f_{12}(\mathbf{a}) = 3$.

The stationary source is **ergodic** if $\Pr\left(\lim_{n\to\infty} \frac{f_n(\mathbf{a})}{n} = \Pr(X_1 = a_1, \ldots, X_t = a_t)\right) = 1$, where $\mathbf{a} = (a_1, a_2, \ldots, a_t)$

Example 14.8 *Let* S *be the source in Example 14.5 of section 14.1. Is* S *ergodic?*

Let $t = 1$ and $\mathbf{a} = (0)$. Then either $f_n(\mathbf{a}) = n$ or $f_n(\mathbf{a}) = 0$. Thus, either $\lim_{n\to\infty} \frac{f_n(\mathbf{a})}{n} = 1$ or $\lim_{n\to\infty} \frac{f_n(\mathbf{a})}{n} = 0$. Now $\Pr(X_1 = 0) = \frac{1}{2}$. Thus $\Pr\left(\lim_{n\to\infty} \frac{f_n(\mathbf{a})}{n} = \Pr(X_1 = 0)\right) \neq 1$. Thus S is stationary, but not ergodic.

Example 14.9 *This example is found in Pierce [Pie79]. The source* S *emits*

i) the sequence $A, B, A, B, A, B \ldots$ *with probability* $\dfrac{1}{3}$

ii) the sequence $B, A, B, A, B, A \ldots$ *with probability* $\dfrac{1}{3}$

iii) and the sequence $E, E, E, E, E, E \ldots$ *with probability* $\dfrac{1}{3}$

Then if $\mathbf{a} = (A)$ *we have* $\frac{f_n(\mathbf{a})}{n} = \frac{1}{2}$ *for the first kind of sequence,* $\frac{f_n(\mathbf{a})}{n} = \frac{1}{2}$ *for the second kind of sequence and* $\frac{f_n(\mathbf{a})}{n} = 0$ *for the third kind of sequence. But* $\Pr(X_1 = A) = \frac{1}{3}$. *Again,* S *is stationary but not ergodic.*

Discussion . If a source is stationary it is time invariant so we can calculate meaningful **time averages** for a *given* output. For a source to be ergodic we need these time averages to be equal to the **ensemble averages**.

These are averages taken over all possible outputs obtained from the given probability distribution of the source. The idea is important, if a bit complicated. Pierce [Pie79] offers the following discussion.

Let's think of a very large number of writers in a given language, say English. For a given message (i.e., for a given writer) the frequency of occurrence of a letter such as S does not vary much along the length of the message. As we analyze a longer and longer piece of a message our estimate of the statistics of a message (e.g., the frequency of occurrence of a letter such as S or of a digram such as AE) and the associated probabilities converges

The point about a source being ergodic is that these statistics or probabilities apply equally well to all possible messages or outputs, i.e., to all possible writers of English in this case.

The ergodic idea applies also in the physical world where we are dealing with electromagnetic sources of various sorts.

From our point of view the main property about ergodic sources is that they satisfy the **asymptotic equipartition property** (AEP) . This is called the **Shannon-McMillan Theorem**. It means that most sequences of length N, with N large, from an ergodic source are "typical" and equiprobable with probability of occurrence equal to 2^{-NH} where H is the entropy of the source. Using this, one can show that the channel capacity theorem of Chapter 12 holds (for a binary symmetric channel) not just for memoryless sources but also for ergodic sources.

Example 14.10 *Every memoryless source is ergodic.*

The proof of this is not immediate and involves proving a special case of "the law of large numbers" in the theory of probability.

In the next section we discuss the examples "par excellence" of ergodic sources which are not memoryless. They are obtained from certain Markov sources.

14.3 Markov Chains and Markov Sources

A source S is said to be a **Markov source** if $\Pr(X_{n+1} = a_{n+1} \mid X_n = a_n, \ldots, X_1 = a_1) = \Pr(X_{n+1} = a_{n+1} \mid X_n = a_n)$, where $a_1, a_2, \ldots, a_n, a_{n+1}$ are elements of the alphabet A of S. In other words, the probability of the event that $X_{n+1} = a_{n+1}$ only depends on the source output immediately preceding X_{n+1}. The **transition probability** p_{ij} is defined as follows.

$$p_{ij} = \Pr\left(X_{n+1} = a_j \mid X_n = a_i\right).$$

Here, if the alphabet A has m letters in it, we have $1 \leq i, j \leq m$. Then the transition probabilities form the **transition matrix** $P = (p_{ij})$.

Each entry in P, being a probability, lies between 0 and 1. The entries in each row of the matrix add up to 1. We regard the elements $a_1, a_2, \ldots, a_m$ of the alphabet A as "states" and p_{ij} as the probability of moving from state i to state j. If we are in a given state i, we either have to remain in state i or move to a new state. Thus the entries in any row of P add up to 1. If we know the initial value X_1 we can find the probability of subsequent values $X_2, X_3, \ldots$ from the transition matrix P.

Example 14.11 *Suppose the source is an English writer. So we have 27 states, say, corresponding to the 26 letters and a space. Then, if A is letter 1, B is letter 2, C is letter 3, etc., we get $p_{2,5} = p_{25}$ the probability of moving from state 2 to state 5 = the probability that the next letter to be written is E given that the present letter is B.*

The matrix P of size 27×27 can be constructed from the statistics of the language.

Example 14.12 *A binary source operates as follows. The next entry equals the present entry with probability p and equals the opposite of the present binary entry with probability $1 - p$.*

Here, the transition matrix P is as follows.

$$P = \begin{array}{c} \\ 0 \\ 1 \end{array} \begin{array}{cc} 0 & 1 \\ \begin{pmatrix} p & 1-p \\ 1-p & p \end{pmatrix} \end{array}.$$

If $p = 0.7$, then P becomes $\begin{pmatrix} .7 & .3 \\ .3 & .7 \end{pmatrix}$.

Question . If the first entry in the output of the source above is 1, what is the probability that the 3rd element is in fact 0?

We want to calculate $\Pr(X_3 = 0 \mid X_1 = 1)$.

It is easier to visualize this with a "**state diagram**", where we have two states named 0 and 1 along with a "**transition diagram**" as indicated below.

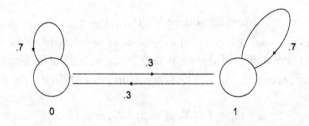

Our question is this. What is the probability of moving from state 1 to state 0 in exactly 2 transitions (denoted by $p_{1,0}^{(2)}$)?

We can move from state 1 to state 1 and then from state 1 to state 0. The probability of this happening is $(0.7)(0.3)$. Or, we can move from state 1 to state 0 and then from state 0 to state 0. The probability of this happening is $(0.3)(0.7)$. Thus the total probability is $(0.7)(0.3) + (0.7)(0.3) = 0.42$

This can all be handled easily from the algebraic point of view. We have

$$P = \begin{pmatrix} p & 1-p \\ 1-p & p \end{pmatrix} = \begin{pmatrix} .7 & .3 \\ .3 & .7 \end{pmatrix}.$$

Now $P^2 = PP$ is a transition matrix. In fact P^n, for any positive integer n is also a transition matrix. This means that each entry of P^n is a probability and that, moreover, each row of P^n adds up to 1.

Significance of P^n. The $(ij)^{th}$ entry of P^n, i.e., $(P^n)_{i,j}$ tells us what the *probability is of moving from state i to state j in exactly n steps or transitions.* Here, $P^2 = \begin{pmatrix} .58 & .42 \\ .42 & .58 \end{pmatrix}$.

So, .58 represents the probability of moving from state 0 to state 0 in exactly 2 steps or transitions. Similarly .42 represents the probability of moving from state 0 to state 1 (and also, the probability of moving from state 1 to state 0) in exactly 2 steps.

Caution. Here p has a different meaning than it had when we discussed the BSC in earlier chapters. There, p was the probability of a mistake being made, so we had $p < \frac{1}{2}$. Here, p is the probability of moving from state 0 to state 0 or state 1 to state 1 so $p = .7$.

In general, if we have some initial probability distribution (α, β) for the 2 states, then the vector $(\alpha, \beta)P^n$ tells us the probability distribution after exactly n steps (transitions). To calculate $(\alpha, \beta)P^n$ it is much shorter, instead of calculating P^n, and then getting $(\alpha, \beta)P^n$ to calculate $((((\alpha, \beta)P)P)P \ldots)$ In this approach we are only multiplying a vector by a matrix at each stage. This is much easier than multiplying an entire matrix by itself at each stage.

We now come to the two main results about Markov chains that show their power and utility.

First, we say that a Markov source (or Markov chain) with transition matrix P is **irreducible** if there exists some power of P, say P^m, such that all entries of P^m are *positive*.

Example 14.13 *Let*

$$P = \begin{pmatrix} \frac{1}{3} & \frac{1}{3} & \frac{1}{3} \\ 0 & \frac{1}{2} & \frac{1}{2} \\ \frac{1}{3} & \frac{1}{3} & \frac{1}{3} \end{pmatrix}$$

Then $P = P^1$ has some zeros. But P^2 has no zeros as is easily checked, and so, P is irreducible.

Here is our first main result.

Theorem 14.14 *Let P be an irreducible Markov source with transition matrix P. Then there exists a unique fixed probability vector $\mathbf{w}$ such that the following holds.*

(a) $\mathbf{w}P = \mathbf{w}$.

(b) *each entry in $\mathbf{w}$ is positive.*

(c) *As n gets large, P^n tends to W where each row of W is $\mathbf{w}$.*

(d) *No matter what the initial probability vector $\mathbf{p}_0$ is, $\mathbf{p}_0 P^n$ tends to $\mathbf{w}$ as n gets large. In other words, the probability of being in state number i tends to the ith component of the fixed probability vector $\mathbf{w}$ regardless of the initial probability distribution.*

This theorem is one of the major results in discrete mathematics. The proof is quite involved. For a discussion we refer the interested reader to Ash [Ash90], Feller [Fel50], and Goldie-Pinch [GP91]. Part (d) is especially important as it shows that **long range predictions** such as weather forecasts can easily be derived from the Markov model.

Example 14.15 *Let P be as above. We want to calculate the fixed probability vector $\mathbf{w}$.*

Let $\mathbf{w} = (x, y, z)$. We have $\mathbf{w}P = \mathbf{w}$ so

$$(x, y, z) \begin{pmatrix} \frac{1}{3} & \frac{1}{3} & \frac{1}{3} \\ 0 & \frac{1}{2} & \frac{1}{2} \\ \frac{1}{3} & \frac{1}{3} & \frac{1}{3} \end{pmatrix} = (x, y, z).$$

Then

$$\frac{1}{3}x + \frac{1}{3}z = x$$
$$\frac{1}{3}x + \frac{1}{2}y + \frac{1}{3}z = y$$
$$\frac{1}{3}x + \frac{1}{2}y + \frac{1}{3}z = z.$$

Since $\mathbf{w}$ is a probability vector we also demand that

$$x + y + z = 1.$$

In solving these equations we see from the second and third equations that $y = z$. Substituting and using the forth equation gives that $\mathbf{w} = \left(\frac{1}{5}, \frac{2}{5}, \frac{2}{5}\right)$.

Now we come to the *coup de grâce*.

Theorem 14.16 *Suppose P is the transition matrix for an irreducible Markov source S. If the fixed probability vector $\mathbf{w}$ is used as the initial distribution then S is an ergodic source.*

The distribution $\mathbf{w}$ is sometimes called the **stationary distribution** as well as the **fixed probability vector**.

14.4 Irreducible Markov Sources, Adjoint Source

If S is irreducible, then the source S is stationary. Let P denote the transition matrix. Thus, from Section 14.1, S has an entropy which can be defined as

$$\lim_{n \to \infty} H(X_{n+1} \mid X_1, X_2, \ldots, X_n) = \lim_{n \to \infty} H(X_{n+1} \mid X_n)$$

since the source is Markov.

Recall that $H(X \mid Y)$ can be written as the average value of $H(X \mid y)$. Thus

$$H(X, Y) = \sum_y H(X \mid y) \Pr(y).$$

Let a_i be the i-th element of the alphabet $A = \{a_1, a_2, \ldots, a_m\}$. We regard the possible elements of the alphabet as the possible states. (Thus, for example, we could model the English language using 26 states for the 26 letters.) Then, $H(X_{n+1} \mid X_n) = \sum_{i=1}^m H(X_{n+1} \mid X_n = a_i) \Pr(X_n = a_i)$. Now $\Pr(X_n = a_i)$ is, (in the limit, as n gets large) w_i, where $\mathbf{w} = (w_1, w_2, \ldots, w_m)$ is the fixed probability vector of P: This follows from Theorem 14.14 in Section 14.3.

Also, $H(X_{n+1} \mid X_n = a_i)$ is easily calculated simply by thinking of the ith state as a source with emission probabilities p_{ij}. Recall that p_{ij} is the probability of moving from state i to state j. The source corresponding to the ith state then has an entropy H_i. So we get the following result.

The entropy H of an irreducible Markov source is given by the following formula

$$H = -\sum_{i=1}^m w_i \sum_j p_{ij} \log p_{ij}.$$

Alternatively, we have, from the above, that

$$H = \sum_{i=1}^m w_i H_i,$$

where H_i is the entropy corresponding to the i-th state. Now if S is an ergodic Markov source with alphabet $A = \{a_1, a_2, \ldots, a_m\}$ and fixed probability vector $\mathbf{w}$ we can ignore the transition matrix P and form the **adjoint source** S^*, which is memoryless, with alphabet A and with w_i defined to be the probability that S^* emits a_i, $1 \leq i \leq m$. This source will then have its own entropy $H^* = H(S^*)$.

Theorem 14.17 (Comparison theorem for entropies) *The entropy of an ergodic Markov source is less than or equal to the entropy of the adjoint source. In symbols we have $H \leq H^*$.*

Example 14.18 *Let the Markov source have transition matrix*

$$P = \begin{pmatrix} \frac{1}{3} & \frac{1}{3} & \frac{1}{3} \\ 0 & \frac{1}{2} & \frac{1}{2} \\ \frac{1}{3} & \frac{1}{3} & \frac{1}{3} \end{pmatrix}.$$

As we have seen, $\mathbf{w} = \left(\frac{1}{5}, \frac{2}{5}, \frac{2}{5} \right)$. *Then*

$$H^* = \frac{1}{5} \log 5 + \frac{2}{5} \log \left(\frac{5}{2} \right) + \frac{2}{5} \log \left(\frac{5}{2} \right)$$

$$= \frac{1}{5} \log 5 + \frac{2}{5} (\log 5 - \log 2) + \frac{2}{5} \log(5 - \log 2)$$

$$= \log 5 - \frac{2}{5} \log 2 - \frac{2}{5} \log 2 = \log 5 - \frac{4}{5} = 1.5222.$$

The source corresponding to the ith state has entropy $H_i = -\sum_j p_{ij} \log(p_{ij})$. Alternatively, $H_i = \sum_j p_{ij} \log \left(\frac{1}{p_{ij}} \right)$. Then

$$H_1 = \left(\frac{1}{3} \log 3 + \frac{1}{3} \log 3 + \frac{1}{3} \log 3 \right) = \log 3 = 1.585$$

$$H_2 = \frac{1}{2} \log 2 + \frac{1}{2} \log 2 = \log 2 = 1$$

$$H_3 = \frac{1}{3} \log 3 + \frac{1}{3} \log 3 + \frac{1}{3} \log 3 = \log 3 = 1.585.$$

Now the entropy of H is

$$w_1 H_1 + w_2 H_2 + w_3 H_3 = \frac{1}{5}(1.585) + \frac{2}{5}(1) + \frac{2}{5}(1.585) = \frac{3}{5}(1.58\overline{5}) + \frac{2}{5}$$

$$= 1.351.$$

Since $1.351 < 1.522$ we have $H < H^*$.

14.5   Cascades and the Data Processing Theorem

Given three random variables X, Y, Z we can define the mutual information $I(X, Y : Z)$ which we can think of as the amount of information that (X, Y) provide about Z (or, as the amount of information that Z provides about (X, Y)).

Using the work in Chapter 10 we can see that for two variables X, Y we have $I(X : Y) = \sum_{x,y} \Pr(x, y) \log \left(\frac{\Pr(y|x)}{\Pr(y)} \right)$. Similarly

$$I(X, Y : Z) = \sum_{x,y,z} \Pr(x, y, z) \log \left(\frac{\Pr(z \mid x, y)}{\Pr(z)} \right).$$

Applying Jensen's inequality, as in McEliece [McE78] it can be shown that $I(X, Y : Z) \geq I(Y : Z)$ with equality if and only if $\Pr(z \mid x, y) = \Pr(z \mid y)$ for all (x, y, z) with $\Pr(x, y, z) > 0$.

We want to study the case of equality. Suppose X, Y, Z form a "Markov triple" as in the diagram.

We have a channel connecting X and Y described by the forward probabilities $\Pr(y \mid x)$ and a channel connecting Y and Z with forward probabilities $\Pr(z \mid y)$. By saying that X, Y, Z form a Markov triple, we mean that $\Pr(z \mid x, y) = \Pr(z \mid y)$. So from the above we have $I(X, Y : Z) \geq I(Y : Z)$ with equality if and only if X, Y, Z is a Markov triple (i.e., Z depends on X only through Y).

In fact, following McEliece [McE78], we have the following result

Theorem 14.19 (Data Processing Theorem) *If (X, Y, Z) is a Markov triple then*

$$I(X : Z) \leq \begin{cases} I(X : Y) \\ I(Y : Z) \end{cases}.$$

This important result says that the "extra processing" involving Z cannot increase mutual information.

14.6 The Redundancy of Languages

This is an important topic for various reasons and is much-studied. For example, in cryptography one needs to know how much **redundancy** is carried by a language such as English to ensure that encryption is secure. Also, the question is of interest in linguistics and in mathematical modeling for testing the assumption that a suitable ergodic source can give a reasonable mathematical model of a language. In biology it is also important as we will see in Chapter 17.

First, let's talk about some definitions. Suppose we are working with some language over an alphabet Γ. For English we can use 26 letters and a space, so $|\Gamma| = 27$.

We regard the language as an ergodic source. (Then, assuming the Shannon-McMillan theorem, the number of typical sequences Y_n of length n is about 2^{nH}, where H denotes the entropy per symbol in Shannon bits). So $|Y_n| = 2^{nH}$

From our work with source coding and block coding –which is easily generalized to the non-binary case– we know that a source with entropy H can be efficiently encoded in such a way that the average length L_n of a typical sequence of length n is given by the formula

$$L_n \log |\Gamma| = nH,$$

since

$$L_n = \frac{nH}{\log |\Gamma|}. \tag{14.1}$$

This follows since the average encoding length per symbol is $\frac{H}{\log |\Gamma|}$ from Shannon's noiseless coding theorem. This average length is a certain fraction of n. If there were no redundancy it would be n. So we define R as follows

$$L_n = n(1 - R). \tag{14.2}$$

Combining equations (14.1) and (14.2) we get our **formula for redundancy**

$$R = 1 - \frac{H}{\log |\Gamma|}$$

where H is the entropy per symbol.

We observe that for the binary alphabet this gives $R = 1 - H$.

Formula 14.1 says that a typical sequence of length n can be recoded using just L_n characters without loss of information about the sequence. Of course, we need to estimate H or R.

The redundancy R is usually measured in percentages. Calculating R is not an exact science. Estimates suggest that, for English, redundancy is about 70%. As a very rough, if not incorrect approximation, this might suggest that only about 30% of a message –suitably chosen– is needed to recover the entire message. Or we can think of having only a "free choice" with 30% of the message with the rest of the message, being determined by statistical patterns and grammatical structure.

(A colleague, Ernest Enns, has passed on an interesting message on the Internet concerning the redundancy of English. As we mentioned earlier redundancy can be "good" or "bad". The piece is entitled "clever student" and "reads" as follows:

Aoccdrnig to a rscheearch at Cmabrigde Uinervtisy, it deosn't mttaer in waht

oredr the ltteers in a wrod are, the only iprmottent thing is that the first and
lsat ltteer be at the rghit pclae. The rset can be a total mses and you can still
raed it wouthit problem. Tihs is bcuseae the human mnid deos not raed every
ltetter by itself, but the wrod as a wlohe.

(amzanig huh?)

The *correct mathematical interpretation of* R is that in an optimal encoding we can
reduce the length of the message from n characters to L_n characters.

As an example, let's suppose $H = 1.4$. Then we get $L_n = \frac{n(1.4)}{\log 27} = n\left(\frac{1.4}{4.76}\right) = n\,(0.2941)$.
This means that say a 100-letter message can be recoded as a message using only around
29.4 characters without losing any information .

How would we go about calculating the redundancy or entropy of English?

The first crude approximation is to think of a 27-symbol alphabet (including the space)
and to regard English as a memoryless source and one in which each symbol is equally likely
to occur. This gives the entropy as $\log 27 = 4.76$ Shannon bits per symbol.

Our next approximation involves taking account of the probabilities of occurrence of the
symbols. The most probable symbol is a space with probability 0.18, followed by the letter
E with probability approximately equal to 0.13.

This approximation will be an **upper bound** for the following reason. In Chapter 10
we saw that $H(X,Y) \le H(X) + H(Y)$ with equality if and only if X, Y are independent.
Thus we get

$$H \le \sum_{i=1}^{27} p_i \log p_i,$$

giving $H \le 4.03$.

Similarly we get upper bounds using diagrams, trigrams etc. and the probabilistic in-
terdependence of the symbols.

Thus, from digrams, $H \le \frac{1}{2}\sum_i\sum_j \Pr(i,j)\log p_{i,j}$, where $p_{i,j}$ is the estimated probability
of the ordered pair (i,j) of symbols and zero-probability pairs are thrown out. This method
gives $H \le 3.32$. One can experiment with trigrams, getting $H \le 3.10$.

Other approaches included estimating the quantity $H(X_{n+1} \mid X_1, \ldots, X_n)$ that was
discussed earlier in this chapter. This estimate can be carried out by estimating the average
number of guesses needed to obtain the $(n+1)^{th}$ letter given the previous n letters.

Also, Shannon proposed a different approach to finding the entropy of English by working
with words and finding the "word-entropy".

The conclusion of all this is that the entropy (per symbol) of English comes out to a
little over 1 Shannon bit (giving a redundancy of around 70% for average English).

14.7   Problems

1. Assume that source S is emitting $X_1, X_2, \ldots$ and that $\lim_{n \to \infty} H(X_n \mid X_1, X_2, \ldots, X_{n-1})$ exists. Show that $\lim_{n \to \infty} H(X_1, X_2, \ldots, X_n)$ exists and that the two limits are equal.

2. Let S be a binary stationary source

   (a) Show that $\Pr(X_1 = 1) = \Pr(X_2 = 1) = \cdots = \Pr(X_n = 1)$ for all n.

   (b) Why does it not follow that S is memoryless?

3. A memoryless source emits symbols in blocks of size 2 according to the following probability distribution:

$$\Pr(0,0) = \frac{1}{4}, \quad \Pr(0,1) = \frac{3}{4}.$$

   (a) Is S stationary?

   (b) Find the entropy of S.

4. Suppose we now regard the source S in question 3 as a source T which emits single binary symbols, not symbols in blocks of size 2.

   (a) Is T stationary?

   (b) Calculate the entropy (*per symbol*) of T.

5. Let S be a memoryless binary source emitting the sequence $X_1, X_2, \ldots$ with the following probability distribution:

$$\Pr(0) = .8, \quad \Pr(1) = .2.$$

   (a) Calculate $H(X_1, X_2)$ from first principles.

   (b) Calculate $H(X_1, X_2)$ using the fact that S is memoryless.

6. For the source S in Problem 5 estimate the number of typical output sequences of length 1000.

7. A binary source S emits symbols 0 and 1 as follows. The first 100 entries are all zero. After that the source is memoryless with probability 0.7 (or 0.3) of emitting 0 (or 1).

   Estimate the number of typical output sequences of length 1000 from this source.

8. Let a Markov source S with 3 states have a transition matrix

$$P = \begin{pmatrix} \frac{1}{4} & 0 & \frac{3}{4} \\ \frac{2}{3} & \frac{1}{6} & \frac{1}{6} \\ \frac{1}{2} & \frac{1}{2} & 0 \end{pmatrix}.$$

Find the entropy of S.

9. Given that the initial probability distribution of the three states is $\left(\frac{6}{14}, \frac{3}{14}, \frac{5}{14}\right)$, calculate the approximate number of output sequences of length n emitted by the source.

10. With reference to the source S in Problem 8 find the entropy of S^*, the adjoint source.

14.8 Solutions

1. Put $v_n = H(X_n \mid X_1, X_2, \ldots, X_{n-1})$. $w_n = H(X_1, X_2, \ldots, X_n)$, $n \geq 1$, with $w_0 = 0$. Using the fact that $H(A, B) = H(A) + H(B \mid A)$ with $A = (X_1, X_2, \ldots, X_{n-1})$ and $B = X_n$, we get

$$w_n = w_{n-1} + v_n.$$

This gives that $w_n = v_1 + v_2 + \cdots + v_n$.

Thus, $\lim_{n \to \infty} \frac{w_n}{n} = \lim_{n \to \infty} \left(\frac{v_1 + v_2 + \cdots + v_n}{n}\right)$. This last limit is the limit of the average of a sequence that tends to a definite limit A, (where $A = \lim_{n \to \infty}(H(X_n \mid X_1, X_2, \ldots, X_{n-1}))$ and so, in the limit, is itself equal to A.

2. Part a) is explained in the text in section 14.1.

   Part b) is trickier. S may not be memoryless because the random variables X_n^*, X_{n+1} may not be independent. For X_n, X_{n+1} to be independent it must be the case that, for example, $\Pr(X_{n+1} = 0 \mid X_n = 1) = \Pr(X_{n+1} = 0)$. But, in Example 14.5 of Section 14.1, $\Pr(X_{n+1} = 0) = \frac{1}{2}$ and $\Pr(X_{n+1} = 0 \mid X_n = 1) = 0$.

3. (a) Yes: think of S as a memoryless source emitting possible letters corresponding to two symbols with probabilities equal to $\frac{1}{4}$ or $\frac{3}{4}$.

   (b) $H(S) = H(0.25) = H(0.75) = 0.8113$.

4. (a) No, T is not stationary. To see this we have that $\Pr(X_n = 0) = 1$ if n is odd but $\Pr(X_n = 0) = \frac{1}{4}$ if n is even. But, from Problem 2, we have that if S is stationary then $\Pr(X_1 = 0) = \Pr(X_2 = 0) \ldots = \Pr(X_n = 0)$ for all n.

   (b) The entropy of T is $\frac{1}{2}$(entropy of S) $= \frac{1}{2}(.8113) = 0.4057$.

5. (a) The possible values corresponding to the ordered pair (X_1, X_2) are $(0, 0)$, $(0, 1)$, $(1, 0)$, $(1, 1)$. Then, the corresponding probabilities are $0.64, 0.16, 0.16, 0.04$. Thus

$$H(X_1, X_2) = -[0.64 \log(0.64) + .32 \log(0.16) + (.04) \log(0.04)].$$

(b)

$$H(X_1, X_2) = H(X_1) + H(X_2) \quad \text{(by independence)}$$
$$= 2H(0.8) = 2(0.7219) = 1.4438.$$

6. $2^{nH(0.8)} = 2^{(1000)(0.7219)} = 2^{722}$

7. $2^{900H(0.8)} = 2^{(900)(0.7219)} = 2^{650}$

8. The fixed probability vector is $\left(\frac{6}{14}, \frac{3}{14}, \frac{5}{14}\right)$. Thinking of the first state as a source with entropy H_1 we have

$$H_1 = -\left[\frac{1}{4}\log\left(\frac{1}{4}\right) + \frac{3}{4}\log\left(\frac{3}{4}\right)\right] = 0.8113.$$

Similarly,

$$H_2 = -\left[\frac{2}{3}\log\left(\frac{2}{3}\right) + \frac{1}{6}\log\left(\frac{1}{6}\right) + \frac{1}{6}\log\left(\frac{1}{6}\right)\right] = 1.2516,$$

$$H_3 = -\left[\frac{1}{2}\log\left(\frac{1}{2}\right) + \frac{1}{2}\log\left(\frac{1}{2}\right)\right] = 1.$$

Then the entropy of S is $H = \frac{6}{14}H_1 + \frac{3}{14}H_2 + \frac{5}{14}H_3 = 0.9730$, say.

9. Because the initial distribution is $\mathbf{w}$, S is ergodic (see theorem 14.16 in Section 14.3). Then, by the Shannon McMillan Theorem we have that most sequences emitted are "typical sequences" where the number of typical sequences is $2^{nH} = 2^{0.9730n}$. Here H is the number indicated in Problem 8.

10. The fixed probability distribution is $\left(\frac{6}{14}, \frac{3}{14}, \frac{5}{14}\right)$. Thus $H(S^*)$ is as follows:

$$H(S^*) = -\left[\frac{6}{14}\log\left(\frac{6}{14}\right) + \frac{3}{14}\log\left(\frac{3}{14}\right) + \frac{5}{14}\log\left(\frac{5}{14}\right)\right] = 1.5305..$$

Chapter 15

Perfect Secrecy: the New Paradigm

Goals, Discussion This chapter links encryption and information theory. We want to study the idea of "perfect secrecy" in cryptography from both abstract and concrete points of view. It is frequently (if incorrectly) asserted that "Shannon proved that the only mechanism for perfect secrecy is the one-time pad." This statement is incorrect on three accounts. First of all, Shannon never asserted such a result. Secondly, the assertion is false. Thirdly, our main result is clearly implicit in Shannon's paper [Sha49]. We will see that for perfect secrecy, our message set can have any size, not just a power of 2, as is the case for the one-time pad.

New, Noteworthy We clarify the idea that for perfect security the key must be as long as the message. We prove that the fundamental idea for perfect security is **not** the one-time pad but rather a latin square of size $n \times n$, for **any** positive integer n. However, we do show (see problems 13 and 14) how the one-time pad and the symmetric algorithm discussed in Chapter 3 can be fitted in to the latin square framework. We also present problems illustrating Shannon's idea that cryptography is a bit like error-correction where an eavesdropper tries to decode the message over a noisy channel.

15.1 Symmetric Key Cryptosystems

We begin with the idea of a *symmetric key cryptosystem*. Here we have a cipher system involving a finite set $\mathbf{M} = \{m_1, m_2, \ldots\}$ of **possible messages**, together with a finite set of **cipher texts** $\mathbf{C} = \{c_1, c_2, \ldots\}$ and a finite number of **keyed enciphering transformations** e_k. The key k in the enciphering transformation is chosen, with various non-zero probabilities, from a finite set $\mathbf{K}$ of keys. It is assumed that each message m has a non-zero probability of transmission. Otherwise we could delete it. Similarly, for any cipher text c in C we assume that there is at least one message that, with non-zero probability, gets enciphered to c.

As usual, an enciphering transformation e_i, associated with key number i, has a deciphering transformation d_i, corresponding to this same key i, such that d_i undoes e_i. Thus, for any message m we have a cipher text $e_i(m)$. Applying d_i, we recover m since $d_i(e_i(m)) = m$.

Let m_1, m_2 be messages. Suppose $e_k(m_1) = e_k(m_2)$. Applying d_k we get $d_k(e_k(m_1)) = d_k(e_k(m_2))$. Therefore, $m_1 = m_2$. We conclude that a given e_k maps different messages (= **plain texts**) to different cipher texts. In other words, e_k is one to one.

Thus, let $\mathbf{M} = \{m_1, \ldots, m_n\}$ be the set of n messages in $\mathbf{M}$. Fix the key k. Then $\{e_k(m_1), \ldots, e_k(m_n)\}$ is a set of n distinct cipher texts in $\mathbf{C}$. Thus, $|\mathbf{C}|$, which is the number of cipher texts in $\mathbf{C}$, is at least equal to the number of messages in $\mathbf{M}$. In symbols

$$|\mathbf{C}| \geq |\mathbf{M}|. \tag{15.1}$$

Let us now suppose that our cryptosystem enjoys **perfect secrecy**. By definition then, for any message-cipher text pair m, c we have that $\Pr(m|c) = \Pr(m)$. This means that the conditional probability that the particular message m was transmitted, given that cipher text c is observed, is in fact equal to the probability that m was transmitted. In other words, we have *independence* of the plain text and cipher text; the cipher text reveals nothing about the plain text.

Let us explore some consequences of this. Let (m, c) be any plain text-cipher text pair. So $\Pr(m|c) = \Pr(m)$. Since $\Pr(m) \neq 0$ we have $\Pr(m|c) \neq 0$. Thus, there is a non-zero probability that c was "caused" by the message m. Thus, given any c in $\mathbf{C}$ and any message m in $\mathbf{M}$ there is at least one key k with $\Pr(k) \neq 0$ and an enciphering transformation e_k with $e_k(m) = c$.

Fix m and vary the key k in $\mathbf{K}$. It follows that the set $\{e_k(m)|k \in \mathbf{K}\}$ contains all cipher texts $\mathbf{C}$ and so is equal to $\mathbf{C}$ since $e_k(m)$ is in $\mathbf{C}$. (We must allow for the possibility that some members of the set $\{e_k(m)\}$ might be equal). We then conclude that

$$|\mathbf{K}| \geq |\mathbf{C}|. \tag{15.2}$$

Combining with formula (15.1) we get

$$|\mathbf{K}| \geq |\mathbf{M}|. \tag{15.3}$$

This says that for perfect secrecy, the total number of (enciphering) keys is at least as big as the possible number of messages.

As an example, let $\mathbf{M}$ denote the set of binary strings of length u, say, chosen at random, and let $\mathbf{K}$ be the set of all keys chosen at random from the set of binary strings of length v. Then, $|\mathbf{M}| = 2^u$, $|\mathbf{K}| = 2^v$. From (15.3), for perfect secrecy, $|\mathbf{K}| \geq |\mathbf{M}|$ so $2^v \geq 2^u$ i.e. $v \geq u$. This example helps to explain the phrase that, for perfect security, the "key is at least as long as the message". We also refer the reader to the discussion of the Vigenère cipher in Chapter 2 for another affirmation of this.

15.2 Perfect Secrecy and Equiprobable Keys

The case of equality in formula (15.3), when $|\mathbf{K}| = |\mathbf{M}|$ gives rise to an elegant mathematical theory, some of which we now want to explore

Since $|\mathbf{C}| \geq |\mathbf{M}|$ from formula (15.1) and $|\mathbf{K}| \geq |\mathbf{C}|$ from formula (15.2) it follows that the assumption $|\mathbf{K}| = |\mathbf{M}|$ leads to the fact that $|\mathbf{K}| = |\mathbf{M}| = |\mathbf{C}| = n$, say. From the above, for each ordered pair (m, c) there is then a unique key k and an enciphering transformation e_k such that $e_k(m) = c$. Choose the pair $(m, c) = (m_1, c)$ with $e_{k_1}(m_1) = c$ in such a way that $\Pr(k_1) \geq \Pr(k)$ for all keys k in $\mathbf{K}$. (Recall that the keys have various probabilities associated with them.) We have the following kind of diagram.

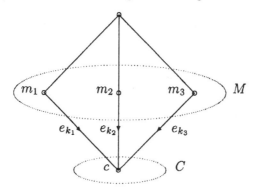

Denote the unique enciphering transformation that maps the message m_i to c by e_{k_i}, $1 \leq i \leq n$. Thus $e_{k_i}(m_i) = c$. Moreover, $\Pr(k_1) \geq \Pr(k_i)$, $1 \leq i \leq n$. From the Bayes formula (see Chapter 10) and the fact that $\Pr(c|m_i) = \Pr(k_i)$, we get

$$\Pr(m_1|c) = \frac{\Pr(m_1)\Pr(k_1)}{\Pr(m_1)\Pr(k_1) + \cdots + \Pr(m_n)\Pr(k_n)}. \tag{15.4}$$

Now, from perfect secrecy,

$$\Pr(m_1|c) = \Pr(m_1). \tag{15.5}$$

Combining formulae (15.4) and (15.5) and dividing by $\Pr(m_1) \neq 0$, we get that

$$\Pr(k_1) = \Pr(k_1)\Pr(m_1) + \cdots + \Pr(k_n)\Pr(m_n).$$

Since $\Pr(m_1) + \cdots + \Pr(m_n) = 1$, we get

$$\Pr(k_1)\Big[\Pr(m_1) + \cdots + \Pr(m_n)\Big] = \Pr(k_1)\Pr(m_1) + \cdots + \Pr(k_n)\Pr(m_n).$$

This gives that

$$\Pr(m_2)\Big(\Pr(k_1) - \Pr(k_2)\Big) + \cdots + \Pr(m_n)\Big(\Pr(k_1) - \Pr(k_n)\Big) = 0.$$

Since $\Pr(k_1) \geq \Pr(k_i)$, $1 \leq i \leq n$, we know that the above is a sum of nonnegative terms summing to zero. It follows that each term in the above sum is zero, and since and $\Pr(m_i) \neq 0$, $2 \leq i \leq n$, we get that

$$\Pr(k_1) = \cdots = \Pr(k_n) = \frac{1}{n}.$$

We have shown the following result.

Theorem 15.1 (The Equiprobable Keys Result) *For perfect secrecy, each key has an equal probability of being chosen, namely $\frac{1}{n}$, assuming $n = |\mathbf{M}| = |\mathbf{C}| = |\mathbf{K}|$.*

Conversely, if $\Pr(k_i) = \frac{1}{n}$, $1 \leq i \leq n$, we see by applying the formula analogous to formula (15.4) to any plain text-cipher text pair (m, c) that $\Pr(m|c) = \Pr(m)$, using the fact that $\sum_{i=1}^{n} \Pr(m_i) = 1$.

To summarize, we have perfect secrecy with $|\mathbf{M}| = |\mathbf{C}| = |\mathbf{K}|$ if and only if $\Pr(k_i) = \frac{1}{n}$, $1 \leq i \leq n$.

15.3   Perfect Secrecy and Latin Squares

Assume we have a cryptosystem with perfect secrecy with $|\mathbf{K}| = |\mathbf{M}| = |\mathbf{C}| = n$, say. Let's denote our message set $\mathbf{M}$ by the set $\{1, \ldots, n\}$ with the messages written in that order. We take $\mathbf{C} = \{1, \ldots, n\}$: our set of enciphering transformation is $\{e_1, \ldots, e_n\}$. We saw earlier that, for fixed i, the set $\{e_i(1), \ldots, e_i(n)\}$ consists of distinct cipher texts. Thus, for a fixed i with $1 \leq i \leq n$, the set $\{e_i(1), \ldots, e_i(n)\}$ is a rearrangement of the sequence $\{1, \ldots, n\}$. We examine the matrix $L = (e_i(j))$, $1 \leq i, j \leq n$. We have seen that each row of L, such as the i-th row above, is a re-arrangement of $\{1, \ldots, n\}$. As seen earlier, for fixed m, the set $\{e_k(m) | k \in \mathbf{K}\}$ equals the set $\mathbf{C}$ of all cipher texts. It follows also that each column of L is some permutation of $\{1, \ldots, n\}$. The conclusion is that L is a **latin square** of size $n \times n$. That is, each entry of L is some number t, $1 \leq t \leq n$. No element t is repeated in any row or any column.

Using the notation above we have the following result.

Theorem 15.2 (Characterization of Perfect Secrecy) *Let* Γ *be a symmetric key system with perfect secrecy with* $|C| = |M| = |K| = n$, *so we may take* **M** *and* **C** *as the set* $\{1, \ldots, n\}$. *Each enciphering transformation* e_k *yields a unique row of an* $n \times n$ *latin square* $L = (e_{ij})$, $1 \le i, j \le n$, *i.e. a permutation of* $\{1, \ldots, n\}$ *and the key is the index of that row. Each key is chosen with probability* $\frac{1}{n}$. *If the message is* j *and the enciphering key is* e_i *we have* $e_i(j) = e_{ij}$, $1 \le i, j \le n$.

Conversely, given any $n \times n$ *latin square* L *we may construct a cryptosystem with perfect secrecy as above.*

Now we see the true nature of a symmetric cryptosystem with perfect secrecy having $|K| = |M|$, not as a one-time pad but as an $n \times n$ latin square. The number of latin squares of order n, denoted by $f(n)$, grows exponentially with n. So there is no shortage of such symmetric cryptosystems!

Example. Let $n = 4$, so our message set is $\{1, 2, 3, 4\}$. Choose the latin square given by

$$L = \begin{pmatrix} 2 & 3 & 4 & 1 \\ 3 & 4 & 1 & 2 \\ 1 & 2 & 3 & 4 \\ 4 & 1 & 2 & 3 \end{pmatrix}.$$

Let's suppose that **A** and **B** are in possession of the secret key 2 and that **A** transmits the cipher text 4 to **B**. **B** seeks the message X such that $e_2(X) = 4$. Now the second row of L, which is the row $(3, 4, 1, 2)$, equals $(e_2(1), e_2(2), e_2(3), e_2(4))$. Therefore we have that $e_2(2) = 4$ so $X = 2$, i.e., the message is 2.

So we see from Theorem 15.2 that perfect secrecy corresponds to message sets of size n, for arbitrary n, not just for $n = 2^t$ (which is the case for a one-time pad). The number of latin squares of order n, as mentioned earlier, grows exponentially with n. One easy specific example that works for all n is the cyclic latin square of order n. For $n = 3$ we get

$$\begin{pmatrix} 1 & 2 & 3 \\ 3 & 1 & 2 \\ 2 & 3 & 1 \end{pmatrix},$$

where each row gets shifted one place to the right and around.

The one-time pad can be fitted into this theory. Indeed, if the key is a binary string of length 2^w then, we must have $n = 2^w$ and a suitable latin square of order n can then be constructed to correspond to the one-time pad. For an example we refer to problem 13. The following example, discussed in chapter 3 can also be fitted in. Our message set and cipher text set is $\{0, 1, \ldots, n-1\}$. Each key k is also a number between 0 and $n-1$. The cipher text for the message x, given by the key k, is $e_k(x) = Rem[x+k, n]$. We refer to problem 14. Note that we have changed the notation from $\{1, 2, \ldots, n\}$ to $\{0, 1, \ldots, n\}$.

Remark. In Theorem 15.2 we make no assumption concerning the probability of the messages. Also there do exist cryptosystems having perfect secrecy with $|\mathbf{K}| > |\mathbf{M}|$.

15.4   The Abstract Approach to Perfect Secrecy

Let $X = (x_1, \ldots, x_n)$ denote the random variable corresponding to a plain text message of n bits: we are taking the alphabet to be binary here but this is not really necessary. Let $Y = (y_1, \ldots, y_n)$ denote the corresponding cipher text of n bits. Then, the mutual information $I(X:Y)$ between X and Y as defined in Chapter 11 is given by the formula

$$I(X:Y) = \max_{\Pr(X)} \left(H(X) - H(X|Y) \right).$$

This is the measure of secrecy here. So **perfect secrecy** is defined by the criterion that $I(X:Y) = 0$. In other words, since we always have $H(X) - H(X|Y) \geq 0$, we get

$$H(X) = H(X|Y).$$

Switching to a more familiar notation where we replace X by M and Y by C we get the following *criterion for perfect secrecy*:

$$H(M) = H(M|C). \tag{15.6}$$

This says that the uncertainty of M given C equals the uncertainty of M so that M, C are independent.

Now let K denote the random variable corresponding to the key space. Using the fact that $H(M) \leq H(M, K)$, we can see that $H(M|C) \leq H(M, K|C) = H(K|C) + H(M|K, C)$. Thus,

$$H(M|C) \leq H(K|C) + H(M|K, C) \tag{15.7}$$

Now K and C determine M uniquely: if we know the key and the cipher text we can recover the message with certainty. Therefore

$$H(M|K, C) = 0 \tag{15.8}$$

So from (15.7) and (15.8), we get

$$H(M|C) \leq H(K|C).$$

But $H(K|C) \leq H(K)$. Thus $H(M|C) \leq H(K)$. Using formula (15.6), we get our final result

$$H(M) \leq H(K).$$

Thus, the entropy of the key space is at least as big as the entropy of the message space. Once again we think of the principle that "perfect secrecy occurs when the key is at least as long as the message".

15.5 Cryptography, Information Theory, Shannon

One of the early important ideas of Claude Shannon was that these two subjects are very closely related, due to the following very basic idea. We simply think of an eavesdropper Eve as attempting to recover the message from the cipher text. In other words, Eve can be thought of as trying to recover the message from a "noisy channel" version of the message, namely the cipher text! We refer to the problems for a further discussion of this.

15.6 Unique Message from Ciphertext, Unicity

Given a cipher text C in a symmetric key cryptosystem it seems reasonable to suppose that the longer that C is, the fewer the number of intelligible text messages M there are corresponding to C. Shannon showed that there exists a critical length U called the **unicity point**, such that for cipher texts longer than this length, there is likely to be just one corresponding plain text. If the length of a cipher text C is much shorter than U then there will be many messages which can in principle encrypt to C thereby increasing security.

Shannon showed that U can be calculated as roughly the point where the message entropy plus the key entropy is less than or equal to the cipher text entropy. This is formally shown in Welsh [Wel88].

Let $H(\Gamma)$ denote the entropy per symbol of the language Γ being used. We estimate the cipher text entropy on the basis that all U-sequences of letters are equally likely to occur as a cipher text. Then our cipher text entropy is approximated by $U \log |\Gamma|$ where $|\Gamma|$ is the number of letters in Γ. The unicity point can be approximated on the basis of the following equality, where $H(K)$ denotes the entropy of the key-space.

$$U \cdot H(\Gamma) + H(K) = U \cdot \log |\Gamma|$$

If all keys are equally likely to occur then $H(K) = \log |K|$ and we get

$$U = \frac{\log |K|}{\log |\Gamma| - H(\Gamma)} \tag{15.9}$$

Example 15.3 *Let us work with substitution ciphers over the English alphabet. Then* $|K| =$ 26!. *By Stirling's expansion,* $\log(26!)$ *is approximately 88 and* $\log 26$ *is about 4.7. Take the entropy of English to be about 1.5 Shannon bits per letter. Then*

$$U \approx \frac{88}{4.7 - 1.5} \approx 28$$

Thus, if a cipher text has length 28 or more we expect there to be just one meaningful plain text.

This is in accordance with Shannon's estimate of a unicity point between 20 and 30, as pointed out in Welsh [Wel88].

We can also write equation (15.9) as follows:

$$U = \frac{\log |K|}{\log |\Gamma| \left(1 - \frac{H(\Gamma)}{\log |\Gamma|}\right)}$$

This gives

$$U = \frac{\log |K|}{R \log |\Gamma|}$$

where R is the redundancy.

For security we want U to be big so that it is desirable that R is small. Note that in the hypothetical case when $R = 0$ we get U to be infinite.

15.7   Problems

1. Let L be a latin square of order 6 as follows.

$$L = \begin{pmatrix} 2 & 3 & 4 & 5 & 6 & 1 \\ 3 & 4 & 5 & 6 & 1 & 2 \\ 4 & 5 & 6 & 1 & 2 & 3 \\ 5 & 6 & 1 & 2 & 3 & 4 \\ 6 & 1 & 2 & 3 & 4 & 5 \\ 1 & 2 & 3 & 4 & 5 & 6 \end{pmatrix}.$$

Suppose **A**, **B** are in possession of the secret key 4 and **A** sends the cipher text 3 to **B**. What message does **B** recover?

2. Let

$$L = \begin{pmatrix} 1 & 2 & 3 & 4 & 5 \\ 2 & 3 & 4 & 5 & 1 \\ 3 & 4 & 5 & 1 & 2 \\ 4 & 5 & 1 & 2 & 3 \\ 5 & 1 & 2 & 3 & 4 \end{pmatrix}.$$

Similar to Problem 1, let the secret key be 4 and the cipher text be 1. What is the message (= plain text)?

3. Let the message space in a symmetric cryptosystem Γ be $\{0,1\}$ with $\Pr(0) = x$, $\Pr(1) = 1 - x$ and let the cipher texts be the set $\mathbf{C} = \{0,1\}$. The key-set has two keys k_1 and k_2. The key k_1 i.e. the enciphering transformation e_{k_1} maps 0 to 0 and 1 to 1. The enciphering transformation e_{k_2} switches 0 and 1. We have $\Pr(k_1) = 1 - p$, $\Pr(k_2) = p$. For what values of p does Γ have perfect security?

Problems 4 to 9 use the setup in problem 3.

4. If Γ has perfect secrecy, what is the corresponding latin square?

5. Calculate $H(M), H(C)$.

6. Calculate $H(K)$.

7. Can you think of a channel corresponding to the setup in Problem 3?

8. (a) Find $H(M|C)$.

   (b) For what value of p is $H(M|C) = H(M)$?

   (c) Explain your answer in (b).

9. (a) What is the maximum value of $H(M) - H(M|C)$ as we vary x.

   (b) What does this number represent?

10. A symmetric cryptosystem Γ having perfect secrecy has $|M| = |K| = |C| = n$, say. What is $H(K)$?

11. Let $m = (m_1, m_2)$ be a message in the form of a binary string of length $2n$ obtained by concatenating two different random binary strings, each of length n. Let $k = (k_1, k_1)$ be a key in the form of a random binary string k_1 of n bits concatenated with itself.

    (a) Find $H(m)$.

    (b) Find $H(k)$.

12. In Problem 11, does the corresponding cryptosystem have perfect security?

13. Show how to obtain a latin square corresponding to a one-time pad based on binary messages of length 2.

14. In a symmetric key cryptosystem let $M = C = \{0, 1, \ldots, n-1\}$ and let the possible keys be $0, 1, \ldots, n-1$, chosen at random. Let key k encipher message x to the cipher text $e_k(x) = Rem[x+k, n]$. Show how to obtain a latin square from this system when $n = 5$.

15.8   Solutions

1. 5.

2. 3.

3. $p = \frac{1}{2}$.

4. $\begin{pmatrix} 0 & 1 \\ 1 & 0 \end{pmatrix}$.

5. $H(x), H(x(1-p) + (1-x)p)$.

6. $H(p)$.

7. The binary symmetric channel with parameter p.

8. (a) We start by noting that $H(M|C) = H(M, C) - H(C)$. Now, (M, C) is a random variable with four possible events, for example "the message is zero, and the cipher text is zero" is one such event. The probabilities are $xp, x(1-p), (1-x)p, (1-x)q$, and $q = 1 - p$. Thus,

$$
\begin{aligned}
H(M, C) &= H\Big(xp, x(1-p), (1-x)p, (1-x)q\Big) \\
&= H(x) + H(p)
\end{aligned}
$$

where the last equality follows from algebraic manipulation. $H(C)$ can be calculated by noting that C is zero with probability $xq + (1-x)p$ and 1 with probability $xp + (1-x)q$. It follows that

$$
\begin{aligned}
H(M|C) &= H(M, C) - H(C) \\
&= H(x) + H(p) - H\Big(xp + (1-x)q\Big).
\end{aligned}
$$

(b) $H\Big(xp + (1-x)q\Big) \geq H(p)$ by convexity, as in section 12.5. Thus $H(M|C) = H(M) = H(x)$ if and only $p = q$, i.e., $p = \frac{1}{2}$.

(c) $H(M|C) = H(M)$ corresponds to perfect secrecy which in turn corresponds to $p = \frac{1}{2}$: also see Problem 3.

9. (a) $1 + p\log p + (1 - p)\log(1 - p)$.

(b) the maximum amount of information concerning M that could be gained by an eavesdropper on learning C. It is also the channel capacity of the BSC with parameter p.

10. $\log n$.

11. (a) $2n$.

(b) n.

12. No, since $H(k) < H(m)$.

13. Our message set M is as follows: $M = \{(0,0), (0,1), (1,0), (1,1)\}$. We abbreviate M to $M = \{a, b, c, d\}$, where $a = (0,0), b = (0,1), c = (1,0), d = (1,1)$. The set C of cipher texts is also equal to $\{a, b, c, d\}$. Each key is one of a, b, c, d, chosen at random. Suppose for example that k is $b = (0,1)$. How do we find $e_b(d)$, say? We have, by addition mod 2 (i.e. XORing), using the one-time pad construction that

$$e_b(d) = (0,1) + (1,1) = (1,0) = c.$$

The rows of our latin square L will be the rows $e_a(M), e_b(M), e_c(M)$ and $e_d(M)$. The rows are indexed by the keys in this way. Then,

$$
\begin{aligned}
e_a(M) &= (0,0) + \{(0,0), (0,1), (1,0), (1,1)\} &= \{a, b, c, d\} \\
e_b(M) &= (0,1) + \{(0,0), (0,1), (1,0), (1,1)\} &= \{b, a, d, c\} \\
e_c(M) &= (1,0) + \{(0,0), (0,1), (1,0), (1,1)\} &= \{c, d, a, b\} \\
e_d(M) &= (1,1) + \{(0,0), (0,1), (1,0), (1,1)\} &= \{d, c, b, a\}
\end{aligned}
$$

where $e_i(M)$ means the ordered set $\{e_i(a), e_i(b), e_i(c), e_i(d)\}$. So the latin square (with rows indexed by a, b, c and d) is as follows:

$$
L = \begin{pmatrix}
a & b & c & d \\
b & a & d & c \\
c & d & a & b \\
d & c & b & a
\end{pmatrix}.
$$

In fact, $L = \begin{pmatrix} A & B \\ B & A \end{pmatrix}$ where A, B are latin squares with $A = \begin{pmatrix} a & b \\ b & a \end{pmatrix}$ and $B =$

$\begin{pmatrix} c & d \\ d & c \end{pmatrix}$. This is because of the one-time pad construction.

14. We have that the message set is $M = \{0, 1, 2, 3, 4\}$. We tabulate our results.

key k	enciphering transform	row of L
0	$e_k(x) = x$	$e_0(M) = \{0, 1, 2, 3, 4\}$
1	$e_k(x) = Rem[x + 1, 5]$	$e_1(M) = \{1, 2, 3, 4, 0\}$
2	$e_k(x) = Rem[x + 2, 5]$	$e_2(M) = \{2, 3, 4, 0, 1\}$
3	$e_k(x) = Rem[x + 3, 5]$	$e_3(M) = \{3, 4, 0, 1, 2\}$
4	$e_k(x) = Rem[x + 4, 5]$	$e_4(M) = \{4, 0, 1, 2, 3\}$

and in this case, $e_i(M)$ means the ordered set $\{e_i(0), e_i(1), e_i(2), e_i(3), e_i(4)\}$. We end up with the latin square

$$L = \begin{pmatrix} 0 & 1 & 2 & 3 & 4 \\ 1 & 2 & 3 & 4 & 0 \\ 2 & 3 & 4 & 0 & 1 \\ 3 & 4 & 0 & 1 & 2 \\ 4 & 0 & 1 & 2 & 3 \end{pmatrix}.$$

Chapter 16

Shift Registers (LFSR) and Stream Ciphers

Goals, Discussion Shift registers are at the heart of cryptography, error-correction and information theory. In cryptography they are the main tools for generating long pseudo-random sequences which can be used as keys in symmetric cryptography.

In information theory, the output of shift registers forms a very good testing ground for fundamental questions involving the entropy of a sequence.

In error correction, as we will see, the linear feedback shift registers (LFSRs) are the basic building blocks for the theory of cyclic linear codes. Accoring to Schneier [Sch96] "most practical stream-cipher designs centre around LFSRs", and "stream ciphers based on shift registers have been the workhorse of military cryptography since the beginning of electronics".

This chapter will provide a thorough analysis of linear feedback shift registers.

New, Noteworthy The theory is quite intricate. Given a recurrence relation generated from the equation $x_m = c_0 x_0 + c_1 x_1 \cdots + c_{m-1} x_{m-1}$ it is crucial to assume that $c_o \neq 0$. If this is not the case, all the usual results break down such as the impossibility of moving from a non-zero state to a zero state. We provide results and counter-examples. A very well-known folklore result states that $2m$ consecutive bits determine the output of a linear feedback shift register with m states. We give a formal proof of this. We also describe the celebrated Berlekamp–Massey algorithm at the end of the chapter.

16.1   Vernam Cipher, Psuedo-Random Key

In the Vernam cipher (= one-time pad), communicating parties **A** and **B** must both be in possession of a common secret key in the form of a random binary string. This key must have the same length as the message and hence, may be very long. The Vernam cipher is a *stream cipher* in which the secret message is encrypted bit by bit. For a block cipher, such as DES, the data encryption standard discussed in Chapter 5, the message is first divided up into blocks of a fixed length before encryption.

It can be difficult to arrange for the two parties to have such a long common random key. However, instead of using a long random secret key, **A** and **B** can use a common pseudo-random binary key. A powerful way for achieving this is to use a *linear feedback shift register* (an LFSR). There are also **nonlinear shift registers**. In this chapter, we deal only with the linear case, and we will use "shift register" and "LFSR" interchangeably. Shift registers are used extensively, both commercially and in industry. They are easily implemented in hardware and are very fast, generating several million bits per second.

The main idea is this. Assume that **A** and **B** are already in possession of a common binary secret key of length $n = 2m$. The first m of these $2m$ bits correspond to the initial state vector of a shift register of length m. The other m bits give the recurrence (i.e. they are the coefficients of the binary recurrence relation associated with the LFSR). These $2m$ bits can generate a much longer sequence of N bits, where N can be as large as $2^m - 1$. These N bits serve as a longer pseudo-random secret key K. Then, each bit of a secret message M of length N is XOR'ed with K (i.e. N and K are added, using binary addition) by **A** to yield a cipher text C, which is transmitted in the open to **B**. **B** then XORs C with K to recover M.

For example, if $2m = 20$, then the pseudo-random key K could be as long as $2^{10} - 1 = 1023$ bits. How does a secret random key of just 20 bits get transformed into a secret random key K of over 1000 bits? The answer is that K is not really random. To quantify this precisely, we need to use the ideas of entropy and information theory discussed earlier in Part II. However, if K has length $2^m - 1$, then K has many of the statistical properties that make K appear to be random.

16.2   Construction of Feedback Shift Registers

Let us describe how a typical LFSR of length m works. We start with m binary registers in a row from left to right named $R_{m-1}, R_{m-2}, \ldots, R_1, R_0$. The contents of each register is either a zero or a one. These registers can be considered storage devices such as flip flops.

Denote the entry or value in the first or right-most register by x_0, the value in the second register by $x_1, \ldots$ and the value in the left-most register by x_{m-1}.

An electronic clock controls proceedings.

$$\boxed{x_{m-1}} \rightarrow \boxed{x_{m-2}} \rightarrow \quad \cdots \quad \rightarrow \boxed{x_1} \rightarrow \boxed{x_0} \rightarrow \quad \text{Output}$$
$$\quad R_{m-1} \qquad\quad R_{m-2} \qquad\qquad\qquad\quad R_1 \qquad\quad R_0$$

Initial binary values for $x_0, x_1, \ldots x_{m-1}$ are in place. At the first clock-pulse, the following happens: The entry x_{m-1} is pushed over to the right by one unit to occupy the register R_{m-2}. Simultaneously, x_{m-2} becomes the new entry in R_{m-3} etc. So each entry gets shifted over by one place to the right, apart from the right-most element which is fed into the **Output Sequence**. For example, suppose $m = 5$ and the initial configuration is as follows:

$$\boxed{1} \rightarrow \boxed{0} \rightarrow \boxed{0} \rightarrow \boxed{1} \rightarrow \boxed{1}$$
$$\ R_4 \qquad\quad R_3 \qquad\quad R_2 \qquad\quad R_1 \qquad\quad R_0$$

So $x_4 = 1$, $x_3 = 0$, $x_2 = 0$, $x_1 = 1$, $x_0 = 1$. After the first clock pulse, we now have the following configuration:

$$\boxed{} \rightarrow \boxed{1} \rightarrow \boxed{0} \rightarrow \boxed{0} \rightarrow \boxed{1}$$
$$\ R_4 \qquad\quad R_3 \qquad\quad R_2 \qquad\quad R_1 \qquad\quad R_0$$

The output sequence at the moment just consists of 1. We can symbolically describe what has happened as follows.

$$(x_4, x_3, x_2, x_1, x_0) \rightarrow (-, x_4, x_3, x_2, x_1).$$

How should we fill in the blank? In other words, what should we put in register R_4? Let us denote this element by x_5. The element x_5 will depend on the initial values x_0, x_1, x_2, x_3, x_4 so x_5 is a function of $x_0, x_1, \ldots, x_4$. We can write $x_5 = f = f(x_0, x_1, x_2, x_3, x_4)$. Figure 16.1 gives a graphic representation of this process.

For a linear feedback shift register ($=$ LFSR), x_5 must be a *linear* function of x_0, x_1, x_2, x_3, x_4. Hence,

$$x_5 = c_4 x_4 + c_3 x_3 + c_2 x_2 + c_1 x_1 + c_0 x_0 \tag{16.1}$$

where $c_0, c_1, \ldots, c_4$ are binary variables and the addition and the multiplication are binary (see the previous section). Note that if c_0 were zero, we could do away completely with the register R_0 holding x_0. Therefore, *in the following discussion, we assume that $c_0 = 1$ unless we state otherwise.*

The sequence of elements occupying the registers at any given time ($\{x_4, x_3, x_2, x_1, x_0\}$ for example) is called the **state vector**. Thus, at each clock pulse, a state vector gets changed to another state vector as in Figure 16.1 and some x_j is adjoined to the output sequence.

For example, assume that $x_5 = x_0 + x_2 + x_4$ and that the initial state vector is $\{10011\}$ (see Figure 16.3). Then $x_5 = 1 + 0 + 1 = 0$ in binary. After one pulse, the state vector is

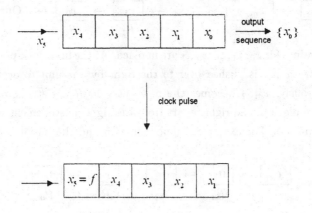

Figure 16.1: One Clock Pulse

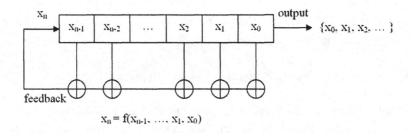

Figure 16.2: A general LFSR

{01001} with output sequence {1} Then, after two clock pulses, the state vector is {10100} with output sequence {11}.

After three clock pulses, the state vector is {01010} with output sequence {110}.

After four pulses, the state vector is {00101} with output sequence {1100}.

The above results were computed from the equations

$$x_5 = x_0 + x_2 + x_4$$
$$x_6 = x_1 + x_3 + x_5$$
$$x_7 = x_2 + x_4 + x_6$$

and so on.

The general situation is as follows. **A (binary) recurrence relation of length m** (corresponding to m registers) is a relationship of the form

$$x_{i+m} = \sum_{j=0}^{m-1} c_j x_{i+j} \tag{16.2}$$

Here $m \geq 0$ is a fixed positive integer equal to the number of registers and i can be any nonnegative integer. All variables x_i are binary and $c_0 = 1$.

For $i = 0$, we get

$$x_m = c_0 x_0 + c_1 x_1 + c_2 x_2 + \cdots + c_{m-1} x_{m-1}$$

Our old formula (16.1) is this result in the case where $m = 5$.

The m given binary numbers $c_0, c_1, \ldots, c_{m-1}$ are called the **recurrence coefficients**.

The recurrence relation in formula (16.2) can be generated by an LFSR (i.e. a shift register) of length m which is easily and economically implemented in hardware. (To avoid trivial cases, we must assume that the initial state vector, namely $(x_{m-1}, x_{m-2}, \ldots, x_1, x_0)$ is not the all-zero state vector. Also, as mentioned earlier, we assume that $c_0 = 1$.) So we have a correspondence between linear recurrence relations and linear shift registers.

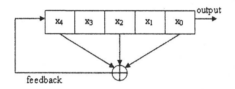

Figure 16.3: The 5-bit LFSR generated by $x_5 = x_0 + x_2 + x_4$

16.3 Periodicity

The **output sequence** will be $\{x_0, x_1, \ldots, x_{m-1}, x_m, \ldots, x_{m+n}\}$, and this makes calculation easy since we start off with $\{x_0, x_1, x_2, \ldots, x_{m-1}\}$ being given. The full mathematical details of an LFSR are complex and intricate. We refer the reader to the books by Golomb [Gol82], and to Rueppel [Rue86]. Beutelspacher [Beu94] and Mollin [Mol00] also give details.

However, we are able to describe many of the main results here. For the most part, we will not distinguish between an LFSR and the corresponding recurrence relation.

Suppose we have a recurrence relation of length m (where the initial state vector is not the all-zero string and the coefficient $c_0 = 1$). At the outset, and after each clock-pulse, the registers hold a binary digit or **bit**, a 0 or a 1. Thus, the m bits in the current state vector give a binary string of length m. The total number of binary strings of length m is 2^m since there are 2 choices for the first bit, 2 for the second, 2 for the third,..., and 2 for the mth bit, giving a total of $(2)(2)(2) \ldots (2) = 2^m$ possible strings. This includes the all-zero string.

So, apart from the all-zero string these are at most $2^m - 1$ distinct state vectors. Thus, after at most $2^m - 1$ clock pulses, some state vector is repeated. Assume that state vector is repeated after exactly t clock pulses. Then all state vectors are repeated after exactly t clock pulses.

To see this, let the state vectors be, successively, $s_0, s_1, s_2, \ldots$ and assume that state vector s_j, is repeated after exactly t clock pulses, so that $s_{j+t} = s_j$. Let the preceding state vector s_{j-1} be given by $s_{j-1} = \{y_{m-1}, y_{m-2}, \ldots, y_2, y_1, y_0\}$. Then we have $s_j = \{y_m, y_{m-1}, y_{m-2}, \ldots, y_2, y_1\}$ and $s_{j+1} = \{y_{m+1}, y_m, y_{m-1}, \ldots, y_3, y_2\}$. From the recurrence relation, we have the following

$$y_m = c_{m-1}y_{m-1} + c_{m-2}y_{m-2} + \cdots + c_1y_1 + c_0y_0. \tag{16.3}$$

$$y_{m+1} = c_{m-1}y_m + c_{m-2}y_{m-1} + \cdots + c_1y_2 + c_0y_1. \tag{16.4}$$

After t pulses, the elements $y_m, y_{m-1}, \ldots, y_2$ and y_1 are repeated, since the state vector s_j is repeated after t clock pulses. Now, from the recurrence (16.3) above we can solve for y_0 in terms of $y_1, y_2, \ldots, y_{m-1}$ and y_m since $c_0 = 1$.

Since $y_m, y_{m-1}, \ldots, y_2$ and y_1 are repeated after t clock pulses, so is y_0. That is,

$$
\begin{aligned}
y_{0+t} &= y_{m+t} + c_{m-1}y_{m-1+t} + c_{m-2}y_{m-2+t} + \cdots + c_1y_{1+t} \text{ by the recurrence and since } c_0 = 1. \\
&= y_m + c_{m-1}y_{m-1} + c_{m-2}y_{m-2} + \cdots + c_1y_1 \\
&= y_0 \text{ again by 16.3 with } c_0 = 1.
\end{aligned}
$$

Thus, s_{j-1} is repeated after t pulses, i.e. $s_{j-1+t} = s_{j-1}$.

Also, from formula (16.4), we see that since $y_m, y_{m-1}, \ldots, y_2$ and y_1 are each repeated after t steps, so is y_{m+1}. Thus, $s_{j+1+t} = s_{j+1}$.

We conclude that *all* state vectors are repeated after exactly t clock pulses.

Result 16.1 *Let Γ_m be a linear shift register of length m with initial state vector*

$$\{x_{m-1}, x_{m-2}, \ldots, x_1, x_0\}$$

and recurrence relation $x_m = \sum_{i=0}^{m-1} c_i x_i$. Provided that $c_0 = 1$, there exists some positive integer t where $1 \leq t < 2^m$ and some state vector s_j such that s_j is repeated after t clock pulses (i.e. such that $s_{j+t} = s_j$). It follows that all state vectors are repeated after t clock pulses; hence, $s_{i+t} = s_i$ for all $i > 0$ and the output sequence $\{x_0, x_1, \ldots, x_{m-1}, x_m, \ldots, x_{m+n}\}$ satisfies

$$x_{i+t} = x_i, \quad i = 1, 2, \ldots$$

Conversely, if there exists an integer t so that $x_{i+t} = x_i$ for every i, then $s_{j+t} = s_j$ for every state vector s_j.

We say that t is a **period** of the recurrence relation. The next result is a logical consequence of the result above:

Result 16.2 *Let t_0 be chosen as small as possible in Result 16.1. Then, if $s_{j+t} = s_j$ for some t then t_0 divides t.*

Proof. Divide t by t_0 to get a remainder u with $0 \le u < t_0$. If u were non-zero, then $s_{j+u} = s_j$, contradicting the minimality of t_0. ∎

The integer t_0 in Result 16.2 is called the **fundamental period**. From the above results, periodicity can be phrased either in terms of state vectors or in terms of the output sequence.

In the proof of Result 16.1, we used the fact that $c_0 \ne 0$. This assumption is needed.

Remark 16.3 *Result 16.1 need not be true if $c_0 = 0$.*

(a) *To construct an example of this, let $m = 3$, let the initial state vector be $s_0 = \{x_2, x_1, x_0\} = \{001\}$ and let the recurrence be generated by $x_3 = x_2 + x_1$.*

Then after one pulse, the state vector is $\{000\} = s_1$.

Then $s_2 = \{000\}$.

It follows that $s_1 = s_{1+1}$, but $s_0 \ne s_{0+1}$ since $s_0 \ne s_1$. So Result 16.1 fails with $t = 1$.

Remark 16.4 *The condition $c_0 \ne 0$ guarantees periodicity, as we have seen in Results 16.1, 16.2. However, even though it is stated in textbooks, the converse does not hold. In other words, we can have periodicity even when $c_0 = 0$. As an example, take $m = 4$ with initial state vector $s_0 = (1011) = (x_3 x_2 x_1 x_0)$ and with the recurrence generated by $x_4 = x_3 + x_1$. Then $s_0, s_1, \ldots, s_6$ all differ and $s_7 = s_0$, from which it follows that $s_{j+7} = s_j$ for all state vectors s_j. Thus, we have periodicity even though $c_0 = 0$.*

A result related to Result 16.1 is the following: (It can be proven by using the same ideas that are used in the proof of Result 16.1.)

Result 16.5 *Given a recurrence relation of length m, any m consecutive terms of the output sequence determines uniquely all past and future elements of the sequence.*

As a counterpoint, we have the following:

Remark 16.6 *The conclusion of Result 16.5 is sometimes false if $c_0 = 0$.*

Example 16.7 *Let $m = 3$ and let the shift register Γ_1 have initial state vector $s_0 = \{x_2 x_1 x_0\} = \{001\}$ and recurrence $x_3 = x_2$. Then $s_1 = s_2 = s_3 = \cdots = \{000\}$.*

Let Γ_2 have the initial state vector $\{011\}$ and recurrence generated also by $x_3 = x_2$, so $s_1 = \{001\}$ and $s_2 = s_3 = s_4 = \cdots = \{000\}$.

The output sequence for Γ_1 is $\{100000\ldots\}$.

The output sequence for Γ_2 *is* $\{110000\ldots\}$.

So we have 2 *different linear shift registers agreeing on* $m = 4$ *consecutive output elements, namely, a sequence of* 4 *consecutive zeros. The assertion concerning future elements in 16.5 is always correct, whether or not* $c_0 = 0$.

16.4   Maximal Periods, Pseudo-Random Sequences

By algebraic means (see Peterson and Weldon [PW72]), the following can be shown.

Result 16.8 *For every positive integer* m *there exists a linear shift register of length* m *with period equal to* $2^m - 1$.

For some theory behind this, see Result 16.14.

When using the LFSR for secure communication between **A** and **B**, each bit of the secret message M (in the form of a binary string that **A** is sending to **B**) is XOR'ed with the key K which is the output of the linear shift register Γ. If $\Gamma = \Gamma_m$ of length m has maximal period equal to $2^m - 1$, then the output sequence of Γ_m is a binary sequence of $2^m - 1$ bits.

Thus, Result 16.8 gives us a way to construct arbitrarily long keys. Although the key K obtained from Γ_m is not random, it does satisfy various statistical tests for randomness if the period of the LFSR Γ_m equals $2^m - 1$.

Before discussing this, it is appropriate to take another look at a linear shift register.

Suppose our initial state vector is $\{x_{m-1}, x_{m-2}, \ldots, x_1, x_0\}$.

After one clock pulse, the new state vector is $\{x_m, x_{m-1}, \ldots, x_2, x_1\}$ where $x_m = c_0 x_0 + c_1 x_1 + c_2 x_2 \ldots + c_{m-1} x_{m-1}$. We can form the $m \times m$ matrix P where

$$
P = \begin{pmatrix}
x_0 & x_1 & \cdots & x_{m-1} \\
x_1 & x_2 & \cdots & x_m \\
\vdots & \vdots & \ddots & \vdots \\
x_{m-1} & x_m & \cdots & x_{2m-1}
\end{pmatrix}
$$

Note that the rows of P give the successive register state vectors *in reverse order* after $0, 1, 2, 3, \ldots, m-1$ clock pulses. Moreover, the first column of P gives the output sequence after m clock pulses, namely $\{x_0, x_1, \ldots, x_{m-2}, x_{m-1}\}$. The j-th column of P gives the last m outputs after $m + j - 1$ clock pulses.

Result 16.9 *Given any finite binary sequence* $(u_0, u_1, \ldots, u_{m-1})$ *there exists a linear shift register whose output sequence contains this sequence.*

Proof. Just take any LFSR Γ, of length m, whose initial state vector is $\{u_{m-1}, \ldots u_1, u_0\}$ Then the output sequence starts with the given sequence. ∎

Frequently, we can find an LFSR of shorter length. For example, if the binary sequence is the output sequence of length $2^n - 1$ obtained from an LFSR of length n and period $2^n - 1$, then n is very small compared to $2^n - 1$. In this case, the length of the shift register is approximately the *log* of the length of the sequence.

In general, the **linear complexity** of a binary sequence is the minimum length of a LFSR generating the sequence as a consecutive subsequence of its output sequence. One algorithm for finding this is the *Berlekamp–Massey algorithm* discussed in Section 16.7.

A **block of length** k is a sequence of the form $\{011 \ldots 110\}$ i.e. it is an all-ones sequence of length k with zeroes at both ends. Similarly a **gap of length** k is a sequence of the form $\{100 \ldots 001\}$ i.e. it is an all-zero sequence of length k with ones at both ends. As an indication of the random nature of the output of an LFSR with maximum period we have the following result.

Result 16.10 *Suppose the linear shift register (= LFSR) of length m has maximum period equal to $2^m - 1$. Then the output sequence of length $2^m - 1$ has the following properties.*

(a) It has exactly $2^{m-1} - 1$ zeros and 2^{m-1} ones.

(b) For any k with $1 \le k \le n - 2$, it contains 2^{n-k-2} blocks of length k and an equal number of gaps of length k.

It suffices to prove this result for the output sequence as defined in Welsh [Wel88] page 130.

16.5 Determining the Output from 2m Bits

Recall that cryptographic stations **A** and **B** use the output of a linear shift register of length m as a common secret key K. It is a remarkable fact that the entire key can be obtained (and the cipher broken) given only $2m$ consecutive bits of K. Hence, if any $2m$ consecutive bits of the cipher text can be correctly decoded to the plain text, the entire cipher text can be decoded.

This is a well known (in the mathematical folklore) result which is fundamental from a theoretical point of view. We have not seen a formal proof of this important result. The proof we give here may be skipped on first reading by the non-expert. The part of the proof involving the subsequent bits of the output sequence is reasonably easy to follow. The part involving the prior bits is more difficult. One of the useful by-products of our proof is a method, used in the problems, for constructing two different linear shift registers of the same length having the same output sequence.

Result 16.11 *Let Γ_m be a linear shift register of length m with initial state vector*

$$\{x_{m-1}, x_{m-2}, \ldots, x_1, x_0\}$$

and recurrence relation $x_m = \sum_{i=0}^{m-1} c_i x_i$. Then the entire output sequence of Γ_m can be determined from any $2m$ consecutive output bits of this sequence, provided that $c_0 = 1$.

Proof. Assume that $2m$ consecutive output bits are known and denoted by $\alpha_0, \alpha_1, \ldots, \alpha_{2m-1}$. Then the recurrence relation for Γ_m can be expressed using the recurrence coefficients by the following matrix equation:

$$\begin{pmatrix} \alpha_0 & \alpha_1 & \cdots & \alpha_{m-1} \\ \alpha_1 & \alpha_2 & \cdots & \alpha_m \\ \vdots & \vdots & \ddots & \vdots \\ \alpha_{m-1} & \alpha_m & \cdots & \alpha_{2m-2} \end{pmatrix} \begin{pmatrix} c_0 \\ c_1 \\ \vdots \\ c_{m-1} \end{pmatrix} = \begin{pmatrix} \alpha_m \\ \alpha_{m+1} \\ \vdots \\ \alpha_{2m-1} \end{pmatrix} \tag{16.5}$$

We will represent this equation by $PC = Z$. If the $m \times m$ matrix P on the left is invertible, then we can solve for the recurrence coefficients by multiplying equation (16.5) on the left by P^{-1}. Once the recurrence coefficients are known, the entire output sequence can be computed.

If P is not invertible, then the equation $PX = Z$ will have more than the one solution $X = C$. Let D be any solution different than C. Then $PD = Z$ and hence, $P(C + D) = PC + PD = Z + Z = \underline{0}$ since the addition is binary. Taking the transpose of this matrix equation, we get $(C + D)^t P^t = \underline{0}^t = \underline{0}$, or

$$UP = \underline{0} \tag{16.6}$$

where $U = (C + D)^t$ is a non-zero row matrix and $P = P^t$ since P is a symmetric matrix.

The vector $\underline{0}$ denotes the all zero row vector of length m.

We wish to show **that the entire output sequence for C equals that of D** given that they agree on some consecutive $2m$ output bits. Now the two output sequences will also agree on the next bit — i.e. bit number $2m + 1$ — if and only if equation (16.5), shifted by one bit, as explained below, is true. Let

$$P_1 = \begin{pmatrix} \alpha_1 & \alpha_2 & \cdots & \alpha_m \\ \alpha_2 & \alpha_3 & \cdots & \alpha_{m-1} \\ \vdots & \vdots & \ddots & \vdots \\ \alpha_m & \alpha_{m+1} & \cdots & \alpha_{2m-1} \end{pmatrix}$$

We want to show that $P_1 C = P_1 D$ where, by assumption, $P_1 C = \begin{pmatrix} \alpha_{m+1} \\ \alpha_{m+2} \\ \vdots \\ \alpha_{2m} \end{pmatrix}$ We have

$PC = Z$. From (16.6), $UP = \underline{0}$. Thus $U(PC) = (UP)C = \underline{0}$. Thus the matrix product of U with each column of P and also with the right column Z is zero. Each column of P_1, apart from the last, is a column of P. Also Z is the same as the last column of P_1. Hence, $UP_1 = \underline{0}$ i.e. $(C+D)^t P_1 = \underline{0}$. Transposing, we get $P_1^t(C+D) = \underline{0}$ i.e. $P_1(C+D) = \underline{0}$, since P_1 is symmetric. This gives $P_1 C = P_1 D$, as required.

So we have shown that if the two sequences agree on any $2m$ consecutive bits, then they agree on all subsequent bits.

What about preceding bits? Here we assume that $P_1 C = P_1 D = \begin{pmatrix} \alpha_{m+1} \\ \alpha_{m+2} \\ \vdots \\ \alpha_{2m} \end{pmatrix}$ and try

to show that $PC = PD$ where $PC = \begin{pmatrix} \alpha_m \\ \alpha_{m+1} \\ \vdots \\ \alpha_{2m-1} \end{pmatrix}$. As before we have $UP_1 = \underline{0}$. Thus,

$UP = \begin{pmatrix} \lambda & 0 & 0 & \cdots & 0 \end{pmatrix}$, since the matrix product of U with all columns of P, save possibly the first, is zero. Also, since the matrix product of U with the last column of P_1 is zero, and this column is PC, we have $U(PC) = \underline{0}$. Now by associativity, $\underline{0} = U(PC) =$

$(UP)C = \begin{pmatrix} \lambda & 0 & 0 & \cdots & 0 \end{pmatrix} \begin{pmatrix} c_0 \\ c_1 \\ \vdots \\ c_{m-1} \end{pmatrix}$

Thus, $\lambda c_0 = 0$. By our fundamental assumption, $c_0 \neq 0$ i.e. $c_0 = 1$. Hence, $\lambda = 0$ and $UP = 0$, giving $(C+D)^t P = 0$. Transposing gives $P^t(C+D) = 0$, i.e. $P(C+D) = 0$ so $PC = PD$. So C and D agree on all preceding bits and the proof is now complete. ∎

We should also mention that the possibility envisaged in the proof of Result 16.11 can occur.

Result 16.12 *There exist two different shift registers of length m, both having the same initial state vector and the same output sequence, and both having a one in the recurrence relation for the coefficient of x_0.*

For a proof we refer to the problems.

Result 16.13 *Result 16.11 need not hold if we only assume that the two sequences have $2m$ (not necessarily consecutive) outputs in common.*

Result 16.14 *Suppose that, in the notation of Result 16.11, $\det P = 0$. Then the sequence $x_0, x_1, \ldots x_{2m-2}$ satisfies a linear recurrence with length less than m. The converse also holds.*

   Proof. This is shown by Trappe and Washington [TW01]. ∎

Remark 16.15 *The relationship between the coefficient of the right-most element in the two sequences in 16.14 is not apparent.*

16.6   The Tap Polynomial and the Period

Next, let the linear shift register Γ_m of length m replace each state vector $(x_{m-1}, x_{m-2}, \ldots, x_1, x_0)$ by the next state vector $(x_m, x_{m-1}, \ldots x_1)$, with $x_m = c_0 x_0 + c_1 x_1 \ldots + c_{m-1} x_{m-1}$. Associate with Γ_m the (binary) **tap polynomial** $h(x) = x^m + c_{m-1} x^{m-1} + c_{m-2} x^{m-2} + \ldots + c_1 x + c_0$, where we assume $c_0 \neq 0$. We say that $h(x)$ is **primitive** if

1. $h(x)$ has no proper non-trivial factors and

2. $h(x)$ does not divide $x^d + 1$ for any d less than $2^m - 1$.

Result 16.16 *Γ_m has period $2^m - 1$ if and only if its tap polynomial $h(x)$ is primitive.*

Example 16.17 *Suppose that the tap polynomial for LFSR Γ is $h(x) = x^8 + x^6 + x^5 + x^3 + 1$. Then $h(x)$ has no factors i.e. is irreducible and does not divide x^{n-1} for any $n < 255$. (It does divide $x^{255} - 1$.) Therefore, the length of the output sequence of Γ is 255 and Γ is of maximum period.*

   The following result is from Peterson and Weldon [PW72]

Result 16.18 *Let Γ be an LFSR with tap polynomial $h(x)$. Suppose that n is the smallest integer such that the polynomial $h(x)$ divides x^{n-1}. Then for any initial state, the fundamental period divides n. Moreover, for some initial state, the fundamental period is exactly n.*

Result 16.19 *Let the linear shift register Γ_m of length m be constructed from the recurrence relation $x_m = c_0 x_0 + c_1 x_1 + \ldots + c_{m-1} x_{m-1}$ with $c_0 = 1$. Assume the initial state vector is not the zero vector. Then no subsequent state vector is the zero vector.*

Proof. Let the initial state vector be $\mathbf{u} = (x_{m-1}, x_{m-2}, \ldots, x_2, x_1, x_0)$. Then the next state vector is $\mathbf{v} = (c_0 x_0 + \ldots + c_{m-1} x_{m-1}, \ldots x_2, x_1)$. Writing of $\mathbf{u}$, $\mathbf{v}$ as columns, we see that $A\mathbf{u} = \mathbf{v}$ where the $m \times m$ matrix A is given by

$$A = \begin{pmatrix} c_{m-1} & c_{m-2} & \cdots & c_1 & c_0 \\ 1 & 0 & \cdots & 0 & 0 \\ 0 & 1 & \cdots & 0 & 0 \\ \vdots & \vdots & \ddots & \vdots & \vdots \\ 0 & 0 & \cdots & 1 & 0 \end{pmatrix}.$$

Now, by expanding along the rightmost column, we find that the determinant of A is c_0 where, by assumption, $c_0 = 1 \neq 0$. Thus, A is non singular. Each state vector can be represented as a column matrix of the form $A^j \mathbf{u}$, $j = 1, 2, 3, \ldots, t - 1$, with t being the fundamental period. Since $\det(A^j) = (\det A)^j = 1$, A^j is non singular. Also, $\mathbf{u} \neq \mathbf{0}$. Thus, each state vector is non-zero. ∎

Result 16.20 *Result 16.19 need not be true if $c_0 = 0$.*

Proof. We refer to the problems. ∎

Instead of using a *linear* shift register generated by the recurrence

$$x_m = c_0 x_0 + c_1 x_1 + \ldots + c_{m-1} x_{m-1}$$

we can also devise a shift register generated by the output sequence $x_m = f(x_{m-1}, x_{m-2}, \ldots, x_1, x_0)$ where f is a suitable **nonlinear** function. An example would be: $x_k = x_{k-1} \cdot x_{k-2} + x_{k-2} \cdot x_{k-3}$. This can improve security. Another technique is to combine the output sequences of different synchronized shift registers to get a single output. For more discussion we refer the reader to Mollin [Mol00].

16.7 Berlekamp–Massey Algorithm

Given the output string of an LFSR, there are many possible LFSRs that will produce that output. But what is the shortest possible LFSR that will produce any given output? That question was answered by J.L.Massey [Mas69] who showed that Berlekamp's [Ber67] algorithm for Reed–Solomon codes also works for finding the shortest LFSR for a given output sequence. Here is some pseudo-code for the algorithm:

Algorithm 16.21 *Here, we take as input, a binary string $s = s_0 s_1 s_2 \ldots s_{n-1}$. We will output the minimum length λ, and the recurrence polynomial $P(X)$.*

Initialize: $P(X) = 1$, $Q(X) = 1$, $\lambda = 0$, $m = -1$

Figure 16.4: Photo by Klaus Peters. Left to right: Richard Guy, John Conway, Elwyn Berlekamp at the re-launching of "Winning Ways" published by A. K. Peters, 2001

For j = 0 to n :
    d = P(X) AND s //This is a bitwise AND.
    if d ≠ 0 : //Check if the current LFSR still works
        T(X) = P(X) //Save the old LFSR
        P(X) = P(X) XOR Q(X)X^{j-m} //Shift Q to the left by (j-m) and XOR
        if 2λ ≤ j : //If the LFSR is too short
            λ = j + 1 − λ
            m = j
            Q(X) = T(X)

This will output (one of) the shortest LFSR(s) to produce that given output string. Here is a sample output that shows the recurrence relation $(P(X))$ after each iteration of j. We find the shortest LFSR that generates the string {101110011110}.

$$
\begin{array}{ll}
\text{Output} & \text{Recurrence} \\
0 & x_0 = 0 \\
10 & x_2 = x_0 \\
110 & x_2 = x_1 + x_0 \\
1110 & x_2 = x_1 \\
11110 & x_2 = x_1 \\
011110 & x_4 = x_3 + x_0 \\
0011110 & x_4 = x_2 + x_0 \\
10011110 & x_4 = x_2 + x_0 \\
110011110 & x_4 = x_2 + x_0 \\
1110011110 & x_4 = x_2 + x_0 \\
01110011110 & x_7 = x_5 + x_3 + x_2 + x_1 \\
101110011110 & x_7 = x_6 + x_5 + x_4 + x_3 + x_1
\end{array}
$$

So our final recurrence is $x_7 = x_6 + x_5 + x_4 + x_3 + x_1$, with initial configuration equal to the right-most 7 input bits $= 0011110$.

In this case the recurrence is unique because of the theorem below.

It may well be that 2 different shift registers will produce the same output. We present an example below. However, in some important situations this cannot happen.

Theorem 16.22 *Let the linear shift register Γ_m of length m with recurrence polynomial $x_m = c_0 x_0 + c_1 x_1 + \cdots + c_{m-1} x_{m-1}$ be the shortest LFSR generating the sequence $S = \{s_0, s_1, s_2, \ldots, s_n\}$, where we assume that $n \geq 2m - 1$. Then Γ_m is unique.*

Sketch of proof Let Γ^* be another LFSR of length m with recurrence polynomial $x_m = d_0 x_0 + d_1 x_1 + \cdots + d_{m-1} x_{m-1}$ generating S. In particular, Γ_m and Γ^* generate the sequence $T = \{s_0, s_1, s_2, \ldots, s_{2m-1}\}$.

Using the notation of the proof of result 16.11 we have that $UP = 0$ where $U = (C+D)^t$ is a non-zero row vector. Thus U annihilates all columns of P, P_1, P_2, etc, as in the proof of result 16.11. Then U causes a dependancy relation among the rows of $P, P_1, P_2, \ldots$. This in turn will yield an LFSR of length $v < m$ that outputs S. But this contradicts the minimality of Γ_m $\square$.

Example 16.23 *It can be shown that Γ_4 and Γ^* with initial state vector given by $\{x_3 x_2 x_1 x_0\} = \{1011\}$ both generate the same sequence. Here Γ_4 has the recurrence $x_4 = x_3 + x_1 + x_0$ and Γ^* has the recurrence $x_4 = x_2 + x_0$. So $C = \begin{pmatrix} 1 \\ 1 \\ 0 \\ 1 \end{pmatrix}$ and $D = \begin{pmatrix} 1 \\ 0 \\ 1 \\ 0 \end{pmatrix}$. Then $U = (C+D)^t =$*

(0111). Now $UP = 0$ implies that $0(\text{Row } 0) + 1(\text{Row } 1) + 1(\text{Row } 2) + 1(\text{Row } 3) = 0$. In

other words, Row 3 = Row 1 + Row 2. This relationship extends to $P_1, P_2, \ldots$, etc. The conclusion is that the output sequence generated by Γ_4 and Γ^* is also generated by a shorter shift register of length 3 with recurrence $x_3 = x_2 + x_1$ (i.e. a "Fibonacci" recurrence) and initial state vector $\{x_2 x_1 x_0\} = \{011\}$. The common output of all 3 shift registers is then $\{x_0 x_1 x_2 x_3 \ldots\} = \{110110110 \cdots\}$.

Remarks

1. The result of Theorem 16.22 is of interest in connection with the Reed–Solomon decoding procedure in Chapter 22

2. The fact that $UP = 0$ leads to a shorter LFSR is also shown in Trappe and Washington [TW01].

16.8 Problems

1. Let the linear shift register Γ of length 5 have the initial state vector $s_0 = \{x_4, x_3, x_2, x_1, x_0\} = \{10011\}$ and a recurrence generated from the relation $x_5 = x_0 + x_2 + x_4$ (see Figure 16.3) so that

$$x_{n+5} = x_n + x_{n+2} + x_{n+4}, \quad n \geq 0.$$

   (a) What is the period of Γ?

   (b) What is the output sequence of Γ?

2. Let the linear shift register Γ of length 4 have the output sequence $x_0 x_1 x_2 1100 \ldots$ and the recurrence relation $x_4 = x_0 + x_2 + x_3$. What are the values of x_0, x_1 and x_2?

3. Let the linear shift register Γ of length 6 have the recurrence relation $x_6 = x_0 + x_3 + x_4$.

   (a) What is the tap polynomial $h(x)$ for Γ?

   (b) Is $h(x)$ a primitive tap polynomial?

4. Let the linear shift register Γ_m have maximal period equal to $2^m - 1$. Show that the output sequence has $2^{m-1} - 1$ zeros and 2^{m-1} ones.

5. Give an example of a linear shift register Γ with non-zero initial state vector such that a subsequent state vector is the all-zero state vector as are all subsequent state vectors. (We will need $c_0 = 0$.)

6. Show that if a linear shift register Γ_m with a given initial state vector $\mathbf{x} \neq \mathbf{0}$ has maximal period equal to $2^m - 1$, then the linear shift register obtained from Γ_m by changing only the initial input $\mathbf{x}$ to any other input $\mathbf{y} \neq \mathbf{0}$ also has period equal to $2^m - 1$.

7. Give an example of two different linear shift registers with the same initial state vector, the same output sequences but with different tap polynomials.

8. Let Γ_4 be a linear shift register with initial state vector $\{1101\}$ which maps $\{x_3x_2x_1x_0\}$ to $\{c_0x_0 + c_1x_1 + c_2x_2 + c_3x_3, x_3, x_2, x_1\}$ at each pulse, where $c_0 = c_3 = 1$ and $c_2 = c_1 = 0$.

   (a) Find the period of Γ_4.

   (b) Find the output sequence of Γ_4.

   (c) Find the associated tap polynomial of Γ_4.

9. Suppose that the linear shift register Γ of length 4 has period 1. Assuming that the initial state vector is not all zeros, what are the possibilities for Γ?

10. (a) What is the shortest LFSR that will yield an output of 10110110?

    (b) What is the shortest LFSR that will yield an output of 101101100?

    (c) What is the shortest LFSR that will yield an output of 010110110?

16.9 Solutions

1. (a) The period is 15 so there are 15 register state vectors starting with $(x_4x_3x_2x_1x_0) = (10011)$ and ending with (00110).

   (b) The output sequence is the following sequence of length 15

$$\{110010100001110\}.$$

2. $(x_0x_1x_2) = (101)$

3. (a) The tap polynomial is $h(x) = x^6 + x^4 + x^3 + 1$.

   (b) Since $x = 1$ is a root of $h(x)$, $h(x)$ has the proper factor $x + 1$ and hence, is not primitive.

4. Any register state vector is a non-zero binary sequence of length m and so is the binary representation of a positive integer N between 1 and $2^m - 1$. To see this just take the ordered sequence of the state vector contents as the binary representation of the integer: the rightmost digit, which is the current output, tells us whether N is odd or even.

   Each non-zero binary m-tuple shows up just once as the machine travels through the $2^m - 1$ state vector registers. So we need only calculate how many odd and even integers N there are. (Note that each bit of the output sequence shows up just once as the current output, namely just after it is the rightmost entry of a register state vector).

5. Let Γ_2 have initial state vector $\{01\}$ with recurrence generated by $x_2 = x_1$. Then the next state vector, and all subsequent state vectors, are the all-zero state vector.

6. The $2^m - 1$ state vectors $A^0\mathbf{x}, A\mathbf{x}, A^2\mathbf{x}, \ldots, A^{2^m-2}\mathbf{x}$ are distinct and give all possible non-zero vectors of length m. Thus $\mathbf{y} = A^i\mathbf{x}$ for some i, with $0 \leq i \leq 2^m - 2$. Now consider the vectors $\mathbf{y}, A\mathbf{y}, A^2\mathbf{y}, A^{2^m-2}\mathbf{y}$. If some two of these, say $A^u\mathbf{y}$ and $A^v\mathbf{y}$ with $1 \leq u, v \leq 2^{m-2}$, and $u \neq v$, were equal, we would have $A^u\mathbf{y} = A^v\mathbf{y}$, so $A^u A^i\mathbf{x} = A^v A^i\mathbf{x}$. Multiplying on the left by $(A^{-1})^i = A^{-i}$ (recall that $\det A = c_0 = 1 \neq 0$) gives $A^u\mathbf{x} = A^v\mathbf{x}$, a contradiction. Thus $A^u\mathbf{y} \neq A^v\mathbf{y}$ for $1 \leq u, v \leq 2^{m-2}$ with $u \neq v$, and so it follows that we have max period since the $A^u\mathbf{y}$ range over all possible states.

7. Let the initial state vector be $\{x_3 x_2 x_1 x_0\} = \{1011\}$. Let Γ_1 be the LFSR generated by the recurrence $x_4 = x_0 + x_2$ and let Γ_2 be the LFSR generated by the recurrence $x_4 = x_3 + x_1 + x_0$. Then the output sequence for both LFSRs has fundamental period 3 and is $110110110\ldots$. The tap polynomial for Γ_1 is $h_1(x) = x^4 + x^2 + 1$ and the tap polynomial for Γ_2 is $h_2(x) = x^4 + x^3 + x + 1$.

8. (a) The period is 15.

   (b) The output sequence is $\{101100100011110\}$.

   (c) The tap polynomial is $h(x) = x^4 + x^3 + 1$. Note that $x^{15} + 1 = (x^4 + x^3 + 1)(x^{11} + x^{10} + x^9 + x^8 + x^6 + x^4 + x^3 + 1)$ and that $h(x)$ does not divide $x^d - 1$ for any divisor d of 15.

9. The only outputs of period 1 are $\{0000\ldots\}$ and $\{1111\ldots\}$. Since the initial state vector is not the all zero vector, it follows that the output must be $\{1111\ldots\}$. This implies that the initial state vector is $\{1111\}$.

10. (a) $x_2 = x_1 + x_0$, Initial configuration $= 10$.

    (b) $x_3 = x_2 + x_1$, Initial configuration $= 100$.

    (c) $x_7 = x_6 + x_5 + x_0$, Initial configuration $= 0110110$.

    Further, the above shortest LFSR are unique.

Chapter 17

The Genetic Code

Figure 17.1: DNA helix ©Marshfield Clinic

Goals, Discussion Our goals in this chapter are modest. We focus on the biological information channel: the transmission of data from DNA in the form of codons into amino acids, the building blocks of proteins. Previous chapters have presented a general discussion of information theory and its varied applications to electronic data but the fundamental concepts apply equally well to any information source or channel. To explain this we briefly discuss appropriate genetic and biological background.

The mathematical treatment of genetics is an old subject with an honorable tradition going back to Mendel. Indeed it was in this very area that Claude Shannon wrote his doctoral dissertation. Recently the connection between mathematics and biology had become closer as analysis of the ever increasing amount of biological data relies upon fundamental mathematical concepts. This is just part of "the biological revolution, the greatest scientific revolution of our times, perhaps of all times" to quote the mathematician Michael C. Reed

(AMS Notices 51, March 2004).

New, Noteworthy   We briefly discuss entropy and compression for DNA and RNA, and In the calculate the channel capacity for the genetic code to a first approximation, using some simplifying assumptions. We then discuss how these assumption could be relaxed using, for example, a suitable Markov model for the process.

17.1   Biology and Information Theory

The compression and transmission of data is not just applicable to electronic media: nature has 'developed' a system that achieves just that. **Deoxyribonucleic Acid** (DNA) provides the encoding of instructions necessary for cellular survival. It may be fair to say that the genetic code is the most important code of all: however, it is still a code.

17.2   History of Genetics

In the late nineteenth and early twentieth centuries many scientists attempted to understand how characteristics were passed from parents to children. Charles Darwin proposed the theory of natural selection in 1859, but it was unable to answer the question of how the inherited traits were 'passed' from one generation to the next. A break-through occurred with the research of an Austrian friar named Gregor Mendel, who showed in 1865 that characteristics of pea plants were passed from generation to generation in a predictable fashion. Sparked by this result many researchers in the early twentieth century attempted to isolate the substance that was responsible for inheritance. This led to early results identifying DNA as a chemical substance responsible for information storage.

However, researchers could not understand how DNA stored information in terms of structure and code. When James Watson and Francis Crick identified the famous double helix in 1953, building on the work of Rosalind Franklin, biology was revolutionized. One of the remaining difficulties was to determine the encoding of the entire human genome. It concerned the entire genetic encoding that is responsible for storing and transferring information to regulate cellular life. The US Department of Energy and the National Institutes of Health sponsored an immense public project to solve the problem. Although planned to last 15 years new techniques in sequencing enabled the first draft release of the human to occur in June 2000. Entitled the **Human Genome Project** it continues to refine and correct the information accumulated on the millions of DNA base pairs that comprise the human genome.

17.3 Structure of DNA

Every living cell on earth stores it's hereditary information in DNA. Double stranded DNA is a polymer consisting of chemical monomers called nucleotides. Nucleotides consist of two major parts: a sugar phosphate group and a nitrogen base. In the case of DNA the sugar is called deoxyribose, and RNA ribose. There are four different **nitrogen bases**: cytosine, guanine, adenine, and thymine. DNA forms a single stranded polymer through **covalent bonding** (sharing of electrons between elements) of the sugar phosphate groups of nucleotides. The double helical structure of DNA (and also double stranded RNA) arises from the **hydrogen bonds** (non-covalent bonds that involve hydrogen) that can form between particular pairs of nitrogenous bases. These bonds form because of the chemical structure of the nitrogen bases. In particular adenine forms hydrogen bonds with thymine, and cytosine with guanine. Such a pairing is commonly referred to as a **base pair**. With only the information from one side of the double helix, *we may reconstitute the other side without any additional information*. This is important during replication (Section 17.5).

There are about 3 billion different base pairs in each complete set of human DNA. If we were to use a fixed length encoding with 2 bits to every base pair, this would take up 750 megabytes. That is about same amount of data that a CD can hold, and is held in every single one of our cells. The estimated 30, 000 genes consist of only a very small percentage ($< 2.5\%$) of all the base pairs in the human genome. The genes are located on exactly 23 **chromosomes**, which are structures that wrap the DNA around specialized proteins. On average chromosomes are 130 million base pairs long. However there is considerable variation since some chromosomes, such as the Y chromosome, contain little information. It is interesting to note that of all the information encoded in each cell of the human body, scientists currently do not understand the function of the majority, or even if such a function exists.

17.4 DNA as an Information Channel

DNA acts as the storage mechanism for hereditary information. However, it is not directly used in catalyzing cellular processes. Instead the information in DNA is transferred into a different form. Both RNA and proteins act as **enzymes** that catalyze reactions in the cell, and it is to these forms that DNA is transferred in complicated cellular processes. RNA transcribed (copied) from the DNA blue-print is quite similar to DNA, but thymine is replaced by the nitrogen base uracil. RNA is single-stranded and can fold up on itself and act as an enzyme. More often, however, the RNA is translated into protein. Proteins are chemicals that are made up of long strands of **amino acids** linked by peptide bonds. There are 20 different types of biologically important amino acids. The mechanism by which the information is transferred from DNA to RNA and protein will be covered in further detail in Section 17.6.

17.5   The Double Helix, Replication

As you read this book, cells are dying and being created all over your body. When you scrape your knee, you lose skin cells (and muscle cells if you are unlucky), which need to be replaced in order for your body to function. Such somatic cells are replaced through a process of cellular division, known as **mitosis**. During this process an entire copy of an organism's genome is copied so that the resulting pair of cells are (nearly) identical.

DNA replication – the part of mitosis when the DNA is actually duplicated, resulting in two copies – is a highly regulated and carefully timed process that relies on stimulus from factors both internal, and external, to the cell. When replicating the DNA double helix is 'unzipped', breaking the weak hydrogen bonds holding the two strands together. Each strand then acts as a 'template' to allow for copying of the other strand. However, because of the complexity of the process transcription errors are sometimes introduced. For example, some bases might not be copied at all, whereas others are copied incorrectly. These errors, or **mutations**, which can occur through other processes besides incorrect copying, may lead to problems. If a mutation in a gene leads to an incorrect protein, it is possible that this protein may not function properly, if at all.

Because of the potentially extreme repercussions and the fact that replication errors may be somewhat common, there are mechanisms in place to 'proof-read' DNA copying. With 'error-correction' the copying fidelity is about one error in every 10^9 nucleotides copied. *There is also another mechanism is place that helps to reduce the number of potentially harmful errors in genes: namely the redundancy of the genetic code.*

17.6   Protein Synthesis and the Genetic code

Proteins are synthesized using the RNA template that is copied from DNA by the ribosome, a primarily RNA-based enzyme. Proteins are synthesized by exploiting complex cellular signaling indicating that certain functions need to be performed in the body. The particulars of all the different signals are not understood in many cases, and researchers continue to work with more simply organisms such as yeast, and *E. coli* to determine the basics of such processes.

To begin the process of protein synthesis, that portion of DNA to be copied into RNA must first be identified by the cell, and exposed by unwrapping it from the proteins with which it is condensed. Proteins are primarily responsible for this identification by binding to **promoter** regions near the gene of interest, and recruiting the machinery that will copy the DNA into RNA in this region. This machinery is primarily responsible for unzipping the hydrogen bonds of the double helix, so that one side of the double helix can be copied into RNA.

This newly created semi-replica of the original DNA strand is known as mRNA or **messenger RNA**. It should be noted that the mRNA is not a replica of the part of the DNA that is used as a template, but rather of the the opposite half. The mRNA is single stranded for two reasons. The first is that single-stranded RNA can (and in some cases, does) fold upon itself to form enzymes. The second reason for copying only one strand relates to the way in which mRNA is translated into protein. However DNA, and RNA are only copied due to the pair-wise bonding of nucleotides. When proteins are made this pair-wise bonding is again used by the cell. The nucleotides present in the mRNA serve as a template for translation into protein. Essentially each triplet of the mRNA acts as a **codon**. The ribosome is responsible for bonding to the single stranded RNA, and bringing special RNA molecules called **transfer RNAs** to bind to the codons. These tRNAs are special because one part will bind to the mRNA while another will bind to an amino acid. Thus each codon has a matching amino acid. Each sequence of three nucleotides on tRNA is called an **anticodon** and binds to a unique codon on the mRNA.

We now have a fixed length, 3 to 1 encoding of codons to amino acids. Every piece of mRNA starts with the codon 'AUG' which will translate to the amino acid methionine (Met). After the tRNA (which maps AUG to Met) attaches to the mRNA in the correct start position (directed by the ribosome), another tRNA attaches to the next codon. The amino acids on the other end of the tRNAs then form a covalent **peptide** bond, and are released by the ribosome from their contact with the mRNA. The process of tRNA bonding to the mRNA, and the subsequent attachment of an amino acid to the growing peptide sequence, continues until a stop codon (either 'UAG', 'UAA, or 'UGA') is reached. Stop codons have no corresponding amino acid, and translation of the mRNA into protein stops.

It is possible for the incorrect tRNA to bind to the mRNA codon but this is rare. Also, although this might lead to a non-functional protein, the process is repeated so many times for a given gene that such errors do not cause interference in cellular operations, unless the error is systematic.

Table 17.1 shows the genetic code, determining which codons in RNA, correspond to particular amino acids. Because there are four nucleotides, and three nucleotides are used for each codon, there are altogether 64 possible triplets that can represent amino acids. As previously noted three are stop codons, and are these not translated into amino acids. The remaining 61 codons code for 20 amino acids, allowing for a degree of redundancy in the genetic code that helps to minimize the effect of mutation errors. After the process begins with the amino acid methionine (Met) it may well be that Met appears again in the process. Thus, mathematically we will think of the genetic code as a channel from a set of size 61 to a set of size 21.

	U		C		A		G	
U	UUU	Phe	UCU	Ser	UAU	Tyr	UGU	Cys
	UUC	Phe	UCC	Ser	UAC	Tyr	UGC	Cys
	UUA	Leu	UCA	Ser	UAA	Stop	UGA	Stop
	UUG	Leu	UCG	Ser	UAG	Stop	UGG	Trp
C	CUU	Leu	CCU	Pro	CAU	His	CGU	Arg
	CUC	Leu	CCC	Pro	CAC	His	CGC	Arg
	CUA	Leu	CCA	Pro	CAA	Gln	CGA	Arg
	CUG	Leu	CCG	Pro	CAG	Gln	CGG	Arg
A	AUU	Ile	ACU	Thr	AAU	Asn	AGU	Ser
	AUC	Ile	ACC	Thr	AAC	Asn	AGC	Ser
	AUA	Ile	ACA	Thr	AAA	Lys	AGA	Arg
	AUG	Met	ACG	Thr	AAG	Lys	AGG	Arg
G	GUU	Val	GCU	Ala	GAU	Asp	GGU	Gly
	GUC	Val	GCC	Ala	GAC	Asp	GGC	Gly
	GUA	Val	GCA	Ala	GAA	Glu	GGA	Gly
	GUG	Val	GCG	Ala	GAG	Glu	GGG	Gly

Table 17.1: The Genetic Code

17.7   Viruses

There is much debate within the scientific community as to whether or not a virus is alive. The question arises because viruses consist only of single or double stranded nucleic acid (DNA or RNA) and a protein coat encapsulating the 'viral genome'. A virus operates by infecting a host cell, and using the machinery from the cell to replicate. Given that the innate ability to reproduce is one of the accepted 'requirements' for something to be living, it is not clear that a virus is living, since without a host cell it has no ability to reproduce.

A virus infects a host cell through recognition of certain proteins that reside on the surface of the cell. For instance the influenza virus will recognize certain proteins that reside on surface of lung cells. When this recognition occurs - it operates rather like a lock and key, where the host cell has the lock, and the virus unfortunately has the correct key to get in - the virus is either completely engulfed by the host cell, or is able to inject it's nucleic acid sequence into the host cell. The viral genome contains certain specialized sequences that allow it to be incorporated into the host genome, by cutting the host DNA and inserting itself. If the genome consists of RNA it might **reverse** copy into what is known as copy DNA (or cDNA), before being inserted into the host genome. It is possible that the viral DNA might lie dormant for a period of time, allowing the cell to grow to a certain site, before beginning reproduction and creating many copies of the viral genome, encased by the protein coat. After a certain point the cell can no longer operate normally and produce more copies of the virus. At this point the cell **lyses** or explodes, releasing all the new viral particles into the host's body, which will infect more host cells.

One more interesting idea that arises from a greater understanding of how viruses operate is the use of viruses in health care. Scientists hope to learn how to create man-made viruses that can target specific cell types, and encapsulate in their genomes some sequences for proteins that an individual might lack. For instance in cystic fibrosis there is a particular gene that is not properly made into protein in the lungs of infected individuals. If one could create a virus that could be targeted to the lungs, and contain a sequence for this missing protein, it would be possible to help individuals who are missing the cystic fibrosis gene. This work is still at a very preliminary stage.

17.8 Entropy and Compression in Genetics

Although the transfer of information from DNA to protein does proceed through the intermediate RNA, we will ignore the intermediate step for our analysis, due to the relative infrequency of errors in copying from DNA to RNA, and translating the RNA into protein (also because when those errors do occur, the final effect on the amount of functional protein is often minimal, unless as noted the errors are systematic for some reason).

We therefore consider possible strings over a four letter letter alphabet with letters A, C, T and G (to represent the nucleotides adenine, cytosine, thymine and guanine respectively). This is a source, and the size of the alphabet means that the upper bound for the entropy of this source is $\log_2 4 = 2$ Shannon bits. The correctness of this bound is dependent upon the simplifying assumption that each of the four letters is equally likely to occur at any point in the string, which in general is not the case for the majority of genomes.

In general it is safe to assume that only one side of the double stranded DNA actually encodes for genes and thus for proteins, although it is possible in some more compacted genomes (such as some bacteria) that both sides of the double helix will encode genes. Because this is not the general case (especially in larger genomes, such as those of mammals) we will ignore it. However there remains redundancy in the message due to statistical patterns among the four alphabet letters. In theory it is possible to use this redundancy for compression. In practice, DNA cannot be encoded to much less than 1.9 bits per symbol. However, in the case of very long sequences that are studied in genetics this can make quite a difference. For further details we refer to a paper titled Biological Sequence Compression Algorithms (http://como.vub.ac.be/Research/CRG/seqcompr-3.pdf). The amount of compression is very dependent on what sequences are being compressed.

This minimal amount of compression that can be performed on DNA sequences contrasts dramatically with the amount of compression that can be obtained for a message in English text. For such text the entropy can be reduced from an initial estimate of $\log 26 \approx 4.7$ Shannon bits to an estimate of less than 1.5 Shannon bits per symbol.

17.9   Channel Capacity of the Genetic Code

Recall that we are mapping the 64 ordered triples of nucleotides to a total of 20 distinct amino-acids, plus the stop instruction. Let X be the random variable corresponding to the codons, and let Y be the random variable representing the amino acids and the stop instruction. Then, X has 64 symbols, while Y has 21 symbols. The channel capacity (see Chapter 12) is defined to be

$$\Lambda = \max_{Pr(X)} I(X:Y) = \max_{Pr(X)} (H(X) - H(X|Y)) \tag{17.1}$$

Let us discuss "channel capacity". The traditional definition of the capacity Λ of a channel is the maximum, taken over all (mathematically possible) input probability distributions X of the mutual information I(X:Y) as above. Here we can proceed in two different ways:

1. Modify this definition to only allow for biologically possible input probability distributions. Indeed, as we do below for a first approximation, we restrict to just one input probability distribution, namely the equiprobable distribution for the codons. Then, no maximization is involved.

2. Stay with the classical definition. Then the number we get below is a *lower bound* on the actual channel capacity since it represents only one particular value of $H(X) - H(X|Y)$.

We start off with a very simple model based on the following two assumptions:

1. All ordered triples of nucleotides (i.e. all codons) are equally likely.

2. There are no channel (transcription or translation) errors.

Then, when calculating the capacity, we only consider one input probability distribution. We have $\Lambda = H(X) - H(X|Y) = H(Y) - H(Y|X)$ since we only allow one input probability distribution, namely the equiprobable one.

Let's try to calculate Λ. Certainly $\Lambda \leq H(X) = \log 64 = 6$ so the capacity is upper-bounded by 6 Shannon bits. We can also calculate $H(X|Y)$ in the usual way, by averaging the quantities $H(X|y)$ over y in Y and obtain Λ in this way.

Fortunately, because of assumption 2, we can get the answer much more quickly by using the fact that

$$H(X) - H(X|Y) = I(X:Y) = I(Y:X) = H(Y) - H(Y|X) \tag{17.2}$$

Now, from assumption 2, $H(Y|X) = 0$. Thus

$$\Lambda = H(Y) = -\sum_{y} Pr(y) \log Pr(y) \tag{17.3}$$

At this stage Λ is upper-bounded by $\log 21 = 4.3920$ Shannon bits, assuming an equiprobable distribution for Y. However, the actual probabilities are easily calculated from assumption 1. For example, $Pr(Phe) = \frac{2}{64}$, since 2 triples map to it. We then get our result.

Theorem 17.1 *Under assumption 1 and 2, the channel capacity of the genetic code is 2.92 Shannon bits.*

Proof.

$$Capacity = H(Y) = H(\frac{2, 6, 3, 1, 4, 6, 4, 4, 4, 2, 2, 2, 2, 2, 2, 2, 2, 3, 1, 6, 4}{64})$$

$$= 2.92$$

∎

This result is not surprising due to the fact that there is much redundancy in the genetic code, mainly arising from the 3rd nucleotide.

Our model is much too simplistic, but out modest goal was to merely illustrate a first pass at understating the channel capacity. To modify assumption 1, we might best proceed on the assumption that the source is ergodic. Following on from this, we might want to assume (as has been done) that we can obtain a useful Markov model using an irreducible Markov chain. We refer to Chapter 14 for discussion of Markov chains.

In his book, Guiasu [Gui77] states that "this capacity seems to be different from the maximum possible quantity $\log 21 = 4.3920$". In practice, the bases A, T, G, C are roughly equiprobable but not independent. Thus, assumption 1 is unrealistic.

Denote the bases A, T, G, C as X_1, X_2, X_3, X_4. In applications it is easier to calculate the transition probabilities $Pr(X_j|X_i)$. We can then form the 4 by 4 Markov transition matrix $M = (p_{ij})$ where $p_{ij} = Pr(X_j|X_i)$. We can also think of p_{ij} as the probability of moving from state X_i to state X_j. Then assuming that the source is stationary we calculate the fixed probability vector $\mathbf{w}$ where $\mathbf{w}P = \mathbf{w}$ and proceed along the lines of Chapter 14

To elaborate on this, we would find the stationary probabilities $Pr(A)$, $Pr(T)$, $Pr(G)$, $Pr(C)$ on solving the following system of equations. (We assume that the transition probabilities have already been experimentally calculated). We have

$$(Pr(A), Pr(T), Pr(G), Pr(C))M = (Pr(A), Pr(T), Pr(G), Pr(C)) \qquad (17.4)$$

Where M is the following transition matrix

$$M = \begin{pmatrix} Pr(A|A) & Pr(T|A) & Pr(G|A) & Pr(C|A) \\ Pr(A|T) & Pr(T|T) & Pr(G|T) & Pr(C|T) \\ Pr(A|G) & Pr(T|G) & Pr(G|G) & Pr(C|G) \\ Pr(A|C) & Pr(T|C) & Pr(G|C) & Pr(C|C) \end{pmatrix} \qquad (17.5)$$

Having found the stationary probability distribution for the variables $(X_1, X_2, X_3, X_4) = (A, T, G, C)$, we can now find the input distribution for the codons to be transmitted over the memoryless channel that constitutes the genetic code. We already know the transition probabilities. Now, by the Markov property we have:

$$Pr(X_i X_j X_k) = Pr(X_i) Pr(X_j | X_i) Pr(X_k | X_j) \qquad (17.6)$$

We also have an expression for the transition probabilities of the codons as follows:

$$Pr(X_a X_b X_c | X_i X_j X_k) = Pr(X_a | X_i X_j X_k) Pr(X_b | X_a) Pr(X_c | X_b) \qquad (17.7)$$

Summing up then we have a source, namely a source of codons. This is a stationary Markov source for which we have calculated the stationary probabilities and the transition probabilities. We can now use this for our calculation of the channel capacity of the genetic code.

Of course we could generalize to n-step Markov processes for better accuracy.

The question of deriving a more realistic assumption 2 to allow for channel transcription errors and also to incorporate mutation errors has led to an enormous volume of biological research which lies beyond the scope of this book.

The long term goal would be to discover a general method for regulating biological information in an error-free way, and reducing genetic disorders by using various channel capacities by analogy with the classical results of Shannon. But success seems a long way off at this stage.

Part III

Mainly Error-Correction

Chapter 18

Error-Correction, Hadamard, Block Designs and p-rank

18.1 General Ideas of Error Correction

Error-correcting and error-detecting codes have their origins in the pioneering work of Hamming and Golay around 1950. The general theory is closely connected to topics in combinatorics and statistics such as block designs, which are also discussed here. On the other hand, we are able to introduce linear codes by the "back door" in order to get an improvement on the (combinatorial) Gilbert-Varshamov bound.

Nowadays the theory is applied in all situations involving communication channels. The channel might involve a telephone conversation, an encrypted electronic message, an internet transaction, a deep space satellite transmission, or a compact disc recording.

The basic principle is to introduce redundancy — "good redundancy" — with a message in order to improve reliability, and has been discussed in Chapter 12. We review some basic ideas and give several examples here. We also develop some classical bounds on the size of codes.

On the practical side, one of the main problems in coding theory is to construct codes getting close to the Shannon capacity bound of Chapter 12.

On the theoretical side, the "main" coding theory problem is discussed in Section 18.9.

The general idea is illustrated in Figure 18.1. As pointed out in Chapter 12, an important special case occurs when

$$(y_1, \cdots, y_n) = (x_1, \cdots, x_k, y_{k+1}, \cdots, y_n) \tag{18.1}$$

where the message is simply lengthened by adjoining various parity checks $y_{k+1}, \cdots, y_n$

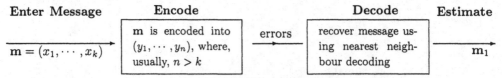

Figure 18.1: Error Correction

which are linear combinations of the message digits $x_1, \cdots, x_k$. This is the case when we use "systematic linear codes" which are covered in Chapter 20. But there do exist other important examples of codes, some of which are discussed here.

Also, Figure 18.1 mainly illustrates the situation for **error correction**. With **error-detection** the receiver can tell if there has been a transmission error (or errors) but cannot correct the errors. Usually, however the receiver can ask for a re-send. This is the case for example **with TCP and CRC protocols** on the internet.

18.2   Error Detection, Error Correction

Let us give some examples. (See Hill [Hil86]).

The International Standard Book Number (ISBN).   This is a 10-digit code word assigned by the publisher. For instance, an ISBN might look like this: ISBN 0-471-12832-8. The first digit refers to the language (0 stands for English) the next 3 refer to the publisher, the next 5 refer to a sequence supplied by the publisher and the last digit is a check digit. In reality, the position of the hyphens is unimportant (there is some variation in conventions on this matter).

The **check digit** x_{10} is chosen so that the **weighted check sum** of the 10 digits gives a remainder of 0 upon division by 11. In symbols, we have

$$Rem\Big[1x_1 + 2x_2 + \cdots + 9x_9 + 10x_{10}, 11\Big] = 0. \tag{18.2}$$

Thus, for the 9 digit number $x_1, \cdots, x_9$ already chosen, we can obtain the check digit x_{10} from the formula

$$Rem[1x_1 + 2x_2 + \cdots + 9x_9, 11] = x_{10}. \tag{18.3}$$

This is because 10 is the additive inverse of -1 (modulo 11) (see Chapter 19). In detail, from equation (18.3) we have

$$x_1 + 2x_2 + \cdots + 9x_9 = x_{10} + \lambda(11). \tag{18.4}$$

Thus

$$x_1 + 2x_2 + \cdots + 9x_9 + 10x_{10} = 11x_{10} + \lambda(11) = 11(x_{10} + \lambda) \qquad (18.5)$$

for some integer λ, and so is divisible by 11.

We claim that the ISBN code detects any single error. For if x_j gets changed to $x_j + a$ we must have, from equation (18.2) that $Rem[ja, 11] = 0$. Thus 11 divides ja which is impossible since 11 is a prime and j, a are integers lying between 1 and 10 (see Chapter 19).

We have already seen that when we use parity checks in the binary case we can also detect a single error.

Generally speaking, with modern technology, multiple errors are rare: one is fairly certain that at most one error has occurred. Still it is important to explore the general theory.

18.3   A Formula for Correction and Detection

We recall from Chapter 12 that the Hamming distance between two code words in a code C of length n over any alphabet is defined to be the number of positions in which they differ. The **minimum distance** of C is defined to be the minimum distance between any 2 code words in C. We have the following two results.

Theorem 18.1 (Error Detection Result) *A code C can detect up to $d - 1$ errors.*

To see this, let the code word $\mathbf{u}$ be transmitted. Assume the received vector $\mathbf{u}_1$ differs from $\mathbf{u}$ is at most $d - 1$ positions. Since the minimum distance between any two words of C is at least d, it follows that $\mathbf{u}_1$ is *not* a code word so the receiver can detect that an error or errors have occurred, upon receipt of $\mathbf{u}_1$.

For a positive integer e, we say that the code is **e-error correcting** or **corrects e errors** if given any string or vector $\mathbf{v}$ of the given length n there is at most one code word $\mathbf{c}$ such that $d(\mathbf{v}, \mathbf{c}) \leq e$, where d denotes the Hamming distance (between $\mathbf{v}$ and $\mathbf{c}$).

We have also the following result.

Theorem 18.2 (Error-Correction Result) *A code C with minimum distance d is e-error correcting if and only if $d \geq 2e + 1$.*

Proof. Suppose that $d \geq 2e + 1$. If a vector $\mathbf{w}$ lies at a distance e or less from two different code words $\mathbf{c}_1$, $\mathbf{c}_2$ then

$$d(\mathbf{c}_1, \mathbf{c}_2) \leq d(\mathbf{c}_1, \mathbf{w}) + d(\mathbf{w}, \mathbf{c}_2) \leq e + e = 2e, \qquad (18.6)$$

which contradicts the assumption that the minimum distance of the code is $d \geq 2e + 1$.

On the other hand, suppose $d \leq 2e$ and let c_1, c_2 be code words at distance d. Put $t = \frac{d}{2}$ if d is even and $t = \frac{d-1}{2}$ if d is odd. Then $t \leq e$ and $d - t \leq e$. We can get from c_1 to c_2 by successively changing coordinates one at a time. Let v be the vector "in the middle". So v is the vector obtained from c_1 after the first t changes. Then $d(c_1, v) = t$ and $d(c_2, v) = d - t \leq e$. Then there are 2 code words c_1, c_2 at distance at most e from v. Therefore C is *not* e-error correcting. ∎

Explanation. *If a code is e-error correcting, it follows that, if at most e transmission errors occur, correct decoding is assured.*

18.4   Hadamard Matrices

Let A be a square matrix of size $n \times n$ in which every entry of A is either 1 or -1. Assume that the dot product of any 2 distinct rows of A is zero. Then A is called a **Hadamard matrix**.

Example 18.3 $A = \begin{pmatrix} 1 & 1 \\ 1 & -1 \end{pmatrix}$ *is a 2×2 Hadamard matrix.*

These matrices are at the heart of many mathematical and scientific questions pertaining to cryptography, error correction, statistics, algebra, and geometry.

Lets start with some geometry.

Theorem 18.4 (Hadamard's Theorem) Let $A = (a_{ij})$ be an $n \times n$ real matrix with $-1 \leq a_{ij} \leq 1$. Then, denoting the determinant of A by $\det(A)$, we get $|\det(A)| \leq n^{\frac{n}{2}}$ where $|\det(A)|$ is the absolute value of $\det(A)$. Equality occurs precisely when A is a Hadamard matrix of size $n \times n$.

Outline of Proof. We have that $|\det(A)|$ is the volume of the box in Euclidean n-space whose sides are the rows of A. Since $-1 \leq a_{ij} \leq 1$, the length of the i-th row is at most

$$(a_{i1}^2 + \cdots + a_{in}^2)^{\frac{1}{2}} \leq \sqrt{n}. \tag{18.7}$$

Then the volume of the box is at most the product of the row lengths with equality exactly when any 2 rows are orthogonal. Then, $|\det(A)| \leq (\sqrt{n})^n = n^{\frac{n}{2}}$. Equality occurs if and only if $a_{ij} = \pm 1$ (so the length of each row is $\sqrt{n}$) and any two rows are orthogonal.

For other geometrical characterizations of Hadamard matrices we refer to [AB76].

What about the existence of Hadamard matrices?

Theorem 18.5 (Existence Criterion) If A is an $n \times n$ Hadamard matrix, then either $n = 1$ or $n = 2$ or 4 divides n.

Outline of Proof. Since the dot product of any 2 distinct rows is zero it follows that the number of positions where they agree equals the number of positions where they disagree. Thus, n is even.

If we change the sign of any column we still have a Hadamard matrix. So we can assume that the first row has all 1s.

Let $n \geq 3$. Let a, b, c, d be the number of columns in which the second and third rows have entries $(+1, +1)$, $(+1, -1)$, $(-1, +1)$, and $(-1, -1)$ respectively. Taking dot products of the 3 rows in pairs, namely, $\mathbf{R_1}.\mathbf{R_2}$, $\mathbf{R_1}.\mathbf{R_3}$, $\mathbf{R_2}.\mathbf{R_3}$:

$$
\begin{aligned}
a + b &= c + d = \tfrac{n}{2}, \\
a + c &= b + d = \tfrac{n}{2}, \\
a + d &= b + c = \tfrac{n}{2}.
\end{aligned}
\tag{18.8}
$$

Solving we get $a = b = c = d = \tfrac{n}{4}$. So 4 divides n.

We now turn to constructions. If $A = (a_{ij})$ and B are Hadamard matrices, not necessarily of the same size, then their **tensor product** in block form, namely,

$$
A \otimes B = \begin{pmatrix}
a_{11}B & a_{12}B & \cdots & a_{1n}B \\
a_{21}B & a_{22}B & \cdots & a_{2n}B \\
\cdots & \cdots & \cdots & \cdots
\end{pmatrix}
\tag{18.9}
$$

is a Hadamard matrix.

Example 18.6 Let $A = H_2 = \begin{pmatrix} 1 & 1 \\ 1 & -1 \end{pmatrix}$, and $B = \begin{pmatrix} 1 & -1 \\ 1 & 1 \end{pmatrix}$. Then

$$
A \otimes B = \begin{pmatrix} B & B \\ B & -B \end{pmatrix} = \begin{pmatrix}
1 & -1 & 1 & -1 \\
1 & 1 & 1 & 1 \\
1 & -1 & -1 & 1 \\
1 & 1 & -1 & -1
\end{pmatrix}.
\tag{18.10}
$$

Thus, we have from this construction that there exists a Hadamard matrix of size $n \times n$ if $n = 2^m$.

But we would like more possibilities. Although Hadamard matrices are firmly grounded in Euclidean geometry (from Hadamard's theorem above) the next construction involves the more exotic world of finite fields, as follows.

Let F be any finite field of odd order. Thus $|F| = q = p^m$ with p any odd prime. For x in F, the **quadratic character** or **Legendre symbol**, denoted by χ, is defined as follows.

$\chi(x) = 0$ if $x = 0$,

$\chi(x) = 1$ if x is a non-zero square,

$\chi(x) = -1$ if x is a non-square.

Example 18.7 *Let $F = GF(5)$ be the field of order 5. The elements of F are 0, 1, 2, 3, 4. The non-zero squares in F are $1^2 = 1$, $2^2 = 4$, $3^2 = 4$, $4^2 = 1$. Thus we have $\chi(0) = 0$, $\chi(1) = \chi(4) = 1$ and $\chi(2) = \chi(3) = -1$, since 2 and 3 are non-squares.*

The **Jacobstahl matrix** $R = (r_{ij})$ is a square matrix of size $q \times q$ which is defined by $r_{ij} = \chi(i - j)$.

Example 18.8 *Let $F = GF(7)$. The elements of F are 0, 1, 2, 3, 4, 5, 6. Index the rows and columns by these elements. Then*

$$R = \begin{pmatrix} 0 & 1 & 1 & -1 & 1 & -1 & -1 \\ -1 & 0 & 1 & 1 & -1 & 1 & -1 \\ -1 & -1 & 0 & 1 & 1 & -1 & 1 \\ 1 & -1 & -1 & 0 & 1 & 1 & -1 \\ -1 & 1 & -1 & -1 & 0 & 1 & 1 \\ 1 & -1 & 1 & -1 & -1 & 0 & 1 \\ 1 & 1 & -1 & 1 & -1 & -1 & 0 \end{pmatrix}. \qquad (18.11)$$

Here we use the fact that the non-zero squares are 1, 2, 4, and the non squares are 3, 5, 6.

Construction of Paley-type Hadamard Matrices. Let q be a prime power with $Rem[q, 4] = 3$ i.e., with $q \bmod 4 = 3$. Let R denote the Jacobstahl matrix of size $q \times q$ and let I be the $q \times q$ identity matrix. Let H be the $(q + 1) \times (q + 1)$ matrix that is obtained by bordering the matrix $R - I$ with a row and column of ones. Then H is a Hadamard matrix which is said to be of **Paley type**.

Example 18.9 *Let $q = 7$. Then*

$$H = \begin{pmatrix} 1 & 1 & \cdots & 1 \\ 1 & & & \\ \vdots & & R - I & \\ 1 & & & \end{pmatrix}, \qquad (18.12)$$

where R is the Jacobstahl matrix obtained previously.

By replacing -1 by 0 in R we get a $(0, 1)$ matrix A which is the *incidence matrix* for a special kind of design called a **Hadamard block design**. This design is "symmetric" since the number of points and blocks is q, as explained below.

18.5 Mariner, Hadamard and Reed–Muller

Imagine a space probe speeding towards a distant planet and transmitting its pictures to far-away Earth at the speed of light. An example to work with is Mariner 69. There, each picture was divided into horizontal scans with each scan line being divided into pixels represented by one of 63 possible gray levels with white corresponding to 0, black to 63 and in-between values of gray being indicated by values between 1 and 62.

Each shade of gray sends a message as a binary string of length 6 with black corresponding to (111111) and white to (000000). However, because of the vast distances, errors will corrupt the transmission so error-correction must be performed.

For this purpose we seek a code C consisting of 64 code-words of reasonably small length such that the Hamming distance between any 2 words is big.

This was achieved for transmission by the Mariner spacecraft in 1969. The construction of C was as follows. We take any Hadamard matrix H of size 32. Next, define B to be the $2n \times n = 64 \times 32$ matrix as follows

$$B = \left(\begin{array}{c} H \\ -H \end{array} \right). \tag{18.13}$$

Define C to be the matrix obtained from B by replacing -1 by 0. We then have the following, using the fact that H is Hadamard.

If $\mathbf{u}$, $\mathbf{v}$ *are code words in* C, *then* $d(\mathbf{u}, \mathbf{v}) \geq 16$.

We can now encode each of the gray-code messages. Each message is encoded into the corresponding row of C. For example, shade of gray number 15 (our messages are 0, 1, $\cdots$, 63) gets encoded as Row 15 of C. Then, from the error correction result above the following holds. If no more than 7 bit errors are made the message is decoded correctly. This represents a significant reliability increase over the unencoded transmission. We refer the reader for discussion of the *coding gain*, to Chapter 13.

In general, we have the following result.

Theorem 18.10 (Hadamard Code Result) *Every Hadamard matrix H of order n gives rise to a binary code of length n with $2n$ code words and minimum distance $d = \frac{n}{2}$.*

18.6 Reed–Muller Codes

These codes were discovered around 1954. First we need the following result.

Theorem 18.11 (Combining Codes) *Let C_1, C_2 be two binary codes of length n with M_1, M_2 code words and minimum distance d_1, d_2 respectively. Then we can define a third* code C_3 of length $2n$, denoted by $C_3 = C_1 \otimes C_2$, with $M_1 M_2$ code words altogether and minimum distance $d_3 = \min(2d_1, d_2)$.

Proof. (See Welsh [Wel88].) Define

$$C_3 = \{(\mathbf{c}_1, \mathbf{c}_1 + \mathbf{c}_2)|\mathbf{c}_1 \text{ in } C_1, \mathbf{c}_2 \text{ in } C_2\}. \tag{18.14}$$

We can now recursively define the binary **m-th order Reed–Muller code** $C(m, n)$ with $0 \le m \le n$ as follows.

$C(m, n)$ has length 2^n.

$C(0, n)$ is $\{(0, \cdots, 0), (1, \cdots, 1)\}$.

$C(n, n)$ is the set of all vectors of length 2^n and

$C(m + 1, n + 1) = C(m + 1, n) \otimes C(m, n)$.

Then, recursively, we have the following codes.

$n = 1$ $C(0, 1) = \{(00), (11)\}$, $C(1, 1) = \{(00), (10), (01), (11)\}$.

$n = 2$ $C(0, 2) = \{(0000), (1111)\}$, $C(1, 2) = C(1, 1) \otimes C(0, 1)$ with code words (0000), $(0011), (1010), (1001), (0101), (0110), (1111), (1100)$.

We then get that $C(m, n)$ is a code of length 2^n, minimum distance 2^{n-m} and having 2^x code words altogether, where

$$x = 1 + \binom{n}{1} + \binom{n}{2} + \cdots + \binom{n}{m} \tag{18.15}$$

∎

The connection with Hadamard codes is as follows.

Reed–Muller and Hadamard. The Reed–Muller code $C(1, n)$ is the binary code obtained from the m-fold tensor product of $H = \begin{pmatrix} 1 & 1 \\ 1 & -1 \end{pmatrix}$ with itself.

18.7   Block Designs

A **block design** $(X, \mathcal{B})$ consists of a finite set X of v points together with a family $\mathcal{B}$ of b subsets called blocks satisfying the following conditions.

(a) Each block contains exactly k points, $k < v$,

(b) Any 2 distinct points are contained in λ blocks, where λ is some constant.

Block designs historically arose in connection with the design of experiments in statistics and they have been much studied because of their connection with combinatorics and coding theory. Several research journals such as *"Designs, Codes and Cryptography"* are devoted to these topics.

A fundamental result in the theory is as follows.

Theorem 18.12 (Fisher's Inequality) *The number of blocks is at least the number of points. In symbols, $b \geq v$.*

Block designs (i.e. "designs") with $b = v$ are called **symmetric designs** and are relatively rare.

Example 18.13 *Let R be the Jacobstahl matrix of Section 18.4. Change each -1 to 0 so we have a $(0,1)$-matrix A of size $q \times q$ with $q \equiv 3(4)$.*

Label the columns of A as $\{0, 1, 2, \cdots, q-1\} = X$. Each row of A yields a subset of X corresponding to those column positions of that row that contain one. Then the set X together with the family of subsets corresponding to the rows of A yields a symmetric design with $b = v = q$ and with $k = \frac{q-1}{2}$, and $\lambda = \frac{1}{4}(q-3)$.

This design is called a **Paley–Hadamard design**. A is called an **incidence matrix** for this design.

In axiom (b) of block designs, if we change 2 to $t \geq 2$ we have a t-**design**. As well as the symmetric designs these t-designs are very rare and have been the subject of much research. The main examples come from finite geometry. If $\lambda = 1$ then we have a **Steiner system** which comes up in connection with the Golay code (Chapter 20). The **inversive plane** over any finite field gives a 3-design which is also a Steiner system. The most-studied symmetric designs are given by the points and hyperplanes of a projective space.

Block designs play a major role in sphere packing (see Thompson [Tho83]) and coding theory. As examples we mention the Golay codes which are intimately related to the projective plane of order 4, the Mathieu t-designs, the Leech lattice and the "Monster" in group theory. The Hamming codes are obtained very easily from the design of points and lines in projective space.

Of all the symmetric designs, those with $\lambda = 1$ are the most-studied because of their mathematical allure and the many open problems surrounding them. They are called **projective planes**. Thus a (finite) projective plane Π is a system of points and lines, with each line having $k = n + 1$ points. The integer n is called the **order of Π**. The total number of points and lines is $n^2 + n + 1$. Any 2 points lie in a unique line; dually, any 2 lines will meet in a unique point. These planes will come up also in subsequent chapters.

Remark   Fisher's inequality has been generalized to finite linear spaces in a well known paper of Erdös and de Bruijn. A very short elegant proof has also been supplied by J.H. Conway. The analogous result for affine planes has been shown independently by P. de Witte and A. A. Bruen. For further references see [Bru73].

18.8   A Problem of Lander, the Bruen–Ott Theorem

Let Π denote a finite projective plane. Arbitrarily label the points as $P_1, P_2, \cdots, P_{n^2+n+1}$: label also the lines as $L_1, L_2, \cdots, L_{n^2+n+1}$. Form an incidence matrix $A = (a_{ij})$ which will be a $(0, 1)$ matrix of size $(n^2 + n + 1) \times (n^2 + n + 1)$ as follows.

$a_{ij} = 1$ if L_i contains P_j;

$a_{ij} = 0$ if L_i does not contain P_j, $1 \leq i, j \leq n^2 + n + 1$.

These incidence matrices have been very useful in error-correction, as follows. We can form a **linear code** C by taking all linear combinations of the rows of A — over some given field F — as code words. There are then several basic questions that can be asked such as

(a) What is the minimum distance of C?

(b) What is the dimension of C?

If the field F is $GF(p)$, for p a prime, the dimension of C is also called the p-rank of C.

There has been a lot of progress with (a). Question (b) was first posed by E. S. Lander in [Lan83]. The main result in the theory is in [BO90] as follows.

Theorem 18.14 (Bruen-Ott Theorem) *Let Π be a projective plane of order n and let p be a prime divisor of n. Then the p-rank of C is at least $n\sqrt{n} + 1$.*

If p^2 does not divide n, then it can be shown that the p-rank of C is equal to $\frac{1}{2}(n^2+n+2)$.

18.9   The Main Coding Theory Problem, Bounds

Let C be a code with **code parameters** (n, M, d). Thus n is the length of each code word, $M = |C|$ and d is the minimum distance. Desirable features of C might include a large value of M, so we can transmit several messages, a large value of d, for reliability, and a small value of n, for fast transmission. These goals are often in conflict with each other. The most common version of the main problem is then the following.

THE MAIN CODING THEORY PROBLEM:

Find the largest code of a given length and given minimum distance (i.e. maximize M) for a given alphabet size q. The maximum value of M is sometimes denoted by $A_q(n, d)$.

In general, this problem is unsolved. This is an understatement. For example, in Hill [Hil86] the author discusses the difficulty of finding $A_2(5, 3)$: it is equal to 4. The existence question for MDS codes discussed below poses an even bigger challenge.

Here is another version of the main problem.

Optimize one of the parameters n, M, d for given values of the other two.

Here we want to obtain some bounds for the construction of codes. Let C be a code of length n. Then each code word is an n-tuple over some alphabet A of size q, say. For example, C might be linear and A might be F, a finite field of order q. We define the **ball** $B_t(\mathbf{c})$ of radius t about $\mathbf{c}$ to be the set of all strings or vectors $\mathbf{x}$ of length n over A such that $d(\mathbf{x}, \mathbf{c}) \leq t$, where d denotes Hamming distance.

Theorem 18.15 (The Ball Theorem)

$$|B_t(\mathbf{c})| = \sum_{i=0}^{t} \binom{n}{i}(q - 1)^i. \tag{18.16}$$

Proof. Suppose we want a string $\mathbf{x}$ with $d(\mathbf{x}, \mathbf{c}) = i$. First we choose the i positions in which $\mathbf{x}$ and $\mathbf{c}$ differ. This can be done in $\binom{n}{i}$ ways. In each of these i positions choose a symbol from A not equal to the symbol that $\mathbf{c}$ has in that position. This can be done in $q - 1$ ways, since $|A| = q$. ∎

Using this we can find *lower bounds* on the size of certain codes.

Theorem 18.16 (The Gilbert–Varshamov Bound) *Given n, q, d, there is a code over an alphabet of size q and minimum distance at least d having at least*

$$q^n \left(\sum_{i=0}^{d-1} \binom{n}{i}(q - 1)^i \right)^{-1} \tag{18.17}$$

code words.

Proof. Start off with any string $\mathbf{c}_1$ of length n over an alphabet A with $|A| = q$. Delete all strings of length n at distance less than d from $\mathbf{c}_1$. Now include in our code any undeleted string $\mathbf{c}_2$ and repeat the process i.e. delete all strings in the ball $B_{d-1}(\mathbf{c}_2)$. Even assuming no overlap in the balls $B_{d-1}(\mathbf{c}_1)$ and $B_{d-1}(\mathbf{c}_2)$ we are still left with

$$q^n - 2 \sum_{i=0}^{d-1} \binom{n}{i}(q - 1)^i \tag{18.18}$$

eligible strings of length n outside the two balls. Proceeding inductively, we get M code words of length n over A at mutual distance at least d, where

$$M \geq \frac{q^n}{\sum_{i=0}^{d-1} \binom{n}{i}(q-1)^i}. \qquad (18.19)$$

∎

Very often it is possible to improve the Gilbert-Varshamov bound by using linear codes. These are discussed in detail in the next chapter. Here, to simplify, we will work in binary although the proof works for any field. We say that a binary code C is **linear** if whenever $\mathbf{x}, \mathbf{y}$ are in C then $\mathbf{x} + \mathbf{y}$ is also in C. For such codes it is easy to see that

$$d(\mathbf{x}, \mathbf{y}) = d(\mathbf{x} + \mathbf{y}, \mathbf{0}) = wt(\mathbf{x} + \mathbf{y}),$$

where $d(-,-)$ denotes Hamming distance and "wt" denotes the **weight** of a vector i.e. the number of non-zero components of that vector.

Theorem 18.17 *There exists a binary linear code satisfying the bound of Theorem 18.16.*

Proof. We proceed inductively constructing linear codes $\{C_0, C_1, C_2, \cdots\}$ having $2^0, 2^1, 2^2, \cdots$ codes words, as follows.

Start off with $C_0 = \{\mathbf{0}\}$. Pick any vector $\mathbf{x}_1 \neq \mathbf{0}$ with $d(\mathbf{x}_1, \mathbf{0}) \geq d$, if possible. Now we have $C_1 = \{\mathbf{0}, \mathbf{x}_1\}$.

Suppose now that

$$|C_1| \left(\sum_{i=0}^{d-1} \binom{n}{i} \right) < q^n$$

where $|C_1|$ is the size of C_1, which is 2. Then there exists a vector $\mathbf{x}_2$ with $d(\mathbf{x}_2, \mathbf{x}_1) \geq d$ and $d(\mathbf{x}_2, \mathbf{0}) \geq d$.

We now construct the smallest linear code C_2 containing C_1 and $\mathbf{x}_2$. This code C_2, being linear, must contain also all sums of the form $\mathbf{x} + \mathbf{c}$, $\mathbf{c}$ in C_1. Thus

$$C_2 = \{\mathbf{0}, \mathbf{x}_1, \mathbf{x}_2, \mathbf{x}_1 + \mathbf{x}_2\}.$$

Now C_2 is a linear code with $|C_2| = 4$. We claim that the distance between any 2 distinct code words in C_2 is at least 4. We need only check the distances between $\mathbf{x}_1 + \mathbf{x}_2$ and the others. But

$$d(\mathbf{x}_1 + \mathbf{x}_2, \mathbf{x}_1) = d(\mathbf{x}_2, \mathbf{x}_1) \geq d.$$

Also,

$$d(\mathbf{x}_1 + \mathbf{x}_2, \mathbf{x}_2) = d(\mathbf{x}_1, \mathbf{x}_2) \geq d.$$

Furthermore,

$$d(\mathbf{x}_1 + \mathbf{x}_2, \mathbf{0}) = d(\mathbf{x}_1, \mathbf{x}_2) \geq d.$$

Now, suppose that

$$|C_2| \sum_{i=0}^{d-1} \binom{n}{i} = 4 \sum_{i=0}^{d-1} \binom{n}{i} < q^n.$$

Then, as in Theorem 18.16, there exists a vector $\mathbf{x}_3$ with $d(\mathbf{x}_3, \mathbf{c}) \geq d$ for any $\mathbf{c}$ in C_2. Now we construct the smallest linear code C_2 containing $\mathbf{x}_3$ and C which will have $2^3 = 8$ code words. Moreover, $d(\mathbf{u}, \mathbf{v}) \geq d$ for any $\mathbf{u} \neq \mathbf{v}$ in C_3.

Eventually, we end up with a binary linear code C_t, where $|C_t| = 2^t$ such that $d(\mathbf{u}, \mathbf{v}) \geq d$ for $\mathbf{u} \neq \mathbf{v}$ in C_t and such that $|C_t| \left(\sum_{i=0}^{d-1} \binom{n}{i} \right) > q^n$. Then C_t is the required linear code, proving the theorem. ∎

Example Let $n = 15, d = 3$. Then $\binom{15}{0} + \binom{15}{1} + \binom{15}{2} = 121$. Now Theorem 18.16 guarantees a code C with $|C| \geq \frac{2^{15}}{121} = 270.8$, i.e., with $|C| \geq 271$ such that any 2 distinct words in C have distance at least 3. However, the argument in Theorem 18.17 guarantees a code C (with the same distance property) with C linear and $|C| \geq 512$. So we get a big improvement on the bound of Theorem 18.16.

We now turn on our attention to *upper bounds*.

Theorem 18.18 (The Hamming Bound) *Suppose that $d \geq 2e + 1$. A q-ary code over A, $|A| = q$, of length n and minimum distance d contains at most*

$$\frac{q^n}{\sum_{i=0}^{e} \binom{n}{i}(q-1)^i} \tag{18.20}$$

code words.

Proof. Since $d \geq 2e + 1$, the balls of radius e about the code words in C must be disjoint as we saw in 9.3. Since the total number of strings of length n over A is q^n we must have

$$|C| \sum_{i=0}^{e} \binom{n}{i}(q-1)^i \leq q^n, \tag{18.21}$$

giving the result. ∎

If equality occurs in the above we have a **perfect code**. This means that every vector lies in one of the balls of radius e about a code word in C.

For $e = 1$ the Hamming codes are perfect. In fact they are the only examples which are linear with $e = 1$.

In general, we have the following result (see Cameron [Cam94]).

Theorem 18.19 (Tietäväinen's Theorem) *Let q be a prime power and let $e > 1$. Then the only perfect e-error-correcting codes of length n are the binary repetition codes (with*

$q = 2, n = 2e + 1$) and the binary and ternary Golay codes (with $q = 2, e = 3, n = 23$ and $q = 3, e = 2, n = 1$ respectively).

Remark The following also holds. Let C be any perfect binary code, linear or not, of length n which is e-error correcting. Then the support of the code words of smallest weight $2e + 1$ in C are the blocks of a t-design with $t = e + 1$, $v = n$, $k = 2e + 1$, $\lambda = 1$. Here, the support of a code word is the set of positions where the code word is non-zero. In this way, we get t-designs associated with the binary Golay codes and Hamming codes

We want now to describe a fundamental upper bound in connection with the main coding theory problem.

Theorem 18.20 (The MDS bound) *Let C be a code of length n over an alphabet of size q with minimum distance d. Then $|C| \leq q^{n-d+1}$*

Proof. We examine any set of $n - d + 1$ coordinate positions, for example, the first $n - d + 1$ positions. Two distinct words of C cannot agree in all these $n - d + 1$ positions, for otherwise they could differ in at most the remaining $d - 1$ positions, contradicting the fact that the minimum distance of C is d. Thus the number of code words is at most the number of $(n - d + 1)$-tuples over an alphabet of size q. This number is at most $\underbrace{(q)(q) \cdots (q)}_{n-d+1}$, i.e.,

$|C| \leq q^{n-d+1}$.   ∎

Theorem 18.21 *In Theorem 18.20 assume that $|C| = q^{n-d+1}$. Then given any set of $n - d + 1$ coordinate positions and any set of $n - d + 1$ symbols from the alphabet there is a unique code word in C having the given symbols in the given positions.*

Proof. This follows from the argument in the proof of Theorem 18.20.   ∎

Theorem 18.22 *Let C be a code of length n over an alphabet of size q satisfying the following condition:*

Given any set of $n - d + 1$ coordinate positions and any set of $n - d + 1$ symbols from the alphabet there is a unique code word having the given symbols in the given position.

Then

(a) $|C| = q^{n-d+1}$.

(b) C has minimum distance d.

Proof. Map each code word $\mathbf{x} = (x_1, x_2, \cdots, x_n)$ in C to, say, its projection on the first $n - d + 1$ positions. So $\mathbf{x}$ gets mapped to $f(\mathbf{x}) = (x_1, x_2, \cdots, x_{n-d+1})$. The given condition implies that f is one to one and onto the set of $(n - d + 1)$-tuples over an alphabet of size q. Thus $|C| = q^{n-d+1}$. This proves (a).

The given condition implies that no 2 distinct code words can agree in as many as $n-d+1$ positions. So the minimum distance of C is $d^* \geq d$. If $d^* > d$, then by Theorem 18.20 we have $|C| \leq q^{n-d^*+1} < q^{n-d+1}$ which contradicts Part (a). ∎

18.10 Problems

1. Let Π be a projective plane of order 2. Let A be any incidence matrix for Π. Regard A as a binary vector space. What is the dimension of A?

2. Show that there exists a binary code C of length 7 containing at least 5 words such that the minimum distance between any 2 distinct words in C is at least 3.

3. Show how the bound in Problem 2 can be improved to give $|\mathcal{C}| \geq 8$.

4. Does there exist a binary code C of length 7 with $|C| = 16$ such that any 2 distinct code words in C are separated by a Hamming distance of 3 or more?

5. Does there exist a binary code C of length 7 with $|C| \geq 17$ such that the distance between any 2 distinct code words in C is at least 3?

18.11 Solutions

1. The rank is $4 = \frac{1}{2}(2^2 + 2 + 2)$. (We can use Theorem 18.14 to get that the rank $\geq 2\sqrt{2} + 1 = 3.8$.)

2. We have $\binom{7}{0} + \binom{7}{1} + \binom{7}{2} = 29$. Then by using the argument in Theorem 18.16 we have a code C with at least $\frac{2^7}{29} = 4.4$ code words, so $|C| \geq 5$ since $|C|$ is an integer.

3. Use the argument in Theorem 18.17 to get a linear code C with $|C| \geq 2^3 = 8$.

4. Yes. We can use the code in Problem 1. This is also a Hamming code as discussed in the next chapter.

5. No, since the Hamming upper bound is violated: we have $d \geq 2(1) + 1$. So pick $e = 1$. If such a code C exists then $(17)\left[\binom{7}{0} + \binom{7}{1}\right] \leq 2^7$ i.e. $144 \leq 128$ which is impossible.

   Note that $16\left[\binom{7}{0} + \binom{7}{1}\right] = 2^7$ and indeed there does exist a code C of length 7 with $|C| = 16$. As in Problem 4 C also has the property that the distance between any 2 distinct code words is at least 3.

Chapter 19

Finite Fields, Linear Algebra, and Number Theory

Goals, Discussion We present a rigorous treatment of modular arithmetic. This leads into linear algebra, which is important for its applicability to coding, especially the class of linear codes.

We also prove several results that have been deferred from earlier chapters. In particular we prove the correctness of the RSA algorithm, and demonstrate that our more efficient choice of d is a valid one. The Euclidean Algorithm is discussed, and we point out how it can be used to find multiplicative inverses modulo N.

The chapter exhibits a construction procedure for finite fields. We also discuss polynomials in terms of this construction procedure, and then mention a few results regarding polynomials that are of use in other contexts. This chapter also has a discussion on complexity issues and a factoring algorithm is described.

19.1 Modular Arithmetic

We use the **integers** Z and the **nonnegative integers** or **natural numbers** N consisting of $\{0, 1, 2, \ldots\}$. An integer divisible only by itself and 1 is called a **prime**. So 19 is a prime number, but $10 = 2 \times 5$ is not a prime.

A useful fact is the following. If two integers a, b are relatively prime, that is, $\gcd(a, b) = 1$, and they both divide another integer c, then their product ab divides c as well. In particular, if a and b are both prime with $a \neq b$, and a divides c, and b divides c, then ab divides c. For example, this holds true with $a = 5$, $b = 3$, and $c = 30$. In the above, $\gcd(a, b)$ denotes the greatest common divisor of a and b. For example, $\gcd(15, 6) = 3$, $\gcd(15, 225) = 15$, and $\gcd(-3, 6) = 3$.

Recall that in Chapter 3, we used the symbol $Rem[u, v]$ to denote the unique remainder that lies between 0 and $v - 1$ when we divide u by v. Thus, $Rem[57, 10] = 7$.

Fix v and think of u as varying, so that we are always dividing by v. Then we can denote $Rem[u, v]$ by $Rem[u]$ or $Remu$. In Chapter 3 we used the following principle, which greatly simplified our calculations.

$$Rem[xy] = Rem[Remx \times Remy] \tag{19.1}$$

This is easily proven as follows. Let $Rem[x] = \alpha$, so $x = \lambda_1 v + \alpha$ with $0 \le \alpha \le v - 1$, and let $Rem[y] = \beta$, so $y = \lambda_2 v + \beta$ with $0 \le \beta \le v - 1$. Then

$$
\begin{aligned}
xy &= (\lambda_1 v + \alpha)(\lambda_2 v + \beta) \\
&= \lambda_1 \lambda_2 v^2 + v(\lambda_1 \beta + \lambda_2 \alpha) + \alpha\beta \\
&= v(\lambda_1 \lambda_2 v + \lambda_1 \beta + \lambda_2 \alpha) + \alpha\beta
\end{aligned}
$$

Thus, when we divide xy by v, the remainder equals the remainder when we divide $\alpha\beta$ by v, establishing formula (19.1).

Before going any further, we need to introduce a few ideas. First, we have the notion of a **group**. A group is a set equipped with an operation, such as addition or multiplication. The operation is **associative**, meaning that the order of adding (or multiplying) is not significant. The operation has a **neutral element** or **unity** which, when added to (or multiplied by) any element of the group, say a, the result is the same element a. Also, each element a has an inverse which, when added to (or multiplied by) the given element, results in the unity element. Groups can be written multiplicatively, or additively, and these properties are listed below.

Multiplicative group	**Additive group**
$(a \cdot b) \cdot c = a \cdot (b \cdot c)$	$(a + b) + c = a + (b + c)$
Unity is 1 so that $a \cdot 1 = a$	Unity is 0 so that $a + 0 = a$
Given a, there exists a^{-1} so that $a \cdot a^{-1} = 1$	Given a, there exists $-a$ so that $a + (-a) = 0$

where a, b, c are group elements. Concrete examples of groups abound. The integers, the rationals, and the real numbers with the usual addition each form an additive group. The rationals and reals, **without zero**, each form a multiplicative group with the usual multiplication.

We will be dealing with mainly **commutative** or **abelian** groups, wherein the order of operation does not matter. That is, written multiplicatively, $a \cdot b = b \cdot a$, or written additively, $a + b = b + a$. Unless otherwise stated, assume we are working with abelian groups.

The number of elements in a finite group is called the **order** of the group. In the case of the real numbers, the order is infinite, but as we shall see, many groups have finite order.

Returning to our main narrative, the set of remainders obtained when we divide any number by v, and consisting of the set $Z_v = \{0, 1, 2, \ldots, v-1\}$ forms a mathematical structure known as a **ring**. This means that we have addition, with neutral element 0, and multiplication with neutral element 1, connected by two distributive laws, namely $x(y+z) = xy + xz$ and $(x+y)z = xz + yz$. In fact, the set under addition is an additive group, but as we shall see, the set under multiplication need not be a multiplicative group. To add or multiply two elements in Z_v, we add or multiply them in the usual way and obtain the remainder upon division by v.

The resulting structure is called the **ring of integers modulo** v, denoted by Z_v. As an example, let $v = 6$. Then $Z_v = Z_6 = \{0, 1, 2, 3, 4, 5\}$. So $3 + 5 = 2$, $(2)(5) = 4$ in this ring.

If $Rem[u, v] = w$, so that $u = \alpha v + w$, $0 \leq w < v - 1$, we also use the notation, called **modular notation** that

$$u \quad (\mathrm{mod}\ v) = w \tag{19.2}$$

or

$$u \equiv w \quad (\mathrm{mod}\ v). \tag{19.3}$$

Equation (19.2) says that w is the remainder when u is divided by v, and equation (19.3) holds if and only if v divides $u - w$. These notations are basically the same, except that $u \equiv w \ (\mathrm{mod}\ v)$ does not imply $u \ (\mathrm{mod}\ v) = w$. For example $17 \equiv 13 \ (\mathrm{mod}\ 4)$, but $17 \ (\mathrm{mod}\ 4) \neq 13$. However, (19.2) implies (19.3), so (19.3) is more general.

With this notation, equation (19.1) can be written as follows

$$xy \quad (\mathrm{mod}\ v) = ((x \quad (\mathrm{mod}\ v))(y \quad (\mathrm{mod}\ v))) \quad (\mathrm{mod}\ v) \tag{19.4}$$

In the ring Z_v, given any element x in Z_v, we can find another element y in $\{0, 1, \ldots, v-2, v-1\}$ so that $(x+y) \ (\mathrm{mod}\ v) = 0$. This element y is unique (with $y = v - x$), and is called the **additive inverse** of x modulo v, and we denote it by $-x$. Thus we have $(x + (-x)) \ (\mathrm{mod}\ v) = 0$. For example, with $v = 6$ and $x = 2$, we get $-x = 4$ so that $(2+4) \ (\mathrm{mod}\ 6) = 0$.

The difficulty in the ring Z_v lies with multiplication and multiplicative inverses. Given x in Z_v we want to find its **multiplicative inverse** y in Z_v. Thus we want to find y in Z_v with $xy \ (\mathrm{mod}\ v) = 1$. It turns out that y can be found if and only if x is **relatively prime** to v, i.e., the **greatest common divisor** of x and v is 1. In symbols, $\gcd(x, v) = 1$, or more simply, $(x, v) = 1$. This means that the only positive integer dividing both x and v is 1.

This important fact concerning inverses will be shown later. It is also true that if x_1 and x_2 are both relatively prime to v, then so is their product $x_1 x_2$. The elements x of

Z_v which have $\gcd(x,v) = 1$ are called the **multiplicative units** or **units** of Z_v. The multiplicative units are "closed under multiplication", i.e., the product of two multiplicative units is also a multiplicative unit. In other words, the multiplicative units form a group (under multiplication). The number of elements in this group is denoted by $\varphi(v)$; φ is called the **Euler phi-function**, so the group of multiplicative units has order $\varphi(v)$.

Let us give some examples. Take $v = 10$. Then $Z_v = Z_{10} = \{0,1,\ldots,8,9\}$. The multiplicative units are 1, 3, 7 and 9. In Z_{10}, the multiplicative inverses of 1, 3, 7, 9 are 1, 7, 3, 9 respectively. This can be seen from the facts that (3×7) (mod 10) $= 1$, and (9×9) (mod 10) $= 1$. Next, let $v = 7$, so $Z_v = Z_7 = \{0,1,\ldots,6\}$. Then **all nonzero elements are multiplicative units** since each of them is relatively prime to 7. This holds since 7 is a prime so, $\varphi(7) = 7 - 1$. This can be generalized as follows:

$$\varphi(p) = p - 1 \tag{19.5}$$

when p is prime. **Thus the group of units in Z_p has order $p - 1$.**

In general it is easy to calculate $\varphi(v)$ using equation (19.5), generalized to prime powers, and using the fact that φ is **multiplicative**. We express these two observations as follows:

$$\varphi(p^t) = p^t - p^{t-1} = p^{t-1}(p - 1) \tag{19.6}$$

where p is prime and t is an integer. Also:

$$\varphi(ab) = \varphi(a)\varphi(b) \tag{19.7}$$

when a and b are relatively prime. The fact that $\varphi(v)$ can be calculated easily from equations (19.6) and (19.7) follows from the **Fundamental Theorem of Arithmetic**, to the effect that any positive integer can be factored uniquely as the product of prime powers.[1]

For example, suppose $v = 36$. We have $\varphi(36) = \varphi(4)\varphi(9) = \varphi(2^2)\varphi(3^2) = (2^2 - 2^1)(3^2 - 3^1) = 12$. So, in Z_{36} there are 12 multiplicative units. They are: 1, 5, 7, 11, 13, 17, 19, 23, 25, 29, 31, 35.

---

[1] What we mean here by uniquely is that it does not depend upon the order of the factors. For example, $2^2 \times 3 \times 5$ and $2 \times 5 \times 3 \times 2$ are the same factorizations of 60.

19.2 A Little Linear Algebra

The **rank** of a matrix is an important concept; it is the maximum number of linearly independent rows of a matrix. For example, the matrix

$$I = \begin{pmatrix} 1 & 0 & 0 \\ 0 & 1 & 0 \\ 0 & 0 & 1 \end{pmatrix}$$

has rank 3, since the rows are linearly independent. However, the matrix

$$A = \begin{pmatrix} 1 & 2 & 2 \\ 1 & 1 & 0 \\ 2 & 3 & 2 \end{pmatrix}$$

has rank 2 since $(1, 2, 2) + (1, 1, 0) = (2, 3, 2)$ but $(1, 2, 2)$ and $(1, 1, 0)$ are linearly independent. We write $\text{rank}(A) = 2$. The rank of a matrix can be found through the process of row reduction.

A useful fact about the rank of a matrix A is that the rank is equal to the rank of its transpose. In symbols, $\text{rank}(A) = \text{rank}(A^t)$. Since the rows of A^t are the columns of A, we could have said that the rank of a matrix is the maximum number of linearly independent columns; **the maximum number of linearly independent columns equals the maximum number of linearly independent rows.** It follows that the rank of a matrix must be less than or equal to the number of rows, and less than or equal to the number of columns. That is, if A is an $n \times m$ matrix, then $\text{rank}(A) \leq n$ and $\text{rank}(A) \leq m$.

Now, the important fact for this section is simply this: A has full rank exactly when it is invertible. That is, when A is an $n \times n$ matrix, $\text{rank}(A) = n$ if and only if there is another matrix A^{-1} such that $AA^{-1} = I$, where I is the identity matrix, which plays the role of 1 for matrix multiplication. This fact is important for solving systems of linear equations, which can be represented in the form $A\mathbf{x} = \mathbf{b}$, where $\mathbf{x}$ is a column matrix of variables, and $\mathbf{b}$ is the column matrix of constant terms; we can solve for $\mathbf{x}$ uniquely when A is invertible. In fact, $\mathbf{x} = A^{-1}\mathbf{b}$, and so rank is an important concept in terms of its relation to inveritibility.

Rank is also important when we deal with the generator matrices of linear codes. In this case, the rank of the generator matrix is exactly the dimension of the linear code.

The Vandermonde Technique   Let V be a Vandermonde matrix, i.e. a matrix of the following form,

$$V = \begin{pmatrix} 1 & 1 & \cdots & 1 \\ \alpha_1 & \alpha_2 & \cdots & \alpha_n \\ \alpha_1^2 & \alpha_2^2 & \cdots & \alpha_n^2 \\ \vdots & \vdots & \ddots & \vdots \\ \alpha_1^{n-1} & \alpha_2^{n-1} & \cdots & \alpha_n^{n-1} \end{pmatrix}$$

where the α_i are in some field F. Fields are discussed in detail later in this chapter, but for now it is enough to say that a field is a special kind of a ring in which all nonzero elements have a multiplicative inverse. Then it can be shown that the determinant of V has a special form. In fact,

$$\det(V) \;=\; \prod_{1 \le i < j \le n} (\alpha_j - \alpha_i) \tag{19.8}$$

What this means is that if we take the product of terms $(\alpha_j - \alpha_i)$ where $i < j$, then we will have the determinant. In particular, $\det(V) = 0$ if and only if at least one of the terms $(\alpha_j - \alpha_i)$ is zero. Now, since $\det(V) \ne 0$ if and only if V is invertible, we know by 19.8 that V is invertible if and only if no two elements α_i, α_j are equal, for $i \ne j$.

This is of particular use in coding theory where Vandermonde matrices are used, since it tells us that the matrix V^{-1} exists exactly when the various α_i are distinct.

Subspaces   We also need the notion of a subspace of a vector space. A subspace is a set of vectors which is closed under addition and closed under scalar multiplication. For example, if we suppose that V is a vector space, and S is a subset of V, then S is a subspace provided that

1. $\mathbf{u} + \mathbf{v}$ is in S whenever $\mathbf{u}$ and $\mathbf{v}$ are in S.

2. $a\mathbf{u}$ is in S whenever $\mathbf{u}$ is in S and a is a scalar.

When we are working in binary, the only nonzero scalar is 1, and multiplication by 1 has no effect. However, the second condition with $a = 0$ guarantees that the zero vector is in the subspace.

For example, consider the set

$$S = \{(000), (101), (110)\}$$

then S is not a subspace of the vector space of all binary vectors of length 3, since the vector $(101) + (110) = (011)$ is not in the set S. However, if we adjoined this vector, then we would have a set which is a subspace.

19.3   Applications to RSA

We use the notation from chapter 3 where $N = pq$ and $p \neq q$ are primes. Then, from equations (19.5) and (19.7) we have $\varphi(N) = (p - 1)(q - 1)$. Since e was chosen with $\gcd(e, \varphi(N)) = 1$, it follows that e is a multiplicative unit in $Z_{\varphi(N)}$. Also, d is chosen so that $(M^e)^d \pmod{N} = M$.

A fundamental fact, holding for any multiplicative group G is this: if we multiply any element g in G by itself t times, where t denotes the order of G (i.e., the number of elements in G), we get the unity element 1. In symbols, $g^t = 1$.

We now apply this principle to obtain two famous results in number theory.

Euler's Theorem
$$x^{\varphi(N)} \pmod{N} = 1 \tag{19.9}$$

whenever $\gcd(x, N) = 1$. As a special case, when N is prime in (19.9) we get the following.

Fermat's Little Theorem
$$x^{p-1} \pmod{p} = 1 \tag{19.10}$$

whenever the integer x is not divisible by p.

Justification of Equations (19.9), (19.10)   Strictly speaking we have only justified (19.9) in the case when x is one of the multiplicative units in Z_N, with $1 \leq x < N$. Suppose x is any integer relatively prime to N. Dividing x by N we get $x = Nu + y$, with $0 < y < N$. Note that if $y = 0$, then x would be divisible by N so that $\gcd(x, N) = N$, contradicting $\gcd(x, N) = 1$. If a number $d > 1$ were to divide y and N then since $x = Nu + y$, d would divide x, so d would divide N and x, contradicting that $\gcd(x, N) = 1$.

We conclude that $\gcd(y, N) = 1$. Now $x^{\varphi(N)} = (Nu + y)^{\varphi(N)}$. When we multiply $Nu + y$ by itself $\varphi(N)$ times, all terms are divisible by N save for $y^{\varphi(N)}$. Then $x^{\varphi(N)}$ equals $y^{\varphi(N)} \pmod{N}$. Now we can think of y as an element of Z_N, since $0 \leq y < N$; moreover, $\gcd(y, N) = 1$. So (19.9) follows, and this in turn implies (19.10).

Question 1   What is the remainder when 2^{86} is divided by 55? This is tantamount to finding x such that $2^{86} \pmod{55} = x$.

We have $55 = 5 \times 11$, so $\varphi(55) = \varphi(5)\varphi(11) = (5-1)(11-1) = 40$. Further, $86 = 2 \times 40 + 6$, and $\gcd(2, 55) = 1$. Then $2^{86} = (2^{40})^2(2^6)$, so

$$
\begin{aligned}
2^{86} &\equiv (2^{40})(2^{40})(2^6) \pmod{55} \\
&\equiv (1)(1)(2^6) \pmod{55} \\
&\equiv 64 \pmod{55} \\
&\equiv 9 \pmod{55}
\end{aligned}
$$

We now prove a result that was mentioned in Chapter 3. We know that $M^e = C + t_0 N$ for some integer t_0, and we claimed that this admits a converse; if $(C + tN)^{\frac{1}{e}}$ is an integer for any integral value of t, with t not necessarily equal to t_0, then $(C + tN)^{\frac{1}{e}} \pmod{N} = M$. This is important from the point of view of an attacker, as guessing t would let the attacker find M.

So, suppose that $x = (C + tN)^{\frac{1}{e}}$ is an integer. We want to show that $x \pmod{N} = M$. Now,

$$x^e = C + tN, \tag{19.11}$$

and also, from the RSA algorithm, we know that

$$(x^e)^d \equiv x \pmod{N}. \tag{19.12}$$

Thus, by equation (19.12) $x^{ed} = x + \lambda N$ for some integer λ. So, by equation (19.11) we get that $x + \lambda N = x^{ed} = (C + tN)^d$. Now, when we multiply $(C + tN)$ by itself d times, we find that all terms are divisible by N except for C^d. It follows that there is a μ so that $x + \lambda N = C^d + \mu N$, and from this, we get that $x + (\lambda - \mu)N = C^d$, which says that $x \equiv C^d \equiv M \pmod{N}$.

19.4   Primitive Roots for Primes and Diffie–Hellman

For the Diffie–Hellman key-exchange we need to find a **generator** for the multiplicative group of units in Z_p where p is a prime (such a generator is also called a **primitive root**). What we mean by a generator is a number g so that the set of all powers of g (mod p) is the non-zero elements of Z_p. That is, $\{1, 2, 3, \ldots, p-1\} = \{g^t, \ t = 1, 2, 3, \ldots, p-1\}$. In this situation what is important is $\varphi(p - 1)$. We have the following result.

The number of primitive roots g with

$$1 \le g \le p - 1 \tag{19.13}$$

is $\varphi(p - 1)$. Thus there is always at least one generator.

Solving Congruence Equations Suppose N_1, N_2 are positive integers with $\gcd(N_1, N_2) = 1$. Given $0 \le a_1 < N_1$ and $0 \le a_2 < N_2$, we can formulate the following result.

Chinese Remainder Theorem Let a_1, a_2 be given with $0 \le a_1 < N_1$ and $0 \le a_2 < N_2$. Then, there is one and only one x with $0 \le x < N_1 N_2$ such that both $x \pmod{N_1} = a_1$ and $x \pmod{N_2} = a_2$. In fact,

$$x \equiv a_1 N_2 m_1 + a_2 N_1 m_2 \pmod{N_1 N_2}$$

where $N_2 m_1 \pmod{N_1} = 1$ and $N_1 m_2 \pmod{N_2} = 1$.

This can be generalized from two relatively prime integers N_1, N_2 to any number of relatively prime integers, as follows.

Let $x \pmod{N_i} = a_i$ for $i = 1, 2, \ldots, k$ be a set of k congruence equations, where the N_i are pairwise relatively prime. Then x has a unique solution modulo $N = N_1 N_2 \cdots N_k$, in fact,

$$x \equiv a_1 M_1 m_1 + a_2 M_2 m_2 + \cdots + a_k M_k m_k \pmod{N}$$

where $M_i = \frac{N}{N_i} = \frac{N_1 N_2 \cdots N_k}{N_i}$ and $M_i m_i \pmod{N_i} = 1$.

Question 2 Solve Question 1 using equation (19.5) and the Chinese Remainder Theorem.

First we want to find $2^{86} \pmod 5$. Now $\varphi(5) = 4$ and $86 = 21 \times 4 + 2$. So

$$
\begin{aligned}
2^{86} &\equiv (2^4)^{41}(2^2) \pmod 5 \\
&\equiv (1^{41})(2^2) \pmod 5 \\
&\equiv 4 \pmod 5
\end{aligned}
$$

Next we find $2^{86} \pmod{11}$. We know that $86 = 8 \times 10 + 6$, and thus

$$
\begin{aligned}
2^{86} &\equiv (2^{10})^8(2^6) \pmod{11} \\
&\equiv (1^8)(2^6) \pmod{11} \\
&\equiv 64 \pmod{11} \\
&\equiv 9 \pmod{11}
\end{aligned}
$$

So we seek the unique x with $0 \le x < 55$ so that $x \pmod 5 = 4$ and $x \pmod{11} = 9$. This forces $x = 9$.

Primes and Primality Testing The famous **prime number theorem** asserts that the number of prime numbers less than the positive integer n is approximately $\frac{n}{\ln n}$. Since this quantity grows at a reasonable rate as n gets large, we can say that there is no shortage of primes! From the prime number theorem, the probability that a number n, represented as a binary string of length m, is prime is a bit bigger than $\frac{log_2 e}{m}$ as n gets large. Recently it has been shown that we can test quickly (in polynomial time) to see whether a given integer is a prime. See [AKS02].

Calculating Inverses and the Euclidean Algorithm   Give two positive integers a, b with $a > b$, we want to do the following.

- Calculate the greatest common divisor d, of a and b, $\gcd(a, b) = d$.

- Express d in the form $d = ax + by$ where x, y are integers, not necessarily positive.

To see how this works, we first divide a by b. This gives the equation

$$a = bq + R, \quad 0 \le R < b \tag{19.14}$$

Note now that any number x dividing a and b must divide R. Thus x divides both b and R. On the other hand, any number dividing b and R must divide a. What this all means is the following: $\gcd(a, b) = \gcd(b, R)$.

Now the numbers b, R are smaller than a, b, respectively, so we proceed until we reach "the bottom." Let's take an example. Find $\gcd(657, 75)$.

$$
\begin{aligned}
657 &= 75 \times 8 + 57 \\
75 &= 57 \times 1 + 18 \\
57 &= 18 \times 3 + 3 \\
18 &= 3 \times 6 + 0
\end{aligned}
$$

We know then that $\gcd(657, 75) = \gcd(75, 57) = \gcd(57, 18) = \gcd(18, 3) = 3$.

Working backward from the bottom we get the following result. There are integers x, y so that

$$d = ax + by. \tag{19.15}$$

In the above example, we have $3 = 657x + 75y$ giving $x = 4, y = -35$, and this can be verified directly, whereas a simpler method for finding x, y will be shown later. Note that, in equation (19.15), if $\gcd(a, b) = 1$, then we have $ax + by = 1$, and so $xa = b(-y) + 1$. So $ax \pmod{b} = 1$, and x is the inverse of a in the group of multiplicative units in Z_b. On the other hand, if $d > 1$, then we cannot find an x, y so that $1 = ax + by$, and so a cannot have a multiplicative inverse in Z_b. In summary, as mentioned earlier, the invertible elements under multiplication in Z_b are exactly those elements in Z_b that are relatively prime to b.

It is important to note, too, that a Euclidean Algorithm exists even for polynomials, since we have a notion of the divisibility of polynomials. In fact, irreducible polynomials, which will be discussed later, play the role of prime numbers for polynomials, and many analogies can be made between integers and polynomials.

19.5   The Extended Euclidean Algorithm

We can use the Euclidean Algorithm to express the greatest common divisor d of two numbers a, b as a **linear combination** of a and b, as in $d = ax + by$. However, using the method, as described above, can be tiresome as we are forced to work backwards after having found "the bottom."

Instead, we present a slightly modified version that keeps track of the numbers x, y as the algorithm runs. It runs as follows.

	x_i	y_i	$657x_i + 75y_i$	
r_1	1	0	657	
r_2	0	1	75	
r_3	1	-8	57	$r_3 = r_1 - 8r_2$
r_4	-1	9	18	$r_4 = r_2 - r_3$
r_5	4	-35	3	$r_5 = r_4 - 3r_3$
r_6	-25	219	0	$r_6 = r_5 - 6r_4$

The initialization of the algorithm is presented in rows r_1 and r_2. At each step, we take a sum of the previous two rows, say r_i and r_{i-1} to find the next row, say r_{i+1} and this sum uses the quotient when the last entry of r_{i-1} is divided by the last entry of r_i. The algorithm stops when the last entry of a row is 0. Compare the above extended process with the previous worked example to see the parallels. It follows that $\gcd(657, 75) = 3$, and $657x + 75y = 3$, where $x = 4$ and $y = -35$.

This algorithm can, of course, be generalized to find the greatest common divisor of any two numbers.

The theory of integers and the integers (mod n) is very similar to the theory of polynomials in one variable over a field. For example, the gcd of two integers u and v is analagous to the gcd of two polynomials $u(x), v(x)$, which is the polynomial of highest degree dividing both $u(x)$ and $v(x)$. This gcd is expressible as a combination (with polynomial coefficients) of $u(x)$ and $v(x)$. $Rem[u, v]$ gets replaced by $Rem[u(x), v(x)]$ which will have a degree lying between 0 and $t - 1$ where t is the degree of $v(x)$. Similarly, the ring Z_n is analagous to the ring of polynomials (mod f) where f is any polynomial of degree n. In coding, $f(x) = x^n - 1$ is often used.

19.6 Proof that the RSA Algorithm Works

We have two distinct primes, p, q and $N = pq$. Also, e is relatively prime to $\varphi(N) = (p-1)(q-1)$. Next, let t be any integer which is divisible by both $p-1$ and $q-1$. So, in particular, t may be $(p-1)(q-1)$, or t may be the least common multiple of $p-1$ and $q-1$. Finally, we have d so that $ed \pmod{t} = 1$, and so $ed = 1 + \lambda t$, for some integer λ.

We want to show that for any integer message M,

$$C^d \equiv (M^e)^d \equiv M \pmod{N} \tag{19.16}$$

Assume first that the message M has $\gcd(M, p) = 1$. Then $M^{ed} = M^{1+\lambda t} = M(M^{p-1})^u$ where u is an integer satisfying $u(p-1) = \lambda t$. It follows by Fermat's Little Theorem that $M^{ed} \equiv M \pmod{p}$. Now, when $\gcd(M, p) \neq 1$, we have that p divides M, as p is prime, and then $M^{ed} \equiv M \pmod{p}$ still holds.

Similarly, we can show that $M^{ed} \pmod{q} = M$. Thus, we have shown that $M^{ed} - M$ is divisible by both p and q, and so it must be divisible by pq, since p, q are prime and therefore relatively prime. That is, equation (19.16) is satisified.

19.7 Constructing Finite Fields

A **field**, F, is a ring with the additional stipulation that every element of the set have a multiplicative inverse. Another way of putting this is that a field is a set F with two operations, addition and multiplication, such that F under addition is an additive group, and F without zero under multiplication is a multiplicative group, and addition and multiplication are linked by the usual distributive laws, $x(y+z) = xy + xz$ and $(x+y)z = xz + yz$. Also, the groups are commutative. Examples of such groups are the rational numbers, the real numbers and the complex numbers. A **finite field**, F, is a field which has only a finite number of elements. The number of elements in F is called the **order** of F. Z_2 is a field, and so is Z_p for any prime p, since the number of multiplicative units in Z_p is $\varphi(p) = p - 1$, which is exactly the number of nonzero elements of Z_p. That is, in Z_p every nonzero element has a multiplicative inverse.

It is a remarkable fact that a field of order q exists if and only if q is a power of a prime, say $q = p^k$ where p is prime, and k is a positive integer. In fact, the field of order p^k is unique (in that all fields of order p^k are essentially the same, or **isomorphic**), and is denoted $GF(p^k)$, where GF stands for "Galois Field" in honor of Galois, the founder of the subject. Further, the smallest number of ones that adds to zero in $GF(q)$ is called the **characteristic** of the field, and this number is guaranteed to be a prime number p. Also going back to Diffie–Hellman, just as in Z_p there is a generator for the field $GF(q)$. That is, there is an element α of $GF(q)$ such that the nonzero elements of the field are the successive powers of α, i.e., $GF(q) = \{0, \alpha^1, \alpha^2, \ldots, \alpha^{q-1}\}$.

When $GF(q)$ has characteristic p, $GF(q)$ will contain $GF(p)$ as a subfield, and ultimately, $GF(q)$ can be built up from $GF(p)$ using the polynomial construction procedure we are about to illustrate. Unfortunately, Z_{p^k} is not a field when $k \geq 2$, as is seen by examining Z_4. The element 2 has no multiplicative inverse in Z_4, and so Z_4 is not a field. However, finite fields turn up again and again in coding theory, so we need to be able to construct fields of order p^k.

An example is in order. Let's construct $GF(4)$ using Z_2 and the polynomial $f(x) = x^2 + x + 1$. Note that f has **degree** 2, since it has an x^2 term, and it does not have a term like x^3 or another higher exponent. Then, our field elements are defined to be all polynomials over Z_2 of degree strictly less than 2, using a "variable" α. Thus $GF(4) = \{0, 1, \alpha, 1 + \alpha\}$. Here, 2 comes from the fact that the degree of f is 2. Also, the polynomial f has a special property; it is **irreducible** over Z_2, meaning that it cannot be factored. This is discussed in more detail later.

Addition is straightforward polynomial addition. For example, in Z_2, $(\alpha) + (1) = (\alpha + 1)$ and, $(\alpha) + (1 + \alpha) = (1)$. Multiplication, however, has a twist. It seems that if multiply α by α we get α^2, which is not a field element. However, we work around this difficulty by decreeing that $f(\alpha) = 0$, i.e. that $\alpha^2 + \alpha + 1 = 0$, or $\alpha^2 = -\alpha - 1 = \alpha + 1$. We can now completely describe the addition and multiplication in $GF(4)$ in the following Cayley tables, setting $\omega = \alpha$ and $\omega^2 = \alpha + 1$.

$+$	0	1	ω	ω^2
0	0	1	ω	ω^2
1	1	0	ω^2	ω
ω	ω	ω^2	0	1
ω^2	ω^2	ω	1	0

$\times$	0	1	ω	ω^2
0	0	0	0	0
1	0	1	ω	ω^2
ω	0	ω	ω^2	1
ω^2	0	ω^2	1	ω

Polynomials   For the sake of completeness, we need a few notions. A **polynomial over** Z_v is a polynomial $f(x) = a_0 + a_1 x + a_2 x^2 + \cdots + a_k x^k$ wherein each a_i is an element of Z_v. That is, a_i is in $\{0, 1, 2, \ldots, v - 1\}$. Polynomial addition and multiplication are as usual, except that the coefficients are added modulo v. That is,

$$(a_0 + a_1 x + a_2 x^2 + \cdots) + (b_0 + b_1 x + b_2 x^2 + \cdots) =$$
$$(a_0 + b_0 \pmod{v}) + (a_1 + b_1 \pmod{v})x + (a_2 + b_2 \pmod{v})x^2 + \cdots \tag{19.17}$$

where each of the above polynomials is finite. Also, the **degree** of f is the highest value of i for which a_i is nonzero. So, if $a_k \neq 0$ in $f(x)$, then the degree of f is k.

A polynomial is **irreducible** provided that it cannot be factored into lower degree terms. For example, $x^2 + x + 1$ is irreducible both as a polynomial over the real numbers, and as a polynomial over Z_2. However, as a polynomial over Z_2, $(x + 1)^2 = x^2 + 1$ is not irreducible. Polynomials which can be factored, that is, polynomials which are not irreducible are **reducible**.

We also need to be able to tell when a polynomial is irreducible. For now, we are satisfied to say that a polynomial $f(x)$ of degree 2 or 3 is irreducible over Z_p provided that for all a in Z_p, $f(a) \neq 0$. That is to say, f has no roots in Z_p. Also, another method to test the irreducibility of a polynomial is to attempt to divide it by all polynomials of lesser degree, and if any divide evenly, then the polynomial is not irreducible.

A useful, if tedious, procedure for finding all irreducible polynomials of degree k is to take all polynomials of degree less than k, and find the products of these which multiply to have degree k. This produces a list of all polynomials of degree k which are *not* irreducible, so any unlisted polynomials of degree k must be irreducible.

Example 19.1 *To find all irreducible polynomials over Z_2 of degree 2, we begin by listing the polynomials of lesser degree. They are x and $x + 1$. Note that we do not list 0 or 1, since these will not provide any useful results. Now, $(x)(x) = x^2$, $(x)(x + 1) = x^2 + x$, and $(x + 1)(x + 1) = x^2 + 1$, and the only unlisted polynomial of degree 2 is $x^2 + x + 1$, so it must be irreducible.*

We also point out that over any field F, there always exists an irreducible polynomial of degree k, for $k \geq 2$.

The General Construction Procedure   We can now describe how to construct finite fields $GF(p^k)$ of order p^k in general.

- We choose a prime p and an irreducible polynomial $f(x)$ over Z_p of degree k.

- The field elements are all polynomials of degree $< k$ over Z_p.

- Addition of the field elements is as described in equation 19.17. That is, we use normal polynomial addition and reduce the coefficients by taking them to be the remainder upon division by p.

- Multiplication is like addition in that the coefficients are reduced modulo p, but with the extra stipulation that $f(\alpha) = 0$. That is, if $f(x) = b_0 + b_1 x + \cdots + b_k x^k$ with $b_k \neq 0$, then

$$\alpha^k = b_k^{-1}(-b_0 - b_1\alpha - \cdots - b_{k-1}\alpha^{k-1}). \qquad (19.18)$$

An Example: Constructing $GF(8)$ We use the irreducible polynomial $f(x) = 1 + x + x^3$. Our field elements are then polynomials of the form $a_0 + a_1\alpha + a_2\alpha^2$, where $a_i = 0$ or 1.

Polynomial in α	Power of ω
0	—
1	$\omega^0 = 1$
α	ω^1
α^2	ω^2
$1 + \alpha$	ω^3
$\alpha + \alpha^2$	ω^4
$1 + \alpha^2$	ω^5
$1 + \alpha + \alpha^2$	ω^6

We have used this presentation rather than the table format since it is less cumbersome. Addition proceeds as usual, and is easily done using the left column, with the added stipulation that the coefficients are reduced modulo 2, so that $(1+\alpha)+\alpha = 1+2\alpha = 1$. Multiplication requires that $f(\alpha) = 0$, i.e. that $\alpha^3 = \alpha+1$. We can easily do multiplication using the right column and the usual exponent laws, namely that $\omega^a \omega^b = \omega^{a+b}$. We also need to know that $\omega^7 = 1$, and we can see this from the fact that $\alpha^3 = \alpha + 1$, as follows,

$$
\begin{aligned}
\omega^7 &= \alpha^7 \\
&= \alpha^{1+3+3} \\
&= \alpha(\alpha+1)(\alpha+1) \\
&= \alpha(\alpha^2 + 1) \\
&= \alpha^3 + \alpha \\
&= (\alpha+1) + \alpha \\
&= 1.
\end{aligned}
$$

A Useful Polynomial for Coding   We mention here a fact that will be useful later when we discuss coding, especially linear codes. The polynomial $x^{q-1} - 1$ over a field of order q is of particular importance. In fact,

$$
\begin{aligned}
x^{q-1} - 1 &= \prod_{i=1}^{q-1}(x - \alpha_i) \\
&= (x - \alpha_1)(x - \alpha_2) \cdots (x - \alpha_{q-1})
\end{aligned}
$$

where $\alpha_1, \alpha_2, \ldots, \alpha_{q-1}$ are the distinct nonzero field elements. This means that a polynomial of the form $(x-b_1)(x-b_2) \cdots (x-b_k)$, divides $x^{q-1}-1$, where $k < q$ and the b_i's are distinct.

For example, over Z_5, we have that

$$
x^4 - 1 = (x - 1)(x - 2)(x - 3)(x - 4).
$$

Another Example: Constructing $GF(16)$ We use the polynomial $f(x) = 1 + x + x^4$, which is irreducible. Our field elements are $\{a_0 + a_1\alpha + a_2\alpha^2 + a_3\alpha^3 | a_i = 0 \text{ or } 1\}$. Then, using $f(\alpha) = 0$, which means that $\alpha^4 = \alpha + 1$, we can describe the field as follows:

Polynomial in α	Power of ω
0	$-$
1	$\omega^0 = 1$
α	ω^1
α^2	ω^2
α^3	ω^3
$1 + \alpha$	ω^4
$\alpha + \alpha^2$	ω^5
$\alpha^2 + \alpha^3$	ω^6
$1 + \alpha + \alpha^3$	ω^7
$1 + \alpha^2$	ω^8
$\alpha + \alpha^3$	ω^9
$1 + \alpha + \alpha^2$	ω^{10}
$\alpha + \alpha^2 + \alpha^3$	ω^{11}
$1 + \alpha + \alpha^2 + \alpha^3$	ω^{12}
$1 + \alpha^2 + \alpha^3$	ω^{13}
$1 + \alpha^3$	ω^{14}

As before, we can use the left column to quickly compute addition, and the right column to quickly compute multiplication, with $\omega^{15} = 1$.

19.8 Pollard's $p - 1$ Factoring Algorithm

Suppose we want to factor some positive integer n which is not a prime. For example, we might be trying to crack RSA, in which case, n is the product of two large primes.

Now, there exists some prime p dividing n. The number $p - 1$ is uniquely expressible as the product of prime powers. We have $p - 1 = p_1^{a_1} p_2^{a_2} \cdots p_s^{a_s}$ where $p_1, p_2, \ldots, p_s$ are the distinct primes dividing $p - 1$. Thus no two of $p_1^{a_1}, p_2^{a_2}, \ldots, p_s^{a_s}$ are equal. We can assume, by relabelling if necessary, that $p_1^{a_1} < p_2^{a_2} < \cdots < p_s^{a_s}$. Now assume that $p_s^{a_s}$ is less than or equal to some small number B. Then $p_i^{a_i}$ is less than or equal to B, for $1 \le i \le s$. The algorithm proceeds as follows:

1. Pick some integer t that is a multiple of all integers less than or equal to B. For instance, take $t = factorial(B) = B!$. Or choose t to be the least common multiple of $\{1, 2, 3, \ldots, B\}$.

2. Choose the integer x randomly, with $2 < x < n - 2$.

3. Calculate $y = x^t \pmod{n}$ by repeated squaring as in Chapter 3.

4. Let d be the greatest common divisor of $x^t - 1$ and n. In symbols, $d = \gcd(x^t - 1, n) = \gcd(y - 1, n)$.

5. We have that d divides n. In fact, we can guarantee that $d > 1$. This means that unless $x^t - 1$ is a multiple of n, d is a proper factor of n.

To see this, proceed as follows. Any 2 of the numbers $p_1^{a_1}, p_2^{a_2}, \ldots, p_s^{a_s}$ are relatively prime and each of them is less than B. By choice of t, it follows that their product, namely $p - 1$, divides t. Thus t is some multiple of $p - 1$, say $t = \lambda(p - 1)$.

Therefore $x^t = x^{(p-1)\lambda}$. By Fermat's Little Theorem $x^{p-1} = 1 \pmod{p}$ so $x^{(p-1)\lambda} = (x^{(p-1)})^\lambda = 1 \pmod{p}$.

Thus $x^t = 1 \pmod{p}$. Therefore p divides $x^t - 1$, and we already know that p divides n. Now because $d = \gcd(x^t - 1, n)$, it follows that p divides d. This means that we have factored n.

The above algorithm is due to Pollard [Pol74]. The method is a predecessor of Lenstra's [Len87] method with elliptic curves.

An example Let $n = 2117$, $B = 7$. Then choose t to be the last common multiple of the integers $2, 3, 4, 5, 6, 7$, so $t = 420$. Choose $x = 2$ (randomly).

Then $2^{420} \pmod{2117} = 1451$. Thus $y = 1451$. So $d = \gcd(y - 1, n) = \gcd(1450, 2117) = 29$. If follows that $n = 29 \cdot 73$.

Remark For the above algorithm to work properly, one must know that $p_s^{a_s} \leq B$. That is, we need some information about the prime factorization of the number $p - 1$. To work around this, one might choose B as a large number, however, for efficiency concerns, we want B to be small.

19.9 Computational Complexity, Turing Machines, Quantum Computing

In previous chapters we discussed public key algorithms such as RSA and pointed out that their security depends on the unproved assumption that, given a cipher text, it is not possible to calculate the underlying message in a reasonable amount of time. This kind of security is called computational security. We want to briefly expand on such ideas in the context of **computational complexity**, which is a much-studied topic in theoretical Computer Science. It also ties in with Information Theory.

Roughly, the complexity of a mathematical problem, algorithm or calculation is a measure of the computational resources needed to solve it. These resources might involve time, space in memory or other considerations. Here we focus on time.

Example 19.2 *Lets take multiplication. Suppose $x = x_1 x_2 \ldots x_n$ is an n digit decimal number. Assume the same for $y = y_1 y_2 \ldots y_n$. To multiply x by y we can use the school method of "long multiplication". So we multiply x by y_1, shift to the left, multiply x by y_2 and so on. Then we sum these up to get x times y.*

*Each multiplication by $y_1, y_2, \ldots$ takes about n **bit operations**. The summing up of the n products account for approximately n^2 bit operations. Thus, this long multiplication algorithm uses $t(n)$ bit operations, where $t(n)$ is $O(n^2)$ as explained below. Note that the actual time taken for the multiplication will depend on the processor being used, so $t(n)$ is just an estimate. This is why we use the big O notation, which is machine independent.*

Big O Notation   We say that $F(n)$ is $O(G(n))$ for functions F, G provided that there exists some positive constant c so that $F(n) \leq cG(n)$ for all n greater than some integer n_o.

In general we need not find (or know) the numbers c, n. All we need to know is that, eventually, some positive multiple of G is larger than F. We think of F as being dominated by G and sometimes write $F \ll G$. An easy example is provided by the log function: here $\log n$ is $O(n)$. To see this, just take $c = 1$ and $n_o = 1$ in the definition.

In the multiplication example above, it is possible to devise a better algorithm that uses only $O\big((n \log n)(\log \log n)\big)$ bit operations.

Input   The complexity of a problem or class of problems depends on the size, n, of the input data. How can this size be defined or measured? Here we fall back on Information Theory and define the **input size** to be the minimum number of Shannon bits of information needed to present the input data.

As an example, consider the following class of problems called PRIME: given an integer N, decide whether N is prime. We can represent N as a sequence of N ones. However it is shorter to represent N as a binary string which will have just $m = 1 + \lceil \log N \rceil$ binary digits corresponding to m Shannon bits (see Chapter 9). Very recently (see [AKS02]) it was shown that the PRIME problem is in the class P of polynomial algorithms as explained below. To discuss this, we need to talk about Turing machines.

Turing Machines   A Turing machine is a theoretical computer which provides the model for all possible computers. This unprovable assumption is known as Church's thesis, and it is generally accepted as being correct. A Turing Machine has an infinite "tape" which is divided into squares. All but a finite number of these squares are blank. So the machine has infinite memory but only a finite part is used at any stage. There is a "head" positioned over the tape and it can read what is on any square on the tape - or write a symbol in any square. The head has a finite number of internal states. Each location on the tape can have a symbol from some finite alphabet A (or a blank) written in it.

In one cycle the machine can write or read a symbol on the tape, move one position left or right, and change the internal state. The **output** of the machine, for a given input, is the characters remaining on the tape when the machine has reached its final state. The time taken from the input to final state is the number of steps (or cycles).

Remarkably, any computation possible on a present-day computer can be performed by a Turing machine. Also, any processor that can be designed can perform only the equivalent of a bounded number of Turing machine steps (see []).

A function f mapping a string X over the finite alphabet A to another string $f(x)$ over A is **computable** by the Turing machine M if for any input x the machine M stops (or halts) with $f(x)$ as its output. The **time complexity** of f with respect to M is a function $t_M(n)$, which is the maximum time taken to compute $f(x)$ over all strings x of length n, $n > 0$.

A function or algorithm f is computable in **polynomial time** or is a **polynomial algorithm** if there exists some Turing machine M that computes f and some polynomial $G(x)$ such that $t_M(n) \leq G(n)$ for all n.

Using the previous terminology we can then say that the algorithm f has complexity $O(G)$. Note also that $t_M(n)$ corresponds to the "worst-case" scenario.

Example 19.3 *Suppose we are given n objects to be sorted in a order. What we want is a procedure or algorithm which accepts as input the unsorted list and outputs a sorted list. It can be shown that there exists an algorithm f which can carry out this procedure with complexity $O(n \log n)$. (Also, this bound is the "best possible" for this problem). Since $\log n \leq n$, we can also say that f is $O(n^2)$, and therefore that f is a polynomial time algorithm.*

Example 19.4 *We consider the PRIME problem. We want an algorithm which accepts as input a positive integer N and outputs "Yes" if N is prime and "No" otherwise. When the output must be "Yes" or "No" the problem is called a **decision problem**. Very recently it has been shown (see [AKS02]) that there exists a polynomial algorithm for the problem PRIME.*

Example 19.5 *We consider the problem of constructing a Huffman code given n source words using the Huffman algorithm in Chapter 11. The algorithm can be shown to have complexity $O(n^2)$.*

A class of decision problems is said to belong to the **class** NP (for **non-deterministic polynomial**) if for any problem for which the answer is "Yes", there is a "certificate" i.e.

an extra piece of information that can be used to verify the correctness of the answer in polynomial time.

For example, consider the question "is the positive integer N non-prime". If the answer is "Yes" a certificate could be a pair of positive integers u, v with $1 < u, v < N$ such that $N = uv$.

It can be shown that, as classes, $P \subset NP$. Whether or not $P = NP$ is a major unsolved problem in theoretical computer science. It should also be pointed out that there do exist decision problems which are not even in NP.

From a practical point of view, a fundamental unsolved problem is whether or not a given function is computable in polynomial time. In other words, we want to know whether there exists an algorithm implementing f in polynomial time. If so, the function is called "feasible" or "tractable" or "computable in a reasonable amount of time" etc, because these algorithms can be effectively implemented on a modern computer.

As an example we mention the factoring problem or the RSA problem discussed in Chapters 2 and 3. It is assumed, but has not been proved, that the factoring problem cannot be solved in polynomial time.

Ideally, to guarantee good security for a public key algorithm we want the encryption function e to be a "one-way function". This means that the public key e is computable in polynomial time whereas the inverse of e, i.e. the private key d corresponding to the decryption index is not computable in polynomial time.

However, to date nobody has been able to prove the existence of such a one-way function.

If we do not restrict ourselves to Turing Machines but allow **quantum computers**, then factoring can be done in polynomial time. This would spell the end of RSA.

We should mention also that various "probabilistic algorithms" have been used with some success such as the **index calculus** method for the Diffie–Hellman problem in Chapter 3. We refer to McCurley [McC90].

Finally it is worth noting, as pointed out in Mollin [Mol00], that for an input n of size 10^6 an algorithm with complexity n takes under a second, where the unit of time for a machine cycle is $\frac{1}{10^6}$ of a second, but an algorithm of complexity n^2 will take days while an algorithm of complexity n^3 will take thousands of years. **Exponential algorithms** with complexity of the order of $2^{f(n)}$, with f a polynomial, are "off the scale" here.

19.10   Problems

1. Use trial division to find the prime factorization of 4023390.

2. Calculate $\varphi(4023390)$.

3. Calculate the remainder when 2^{605} is divided by 783.

4. Use the Euclidean Algorithm to find the greatest common divisor, d, of 2925 and 3055, and find x, y so that $d = 2925x + 3055y$.

5. Repeat the above question for 1547 and 1265.

6. Show that if $d = \gcd(a, b)$, and $d = ax + by$, then there are integers x_0, y_0 with $x_0 \neq x$ and $y_0 \neq y$ so that $d = ax_0 + by_0$.

7. Use the Chinese Remainder Theorem to find $x < 7 \times 11 \times 13$ so that $x \equiv 4 \pmod{7}$, $x \equiv 3 \pmod{11}$, and $x \equiv 8 \pmod{13}$.

8. Is the requirement that the moduli be pairwise relatively prime in the Chinese Remainder Theorem necessary? What happens if we remove that restriction?

9. Find all irreducible second degree polynomials over Z_3.

10. Construct $GF(9)$, writing out the Cayley tables.

11. Show that if a, b are elements of a field F, then $ab = 0$ implies that $a = 0$ or $b = 0$.

12. What happens if we try to construct $GF(9)$ using the reducible polynomial $(x+1)(x+2) = x^2 + 2$. Why is the resulting structure not a field? What is a better term for the resulting structure?

13. We work over Z_7 for this question.

   • Does $(x - 3)(x - 6) = x^2 + 5x + 4$ divide $x^6 - 1$?
   • Does $(x - 2)^2 = x^2 - 4x + 4$ divide $x^6 - 1$?

   What is a quick way to come to these conclusions?

14. Verify that over Z_5, the equation $x^4 - 1 = (x - 1)(x - 2)(x - 3)(x - 4)$ is true. That is, expand the right hand side, and show that it is the left hand side.

19.11 Solutions

1. $4023390 = 2 \times 3 \times 5 \times 7^3 \times 17 \times 23$.

2. We refer to equations (19.6) and (19.7).

$$
\begin{aligned}
\varphi(4023390) &= \varphi(2)\varphi(3)\varphi(5)\varphi(7^3)\varphi(17)\varphi(23) \\
&= (2 - 1)(3 - 1)(5 - 1)(7^3 - 7^2)(17 - 1)(23 - 1) \\
&= (1)(2)(4)(7^3 - 7^2)(16)(22) \\
&= 827904
\end{aligned}
$$

3. We find 2^{605} (mod 783). Now, $783 = 3^3 \times 29$, so $\varphi(783) = \varphi(3^3 \times 29) = \varphi(3^3)\varphi(29) = 3^2(3-1)(28) = 504$, so $2^{605} \equiv 2^{101}$ (mod 783). We now use the repeated squaring technique of Chapter 3 to find that the answer is 50.

4. $d = 65$, $x = 23$, $y = -22$.

5. $d = 1$, $x = -157$, $y = 192$.

6. We can use $x_0 = x + \frac{b}{d}t$ and $y_0 = y - \frac{a}{d}t$ for any integer t, so that $ax_0 + by_0 = a(x + \frac{b}{d}t) + b(y - \frac{a}{d}t) = ax + by = d$.

7. We find $N = 7 \times 11 \times 13$, and then $M_1 = 143$, $M_2 = 91$ and $M_3 = 77$. Using the Euclidean Algorithm, we calculate that $m_1 = 5$, $m_2 = 4$ and $m_3 = 12$. This leads to finding that $x = 333$.

8. The pairwise relatively prime condition implies that the set of congruences is actually solvable. Consider $x \equiv 1$ (mod 2), and $x \equiv 2$ (mod 4). Then no such x exists, as $x \equiv 2$ (mod 4) implies that x is even, whereas $x \equiv 1$ (mod 2) forces that x is odd.

   However, it is possible to have a solvable set of congruence equations in which the moduli are not pairwise relatively prime. To deal with this situation, one can reduce the congruence equations into a set of pairwise relatively prime moduli, and then apply the Chinese Remainder Theorem.

9. The six irreducible polynomials are x^2+1, x^2+x+2, x^2+2x+2, $2x^2+2$, $2x^2+2x+1$, and $2x^2 + x + 1$.

   To find these polynomials, one could list the first degree polynomials and find all products of exactly two of them. The resulting list will be all polynomials which are not irreducible, so any unlisted polynomials are irreducible.

   Also, one could list all second degree polynomials and test whether each polynomial has a root. If a polynomial has no root, then it is irreducible (but this test only works for polynomials of degree 2 or 3).

10. We use the polynomial $f(x) = x^2 + 1$ to find the field, so $\alpha^2 + 1 = 0$, that is, $\alpha^2 = 2$. This is sufficient to let us completely describe the addition and multiplication of the field.

11. Assume that $b \neq 0$. Then b has an inverse, say b^{-1}. Multiplying $ab = 0$ by b^{-1}, we get $(ab)b^{-1} = 0b^{-1} = 0$, so $a = 0$. It follows that either $a = 0$ or $b = 0$.

12. Using $f(x) = x^2 + 2$, we set $f(\alpha) = 0$, i.e., $\alpha^2 + 1 = 0$. So, we find that the field elements $\alpha+1$ and $\alpha+2$ multiply to $(\alpha+1)(\alpha+2) = \alpha+2 = 0$. That is, the condition proven to hold for all fields in the previous problem is violated. This kind of structure is more properly called a ring.

13. Using long division, we find:

- $(x^2 + 5x + 4)g(x) = x^6 - 1$, where $g(x) = x^4 + 2x^3 + 6x + 5$, so that it does divide evenly.

- $(x - 4x + 4)h(x) + (3x + 1) = x^6 - 1$ where $h(x) = x^4 + 4x^3 + 6x^2 + 5x + 2$, so that it does not divide evenly.

We can conclude these quickly, since we know that

$$x^6 - 1 = (x - 1)(x - 2)(x - 3)(x - 4)(x - 5)(x - 6), \tag{19.19}$$

and so it follows that $(x - 3)(x - 6) = x^2 + 5x + 4$ must divide evenly. Further, we know that the factorization of $x^6 - 1$ over $GF(7)$ is unique, so that if $(x - 2)^2$ were to divide $x^6 - 1$, it would also have to divide the right hand side of 19.19. However, this is impossible, and thus we can conclude that $(x - 2)^2$ does not divide $x^6 - 1$ evenly, without having to do the long division.

14. We take $[(x - 1)(x - 2)][(x - 3)(x - 4)]$ and expand inside the square braces to find $[(x - 1)(x - 2)][(x - 3)(x - 4)] = [x^2 - 3x + 2][x^2 - 7x + 12]$. Adjusting the coefficients modulo 5, we get $[x^2 - 3x + 2][x^2 - 7x + 12] = [x^2 + 2x + 2][x^2 + 3x + 2]$. Expanding again, we have that this is equal to $(x^4 + 3x^3 + 2x^2) + (2x^3 + 6x^2 + 4x) + (2x^2 + 6x + 4)$. Collecting like terms and reducing modulo 5, we get $x^4 + 4 = x^4 - 1$.

Chapter 20

Introduction to Linear Codes

Goals, Discussion We introduce the basics of linear codes and give several examples including the "perfect" Hamming and Golay codes. The McEleice cryptographic protocol is described as an application of the theory of linear codes.

New, Noteworthy We include a discussion of the "football pools" problem. Basic ideas from Shannon's fundamental theorem are revisited using Hamming codes over a binary symmetric channel. We discuss some of the fascinating history of the perfect ternary Golay code of length 11, which was in fact discovered independently and published in the Finnish soccer magazine Veikkaaja a year and a half before Golay published the same code.

20.1 Repetition Codes and Parity Checks

We want to revisit some of the general ideas of Chapter 19 and to amplify on them. Mathematical error correction is concerned with the errors which occur when information is transmitted from one place to another. For example, the binary message 010101 may be received as 011101 because of a problem in the transmission process. These problems are referred to as "noise" and may be due to electromagnetic radiation, thermal radiation (heat), cross talk, deterioration of storage devices such as hard disks, or even operator (human) error. The detection of errors is important because it sometimes allows for correction by transmitting the information a second time. The correction of the received message to the correct message is even more desirable, but usually harder to accomplish. Both of these objectives are realized by transmitting the information in a redundant form. For example, the binary message 010111 could be repeated three times by transmitting 010111010111010111. Then if only one error was made in the 18 digits transmitted, it would be easy to pick the correct message, since the corrrect message would occur twice and the wrong message would occur only once.

Repeating the message a large number of times would almost certainly guarantee the received message to be corrected to the sent message, but this process is inefficient and may be too time consuming. For example, the digital encoding of music on a compact disc is decoded by a digital-to-analog converter, but early CD players did not read the data on the disc fast enough to be able to read multiple copies of each note encoded on the disc. Nor would the disc have been able to contain 75 minutes of music if everything were recorded multiple times on the disc. Coding theory is concerned with finding the most efficient methods of transmitting information so that some error detection and usually some error correction is possible. Part of the secret of efficient coding is to carefully choose which sequences of numbers are used in representing the transmitted information.

We now give some examples of linear codes.

Example 20.1 *The repetition code consists of repeating each transmitted digit n times. The map from elements in a field F to n-tuples over F given by $f_n : x \to (x, x, \ldots, x)$ expresses the encoding. If n is odd these will give rise to perfect codes.*

As indicated above, this is not a very efficient approach to coding.

Example 20.2 *A more commonly used coding is called parity-check coding. This is used in places where errors are not likely to occur and the detection of an error is all that is required. This code does not supply enough information to determine what the original message might be if there is an error in transmission. The map from vectors of length $n - 1$ to vectors of length n over F given by $\psi_n : (x_1, x_2, \ldots, x_{n-1}) \to (x_1, x_2, \ldots, x_{n-1}, -x_1 - \cdots - x_{n-1})$ defines this code. It can also be expressed in matrix form by*

$$(x_1, x_2, \ldots, x_{n-1}) \to (x_1, x_2, \ldots, x_{n-1}) \begin{pmatrix} 1 & & & -1 \\ & 1 & & -1 \\ & & \ddots & \vdots \\ & & & 1 & -1 \end{pmatrix}.$$

Note that, if we are working in binary, the effect of this is to adjoin a 1 at the end if the message $(x_1, x_2, \ldots, x_{n-1})$ has an odd number of 1s and to adjoin a 0 otherwise. So this is the same parity check idea that we used previously.

Example 20.3 *Suppose that we want to encode messages which consist of 4 binary digits. The code map f maps a vector $\mathbf{u}$ to $\mathbf{u}G$ where $\mathbf{u}G$ is considered to be a vector of length 7 over the binary field. We transmit the 7 binary digits which correspond to the seven components* *of* **u***G. This gives us a linear code C. The matrix G is*

$$
\begin{pmatrix}
1 & 0 & 0 & 0 & 0 & 1 & 1 \\
0 & 1 & 0 & 0 & 1 & 0 & 1 \\
0 & 0 & 1 & 0 & 1 & 1 & 0 \\
0 & 0 & 0 & 1 & 1 & 1 & 1
\end{pmatrix}
$$

Using this matrix, we find all of the code words C:

$(0,0,0,0)G = (0,0,0,0,0,0,0),\quad (0,0,0,1)G = (0,0,0,1,1,1,1)$

$(0,0,1,0)G = (0,0,1,0,1,1,0)\quad (0,0,1,1)G = (0,0,1,1,0,0,1)$

$(0,1,0,0)G = (0,1,0,0,1,0,1)\quad (0,1,0,1)G = (0,1,0,1,0,1,0)$

$(0,1,1,0)G = (0,1,1,0,0,1,1)\quad (0,1,1,1)G = (0,1,1,1,1,0,0)$

$(1,0,0,0)G = (1,0,0,0,0,1,1)\quad (1,0,0,1)G = (1,0,0,1,1,0,0)$

$(1,0,1,0)G = (1,0,1,0,1,0,1)\quad (1,0,1,1)G = (1,0,1,1,0,1,0)$

$(1,1,0,0)G = (1,1,0,0,1,1,0)\quad (1,1,0,1)G = (1,1,0,1,0,0,1)$

$(1,1,1,0)G = (1,1,1,0,0,0,0)\quad (1,1,1,1)G = (1,1,1,1,1,1,1)$

20.2 Details of Linear Codes

Let $\mathbf{u} = (u_1, u_2, \ldots, u_k)$ be any element of the k-dimensional vector space over the finite field $GF(q)$ of order q. Thus $\mathbf{u}$ is a vector of length k over some finite field F of order q, $q = p^m$. For example, if $q = 2$ then we are talking about binary strings of length k. So $\mathbf{u}$ is a **message** of length k.

Let G be a matrix with k rows and n columns over F, of rank k. This implies that $k \leq n$. The fact that G has rank k means that the k rows of G are linearly independent. We can regard these rows as a set of linearly independent vectors of length n over $GF(q)$. First we break up our message into blocks of length k. Then a block $\mathbf{u}$ gets encoded to the code word $\mathbf{u}G$. This is our **encoding algorithm**.

Thus $\mathbf{u}$ has length k and $\mathbf{u}G$ has length n.

The set $C = \{\mathbf{u}G \mid u \text{ in } F_q^k\}$ is called a **linear code of dimension** k **over** F, because C is a linear subspace of F_q^n. F_q^n means the set of vectors of length n, with entries taken from $GF(q)$. To see this, let $\mathbf{v}_1, \mathbf{v}_2$ be in C. So $\mathbf{u}_1 G = \mathbf{v}_1$ and $\mathbf{u}_2 G = \mathbf{v}_2$ Then $\alpha \mathbf{v}_1 + \beta \mathbf{v}_2$ is in C since

$$(\alpha \mathbf{u}_1 + \beta \mathbf{u}_2)\, G = \alpha(\mathbf{u}_1 G) + \beta(\mathbf{u}_2 G)$$
$$= \alpha \mathbf{v}_1 + \beta \mathbf{v}_2.$$

The matrix G is called a **generator matrix** for C. We say that C is **a linear** (n, k) **code**.

Properties of the encoding .

(a) the mapping $\mathbf{u} \to \mathbf{u}G$ is one to one. Thus, given $\mathbf{u}G$ (= $\mathbf{v}$) there is a unique $\mathbf{u}$ such that $\mathbf{u}G = \mathbf{v}$. We find $\mathbf{u}$ by solving the appropriate system of linear equations.

(b) usually $d(\mathbf{u}G, \mathbf{v}G) \geq d(\mathbf{u}, \mathbf{v})$ where d denotes the Hamming distance.

Example 20.4 *Let G be as in Example 20.3. Let $\mathbf{u} = (0010)$, $\mathbf{v} = (0100)$. Then $d(\mathbf{u}, \mathbf{v}) = 2$. But $\mathbf{u}G = (0010110)$, $\mathbf{v}G = (0100101)$. Then $d(\mathbf{u}G, \mathbf{v}G) = 4$.*

How linear encoding works . The data stream (= message stream) is blocked off into segments of length k. Then each segment $\mathbf{u}$ of length k is encoded to $\mathbf{u}G$. The code word $\mathbf{u}G$ is then transmitted. The receiver has a list of the code words in C and decodes the received vector to the nearest code word in C. From this code word the receiver calculates the message since the encoding function $\mathbf{u} \to \mathbf{u}G$ is one-to-one.

The fact that C is linear has several advantages. For example, if $\mathbf{v}_1, \mathbf{v}_2$ are in C then $d(\mathbf{v}_1, \mathbf{v}_2) = d(\mathbf{v}_1 - \mathbf{v}_2, \mathbf{0})$ where $\mathbf{0}$ is the all-zero code word. Then $d(\mathbf{v}_1 - \mathbf{v}_2, \mathbf{0})$ is the number of non-zero elements in $\mathbf{v}_1 - \mathbf{v}_2$. This is called the **weight** of $\mathbf{v}_1 - \mathbf{v}_2$. Now $\mathbf{v}_1 - \mathbf{v}_2$ is in C. We have shown the following.

The minimum distance between any 2 code words in a linear code C equals the minimum weight d of any non-zero code word in C.

This is useful, because instead of testing all pairs of words in C we need only measure the weight of words in C.

Example 20.5 *The minimum distance between any 2 distinct code words in the code in example 20.3, i.e. the minimum distance of the code C, is 3. This follows from the fact that the minimum weight of any code word in C is 3.*

We now describe another approach to linear codes, as follows. Two vectors of length n are said to be **orthogonal** if their dot product is zero.

Example 20.6 *The binary vectors (1100) and (1111) are orthogonal since $(1)(1) + (1)(1) + (0)(1) + (0)(1) = 1 + 1 = 0$ (in binary).*

Example 20.7 *(1100) is orthogonal to itself (in binary). Thus, if C is the linear code consisting of $\{(1100), (0000)\}$ we have $C \subset C^{\perp}$.*

This is in sharp contrast to the Euclidean situation over the reals. However, the dimension formula still works: it can be shown by solving a system of linear equations.

Dimension formula . *Let C be a linear (n, k) code. Then $C^\perp$, the set of all vectors that are orthogonal to each vector in C, is a linear $(n, n - k)$ code.*

To check if a vector $\mathbf{x}$ is in $C^\perp$ we need only check that $\mathbf{x}$ is orthogonal to each row of a generator matrix G. The code $C^\perp$ is called **the dual code** or **the orthogonal** code of C.

We now examine $\left(C^\perp\right)^\perp$, the set of all vectors orthogonal to all vectors in $C^\perp$. By definition, C is contained in $\left(C^\perp\right)^\perp$. Also, $\dim\left(C^\perp\right)^\perp = n - (n - k) = k = \dim C$.

Double dual formula $\left(C^\perp\right)^\perp = C$.

Thus, C can equally well be described as the set of all vectors orthogonal to $\left(C^\perp\right)$.

A linear (n, k) code C is said to be in **systematic form** if a generator matrix for C is of the form $G = [I_k \mid A]$ where I is the $k \times k$ identity matrix and A is some matrix of size $k \times (n - k)$. In such a case the first k entries of $\mathbf{u}G$ are just the k entries of the message $\mathbf{u}$ and are called the **information** or **message** digits. The remaining $n - k$ digits are called **parity bits** or **parity digits**.

We can now establish the following result.

Algebraic representation of $C^\perp$ Let $G = (I_k \mid A)$ be a generator matrix for a linear (n, k) code C. Then $H = (-A^t \mid I_{n-k})$ is a generator matrix for $C^\perp$, where A^t is the transpose matrix of A. The code C can be described as the set of all vectors $\mathbf{v}$ whose dot product with each row of H is zero. In other words $C = \{\mathbf{v} \mid \mathbf{v}H^t = 0\}$.

To see that H is a generator matrix for $C^\perp$ we need only check that the corresponding dot products are zero. H is called a **parity check matrix** for C (or G).

Example 20.8 *Let G be the matrix in Example 20.3 and C the corresponding binary $(7, 4)$ code. Then C is in systematic form. A parity check matrix H is given by the matrix*

$$H = \begin{pmatrix} 0 & 1 & 1 & 1 & 1 & 0 & 0 \\ 1 & 0 & 1 & 1 & 0 & 1 & 0 \\ 1 & 1 & 0 & 1 & 0 & 0 & 1 \end{pmatrix}.$$

Example 20.9 *The repetition code of length n in Example 20.1 is dual to the parity check code C_n in Example 20.2 over any field.*

To see this we can check that the relevant dot products are zero and that the dimensions add up.

Let G be any generator matrix for an (n, k) linear code C. Then by **row reduction** we can reduce G to systematic form. The resulting code C_1 may not be the same as C (since we may permute columns in the process) but is **equivalent** to C in that distances are preserved.

20.3   Parity Checks, the Syndrome, Weights

Let C be an (n, k) linear code over F. We can show the following result.

Theorem 20.10 C *has a code word of weight t if and only if some t columns of H are linearly dependent, where H is a parity check matrix for C.*

For, suppose $\mathbf{c}$ in C has weight t. Now $\mathbf{c}H^t = 0$ so we get a dependence relation among t rows of H^t i.e. among the corresponding t columns of H. The converse also holds.

Take an easy example. Suppose the sum of the first 2 columns of H is zero. Then the dot product of $(110\dots0)$ with each row of H is zero. So $(1100\dots0)$ is in C — and has weight 2.

Next, let $\mathbf{w}$ be any vector in of length n. The **syndrome** of $\mathbf{w}$ is defined to be $\mathbf{w}H^t$, which is a row vector over F. Using our previous discussions we have part of the following result.

Theorem 20.11 *(a) The vector $\mathbf{w}$ is in C if and only if the syndrome of $\mathbf{w}$ is the zero vector.*

(b) The syndromes of $\mathbf{w}$ and $\mathbf{w} + \mathbf{c}$ are equal for any code word $\mathbf{c}$ in C.

(c) If $\mathbf{w}_1, \mathbf{w}_2$ have the same syndrome then $\mathbf{w}_1 = \mathbf{w}_2 + \mathbf{c}$ for some code word $\mathbf{c}$ in C. Thus the set of all vectors having a given syndrome forms a coset.

To see (c) observe that if $\mathbf{w}_1 H^t = \mathbf{w}_2 H^t$ then $(\mathbf{w}_1 - \mathbf{w}_2) H^t = 0$, so $\mathbf{w}_1 - \mathbf{w}_2$ is in C giving $\mathbf{w}_1 = \mathbf{w}_2 + \mathbf{c}$. Then the set of all vectors $\mathbf{w}_1$ having the same syndrome as $\mathbf{w}_2$ is the coset $\mathbf{w}_2 + C$.

We can use the syndrome idea for decoding. Let C be any linear (n, k) code over F where F is any finite field. Then the set of all vectors of the form $\mathbf{c} + \mathbf{x}$, $\mathbf{c}$ in C, which is denoted by $C + \mathbf{x}$ is called a **coset of** C.

Example 20.12 *Let $\mathbf{x}$ be any code word $\mathbf{c}$. Then $C + \mathbf{x} = C + \mathbf{c} = C$. The reason is that C, being a subspace, is closed under addition.*

Example 20.13 *For any $\mathbf{x}$, the coset $C + \mathbf{x}$ equals the coset $C + \mathbf{x}_1$ where $\mathbf{x}_1$ is any vector in the coset $C + \mathbf{x}$.*

To see this we have $C + \mathbf{x}_1 = C + (\mathbf{c} + \mathbf{x}) = (C + \mathbf{c}) + \mathbf{x} = C + \mathbf{x}$.

From Theorem 20.11 above we see that two vectors have the same syndrome if and only if they lie in the same coset of C. A **coset leader** for a coset $C + \mathbf{x}$ is a vector $\mathbf{c}_0 + \mathbf{e}$ of minimal weight in $C + \mathbf{x}$. Of course, there may be several coset leaders in $C + \mathbf{x}$.

The syndrome decoding procedure works on the assumption that a *small number of errors is more likely than a large number of errors*. Suppose that $\mathbf{y}$ is received. We calculate

$\mathbf{y}H^t = S(\mathbf{y})$, the syndrome of $\mathbf{y}$. If $S(\mathbf{y}) = \mathbf{0}$ then, since $\mathbf{y}$ is in C, we assume that no errors occurred and that $\mathbf{y}$ was transmitted. We note in passing that for **any** c in C, $S(\mathbf{c}) = \mathbf{0}$.

Now suppose $S(\mathbf{y}) = \mathbf{z} \neq \mathbf{0}$. The set of all vectors of length n having syndrome $\mathbf{z}$ consists of a coset $C + \mathbf{x}$, and this coset contains $\mathbf{y}$. Thus $\mathbf{y}$ is one of the vectors $\{\mathbf{c}_1 + \mathbf{x}, \mathbf{c}_2 + \mathbf{x}, \ldots, \mathbf{c}_t + \mathbf{x}\}$ where $t = q^k$ is the total number of code words in C. We can think of $\mathbf{y}$ as having been obtained by transmitting some $\mathbf{c}_i$ which gets distorted by the error pattern $\mathbf{x}$ in transmission, so $\mathbf{y} = \mathbf{c}_i + \mathbf{x}$.

Now let $\mathbf{e}$ be the coset leader in the coset $C + \mathbf{x}$. So $C + \mathbf{x} = C + \mathbf{e}$ as in example 20.13 and $wt(\mathbf{e}) \leq wt(\mathbf{x})$, with wt denoting weight, by definition of the coset leader.

Then $C + \mathbf{x} = C + \mathbf{e} = \{\mathbf{c}_1 + \mathbf{e}, \mathbf{c}_2 + \mathbf{e}, \ldots, \mathbf{c}_t + \mathbf{e}\}$. So $\mathbf{y}$ is also equal to $\mathbf{c}_j + \mathbf{e}$ and $\mathbf{y}$ could have been obtained by transmitting $\mathbf{c}_j$ which gets distorted to $\mathbf{c}_j + \mathbf{e}$ with error pattern $\mathbf{e}$. If $wt(\mathbf{e}) < wt(\mathbf{x})$ this is more likely than having $\mathbf{c}_i$ transmitted as above, with error pattern $\mathbf{x}$. Now to our punch line. We assume $\mathbf{y} = \mathbf{c}_j + \mathbf{e}$ so we decode $\mathbf{y}$ as $\mathbf{c}_j = \mathbf{y} - \mathbf{e}$. This explains the following decoding algorithm.

Syndrome Decoding Algorithm

(a) For a received vector $\mathbf{y}$ calculate $\mathbf{y}H^t$, the syndrome $S(\mathbf{y})$ of $\mathbf{y}$.

(b) Find a coset leader $\mathbf{w}$ in that coset having the property that each coset vector has $S(\mathbf{y})$ as its syndrome.

(c) Decode $\mathbf{y}$ as $\mathbf{y} - \mathbf{w}$. If there are several coset leaders make an arbitrary choice.

Example 20.14 *Working in binary, let the $(4, 2)$ linear code C have systematic generator matrix* $G = \begin{pmatrix} 1 & 0 & 0 & 1 \\ 0 & 1 & 1 & 1 \end{pmatrix}$. *Then the code words in C are $\{(0000), (1001), (0111), (1110)\}$.*

Now $H = \begin{pmatrix} 0 & 1 & 1 & 0 \\ 1 & 1 & 0 & 1 \end{pmatrix}$. *Suppose the code word $\mathbf{y} = (1101)$ is received. Now $S(\mathbf{y}) =$*

$$\mathbf{y}H^t = (1101) \begin{pmatrix} 0 & 1 \\ 1 & 1 \\ 1 & 0 \\ 0 & 1 \end{pmatrix} = (11).$$

The coset $C + (0100) = \{(0100), (1101), (0011), (1010)\}$ has the property that the syndrome of each of its members is $(1, 1)$. The unique coset leader is $\mathbf{w} = (0100)$. So we decode $\mathbf{y}$ to $\mathbf{y} - \mathbf{w} = \mathbf{y} + \mathbf{w} = (1101) + 0100 = (1001)$.

For the record, we write down the cosets of C here with the coset leaders and the syndromes.

Coset	Coset Leader	Syndrome
(0000) (1001) (0111) (1110)	(0000)	(00)
(1000) (0001) (1111) (0110)	(1000) [or (0001)]	(01)
(0100) (1101) (0011) (1010)	(0100)	(11)
(0010) (1011) (0101) (1100)	(0010)	(10)

20.4   Hamming Codes, an Inequality

Initially we work in binary. Pick any positive integer $m \geq 3$. Form the matrix H_m of size $m \times s$, $s = 2^m - 1$ whose columns are all possible *non-zero* binary m-tuples. Using this matrix H_m as a parity check matrix, form the $(2^m - 1 - m) \times (2^m - 1)$ binary linear code C_m. This code C_m is called the **Hamming code** of length $2^m - 1$. Of course the words in C_m will depend on our ordering of the columns of H_m.

Theorem 20.15 C_m *has length* $2^m - 1$, *dimension* $2^m - 1 - m$ *and minimum weight* $d = 3$. *Moreover* C_m *is a perfect code.*

Proof. Since the columns in H contain the standard basis vectors $\begin{pmatrix} 1 \\ 0 \\ \vdots \\ 0 \end{pmatrix}, \begin{pmatrix} 0 \\ 1 \\ \vdots \\ 0 \end{pmatrix}, \ldots, \begin{pmatrix} 0 \\ 0 \\ \vdots \\ 1 \end{pmatrix}$

of length m the rank of H is m. Since the columns of H are distinct it follows from Theorem 20.10 of Section 20.3 that $d = 3$. C_m is perfect since by calculation the Hamming balls of radius 1 about the code words in C_m are pairwise disjoint and fill up the space i.e., the sphere-packing criterion is satisfied as described in Chapter 12.

Decoding is easy for the Hamming codes. Suppose that we index the columns of H in accordance with the binary representations of $1, 2, 3, \ldots, 2^m - 1$. So for $m = 3$ we get

$$H = \begin{pmatrix} 0 & 0 & 0 & 1 & 1 & 1 & 1 \\ 0 & 1 & 1 & 0 & 0 & 1 & 1 \\ 1 & 0 & 1 & 0 & 1 & 0 & 1 \end{pmatrix}.$$

Now suppose $\mathbf{u}$ is transmitted and $\mathbf{v} = \mathbf{u} + \mathbf{e}$ is received, where the error vector $\mathbf{e}$ is either the zero vector or else has exactly one 1 in it. Now the syndrome is $\mathbf{v}H^t = (\mathbf{u} + \mathbf{e})H^t = \mathbf{e}H^t$. We are getting the dot product of $\mathbf{e}$ with the rows of H. Suppose H is as above and

$e = (0001000)$. Then (the transpose) of vH^t will be $\begin{pmatrix} 1 \\ 0 \\ 0 \end{pmatrix}$, corresponding to the fourth column of H. We then know that the error is in the fourth slot. So we correct the received word v by adding 1 in the fourth slot, since this operation will change 1 to 0 or 0 to 1. ■

Example 20.16 *Let $m = 3$ and let H be as above. Let the received word be $v = (1010011)$.*

The transpose of the syndrome is $vH^t = \begin{pmatrix} 0 \\ 1 \\ 1 \end{pmatrix}$. So there is an error in position number three. Thus v gets decoded to $v + (0010000) = (1000011)$.

We now obtain an inequality involving the minimum distance and dimension of a linear code. For convenience, we restrict our attention to the binary case.

Theorem 20.17 *If a binary linear code C of length n and dimension k has minimum distance $d \geq 2t + 1$ then $2^{n-k} \geq \binom{n}{0} + \binom{n}{1} + \binom{n}{2} + \cdots + \binom{n}{t}$.*

Proof. Let u, v be distinct vectors, each of which has weight at most t. If the syndrome of u were equal to the syndrome of v then the syndrome of $u + v$ is 0, implying that $u + v = c$ in C. Then c would be a non-zero code word in C whose weight is at most $t + t = 2t < d$. This is impossible. Thus, since there are exactly $\binom{n}{i}$ binary vectors of length i, $0 \leq i \leq t$, the number of distinct syndromes is at least $\binom{n}{0} + \binom{n}{1} + \binom{n}{2} + \cdots + \binom{n}{t}$. Now any syndrome is a binary string of length $n - k$. There are exactly 2^{n-k} such strings. Therefore, $2^{n-k} \geq \binom{n}{0} + \binom{n}{1} + \binom{n}{2} + \cdots + \binom{n}{t}$. ■

20.5 Perfect Codes, Errors and the BSC

Suppose C is an (n, k) linear code over a field $F = GF(q)$ of order q. Then the number M of code words in C is q^k. The **rate of** C is defined to be $\frac{\log_q M}{n} = \frac{k \log q}{n}$. Note that when $q = 2$ this is in accordance with our previous definition in Part II of this book.

Now suppose also that C is perfect and can correct up to t errors. *Then, if at most t errors occur, all transmitted code words will be decoded correctly* because **every** vector lies within the (Hamming) ball of some code word in C of radius t. However, if more than t errors occur in a transmission, then **there is certain to be a decoding error**. For, if c is transmitted and w is received with $d(c, w) > t$ then w will be decoded to the unique code word c_1 with $d(c_1, w) \leq t$. So $c_1 \neq c$ and we have a decoding error.

For example, let's look at the case when $C = C_m$ is a Hamming code (which is perfect), with $t = 1$. The rate of C_m is $\frac{2^m - 1 - m}{2^m - 1}$ so **the rate tends to 1 as m gets large**. Suppose we transmit over the BSC with **parameter** p. From the above, there is a decoding error if there is more than a one-bit transmission error.

The code length is $n = 2^m - 1$. We have $\Pr(0 \text{ or exactly 1 errors}) = q^n + \binom{n}{1}q^{n-1}p^1$, $q = 1 - p$. Thus the probability of a decoding error is $1 - \left[q^n + \binom{n}{1}q^{n-1}p^1\right]$. This tends to 1 as n gets large as pointed out by Jones and Jones [JJ00].

This gives a nice vindication of Shannon's channel capacity theorem. Here, the rate tends to 1 when n is large. In particular for large n the rate is greater than the capacity function $1 - H(p)$. Shannon's result only guarantees accuracy of decoding *when the rate is less than the capacity*. Here, as we have seen, the error probability tends to 1!

20.6   Generalizations of Binary Hamming Codes

Instead of working in binary we can work over $GF(q)$. Each column of H is defined to be a non-zero vector of length m, and no two columns are allowed to be scalar multiples of each other. We then get a matrix H of size $m \times \left(\frac{q^m-1}{q-1}\right)$ of rank k. H is a parity check matrix for a Hamming code $C_m(q)$ which is a perfect 1-error correcting code over $GF(q)$. The columns of H are the points of a projective space of (projective) dimension $m - 1$. For $m = 3$, $q = 2$ we get the famous 7 point Fano projective plane.

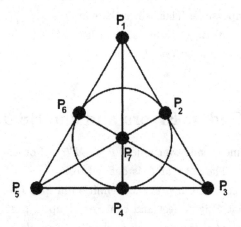

Figure 20.1: Fano's Geometry with 7 lines and 7 vertices

The following matrix covers the case when $q = 3$, $m = 3$.

$$H = \begin{pmatrix} 1 & 1 & 0 & 1 & 2 & 0 & 2 & 2 & 1 & 2 & 1 & 0 & 0 \\ 1 & 0 & 1 & 1 & 1 & 2 & 0 & 1 & 2 & 2 & 0 & 1 & 0 \\ 0 & 1 & 1 & 1 & 0 & 1 & 1 & 1 & 1 & 1 & 0 & 0 & 1 \end{pmatrix}.$$

20.7 The Football Pools Problem, Extended Hamming Codes

We present here a surprising application of Hamming codes to the topic of Football pools, a topic which does not appear to be directly related to the errors that occur in the transmission of messages. Suppose that a **football pool** consists of m soccer matches, each of which can end in either a draw(0), a loss(1) or a win(2) for the home team. Any outcome for all m games can be considered a vector over $GF(3)$ of length m. The number of possible outcomes of all m games is 3^m. A first or second prize is awarded to any person who guesses at least $m-1$ games correctly. We want to solve the following problem. *What is the fewest number of guesses required to be certain of winning at least a second prize?*

Winning at least a second prize corresponds to choosing a ternary m-tuple that differs from the winning m-tuple in at most one slot. We choose as our set of guesses all of the code words in the ternary Hamming code of length $m = 3^n + 3^{n-1} + \cdots + 3 + 1$. From the nature of the Hamming code one can show that this set of guesses is optimal for guaranteeing at least a second prize. Since a Hamming code is a perfect code in which every vector lies in exactly one Hamming ball of radius 1 centered at a code word, we know that the code words are sufficiently close to every vector. Let C_m be the Hamming code which comes from the n-dimensional projective geometry over $GF(3)$. Then the number of rows in a parity-check matrix is $n+1$ and the number of columns is the number of points in the geometry, which is $m = 3^n + \cdots + 3 + 1$. Hence the number of rows in a generator matrix for C_m is $m - n - 1$. Thus, the number of code words is 3^{m-n-1}. Each value of n determines a possible value of m. For example, if $n = 1$, then $m = 4$. If $n = 2$, then $m = 13$. In this case, the number of guesses required to win at least a second prize in a pool of 13 games is $3^{m-n-1} = 3^{10}$.

If the value of m does not equal the number of points in some projective geometry over $GF(3)$, then a different approach is required.

Extended Hamming Codes Let H be a parity-check matrix for a Hamming code. We create a new matrix H^* by first adding a row of ones at the top of H. Then, we add a column of zeros to the left side of H. Next, we change the number in the upper left hand corner of this new matrix to -1. This new matrix H^* is a parity-check matrix for a linear code C, called an **extended Hamming code**. The words in C are the words in the Hamming code with a new first digit which is the sum of all the other digits in the code. Extended Hamming codes are not perfect codes, but they do correct any one error in transmission and are capable of detecting any two errors.

Example 20.18 *The following matrix is a parity-check matrix for the extended Hamming* code arising from the Fano geometry.

$$H^* = \begin{pmatrix} 1 & 1 & 1 & 1 & 1 & 1 & 1 & 1 \\ 0 & 0 & 1 & 1 & 1 & 1 & 0 & 0 \\ 0 & 1 & 0 & 1 & 1 & 0 & 1 & 0 \\ 0 & 1 & 1 & 0 & 1 & 0 & 0 & 1 \end{pmatrix}.$$

20.8 Golay Codes

In this section we will discuss all perfect codes which can correct more than one error. The easy examples are the repetition codes over $GF(2)$. The repetition code of length $2k + 1$ corrects any k errors. It is perfect over $GF(2)$ since any $(2k + 1)$-tuple will have either $k + 1$ zeros or $k + 1$ ones and hence, will be corrected to one of the two possible code words: $(000 \ldots 0)$ and $(111 \ldots 1)$.

Golay searched for perfect linear codes. In addition to finding the repetition codes and some of the Hamming codes, he also found some non-trivial examples. He looked for solutions to the sphere packing condition which must be satisfied by a perfect code, namely the following.

$$\sum_{i=0}^{t} \binom{n}{i} (q - 1)^i = q^{n-j} \tag{20.1}$$

One solution is $q = 3$, $n = 11$, $j = 6$ and $t = 2$. Golay created the following parity-check matrix for the corresponding ternary code G_{11}.

$$H = \begin{pmatrix} 1 & 1 & 1 & 2 & 2 & 0 & 1 & 0 & 0 & 0 & 0 \\ 1 & 1 & 2 & 1 & 0 & 2 & 0 & 1 & 0 & 0 & 0 \\ 1 & 2 & 1 & 0 & 1 & 2 & 0 & 0 & 1 & 0 & 0 \\ 1 & 2 & 0 & 1 & 2 & 1 & 0 & 0 & 0 & 1 & 0 \\ 1 & 0 & 2 & 2 & 1 & 1 & 0 & 0 & 0 & 0 & 1 \end{pmatrix}$$

The five rows of H are linearly independant over $GF(3)$, the field of order 3. Here we refer to Jones and Jones [JJ00]. They comment as follows. "With considerable patience one can show that there are no sets of four or fewer linearly independant columns whereas there is a set of five linearly dependant columns (for instance, $c_2 - c_7 - c_8 + c_9 + c_{10} = 0$)."

Thus, $H^\perp = G_{11}$ has minimum distance 5 and can correct up to 2 errors so $t = 2$. Then, since $\sum_{i=0}^{2} \binom{n}{i}(q - 1)^i = 243 - 3^5$, we conclude that G_{11} is perfect.

There are many other ways of representing this Golay code including quadratic residue codes.

Another solution, $q = 2$, $n = 23$, $j = 12$ and $t = 3$ leads to the binary Golay code G_{23} with a parity-check matrix $H = [A^t \mid I]$ where

$$A^t = \begin{pmatrix} 1 & 0 & 0 & 1 & 1 & 1 & 0 & 0 & 0 & 1 & 1 & 1 \\ 1 & 0 & 1 & 0 & 1 & 1 & 0 & 1 & 1 & 0 & 0 & 1 \\ 1 & 0 & 1 & 1 & 0 & 1 & 1 & 0 & 1 & 0 & 1 & 0 \\ 1 & 0 & 1 & 1 & 1 & 0 & 1 & 1 & 0 & 1 & 0 & 0 \\ 1 & 1 & 0 & 0 & 1 & 1 & 1 & 0 & 1 & 1 & 0 & 0 \\ 1 & 1 & 0 & 1 & 0 & 1 & 1 & 1 & 0 & 0 & 0 & 1 \\ 1 & 1 & 0 & 1 & 1 & 0 & 0 & 1 & 1 & 0 & 1 & 0 \\ 1 & 1 & 1 & 0 & 0 & 1 & 0 & 1 & 0 & 1 & 1 & 0 \\ 1 & 1 & 1 & 0 & 1 & 0 & 1 & 0 & 0 & 0 & 1 & 1 \\ 1 & 1 & 1 & 1 & 0 & 0 & 0 & 0 & 1 & 1 & 0 & 1 \\ 0 & 1 & 1 & 1 & 1 & 1 & 1 & 1 & 1 & 1 & 1 & 1 \end{pmatrix}$$

In addition to being extremely useful codes, Golay codes are related to other mathematical structures such as simple groups, lattices, block designs and Steiner systems. These connections give insight into how Golay chose a correct parity-check matrices for these codes. The reader is referred to Anderson [And89] for details of the block design on 24 points with block size 8, denoted by $S(5, 8, 24)$.

20.9 McEliece Cryptosystem

In order to correct the errors in a linear code, it is necessary to know something about how the code is constructed, such as the parity-check matrix. It is possible to create codes that are so complicated that correcting a message with many errors is infeasible without this knowledge. This fact can be used to enhance the security of a cryptosystem. The McEliece cryptosystem is based on the sender purposely putting errors into the message, but not so many that the message cannot be decoded if the receiver has knowledge of the code. Eavesdroppers, without this knowledge, have little hope of correcting the message and therefore cannot determine what the message is.

The codes which are often used for this purpose are Goppa codes, defined over $GF(2)$. The word length is $n = 2^m$ and the distance between code words is $2t + 1$ for some integer t. A generator matrix G for the code is of size $k \times n$ where $k = n - mt$. For example, if $m = 9$ and $t = 30$, the generator matrix is 242×512, the word length is 512 and the code can correct as many as 30 errors.

The encryption works as follows. A message M of length k is multiplied on the right by the product $G^+ = SGP$ where S is an invertible $k \times k$ matrix, G is a $k \times n$ generator matrix for the code, and P is a $n \times n$ permutation matrix. (A permutation matrix over $GF(2)$ is characterized by the fact that each row and each column of P contains exactly one 1.) **The receiver keeps these matrices secret.** Only G^+, the product of the three matrices, is made public. The sender then adds an error E of weight at most t to the product $M \cdot G^+$. (Making the weight of the error E close to t will make the message harder to correct.) So the transmitted message is $M_1 = MG^+ + E$.

The receiver multiplies M_1 by P^{-1} to get $M_1' = M_1 P^{-1} = MSG + EP^{-1}$. Since P is a permutation matrix, it has the effect of permuting the components of any vector. Hence, EP^{-1} has at most t ones, just as E does. The vector M_1' is decoded, using knowledge of the code, by the receiver producing MSG which is a code word. Assuming that the matrix G is in systematic form, i.e. has the form $G = [I|A]$, MS gives the first k components of MSG. Multiplying MS on the right by S^{-1} produces the intended message M.

One problem with this encryption (which offers only computational security) is that the public key G^+ must be a very large matrix.

20.10   Historical Remarks

Soccer Pools   Suppose we have a soccer pool with 12 matches with the outcomes being a home win, home loss, or draw for each game. Let's assume that even a novice is guaranteed to know one "sure thing." We are now down to a pool with 11 matches. So now our problem is to buy as few tickets as possible in such a way that we differ by at most two from the real outcome. In other words we want to be sure that we are off by at most two results out of the eleven matches.

This is reminiscent of the football pools problem using Hamming codes except that now we allow ourselves to be wrong on two matches instead of just one match.

The most economical wqy of purchasing the tickets is by buying exactly those tickets – regarded as ternary 11-tuples that are in the Golay code G_{11}. Indeed, Juhani Virtakallio published the code G_{11} in issues 27, 28 and 33 of the Finnish soccer magazine Veikkaaja, one and a half years before Golay published G_{11}. The history of this is described in H. Hamalainen and S. Rankinen [HR91], and in the following (unpublished) paper by I. Honkala, *On the early history of the ternary Golay code*, which is an appendix to the paper above in the Journal of Combinatorial Theory.

Quoting from Bary quoting from Honkola, Virtakallio puts it laconically in this way.

> The following system with 729 columns [= codewords] was born in my brain during a period of depression in football-pool prizes. Because the prizes were too small at that time to compensate the investments that would have been required if the system had been used week after week, the system remained unpublished and was forgotten among other systems. When during the last winter the football-pool prizes reached a peak, there was talk with the editors about publishing the system but they could not fit the 729 columns into the magazine. Only now, when I discovered a method to obtain the required saving

of space, does this system get a chance to enrich the possibilities of players, and perhaps the players themselves.

If the match chosen to be sure the match has has been forecast correctly, the system guarantees at least 10 correct results. In the model we only present how to forecast the 11 other matches, the sure match has not been written down.

Golay In his famous and remarkable half-page paper in 1949 [Gol49] Golay describes not just the two Golay codes but also the (perfect) binary Hamming codes and the perfect binary repetition codes of odd length.

20.11 Problems

1. Find all the code words in the linear code C, over the field of order 3, if a generator matrix for C is $G = [I|A]$ where $A = \begin{pmatrix} 1 & 2 \\ 2 & 1 \end{pmatrix}$. Also, find all the code words in the dual code $C^{\perp}$.

2. If $G = \begin{pmatrix} 2 & 4 & 2 & 3 \\ 1 & 2 & 0 & 1 \end{pmatrix}$ is a generator matrix for the linear code C over the field $GF(5)$, find a generator matrix G' in the form $G' = [I|A]$ for an equivalent code C'.

3. Suppose that the minimum weight of any non-zero code word in the linear code C is 8. How many errors can the code correct?

4. What is the rank of a generator matrix G if G encodes 3 digit messages in binary?

5. The following vectors were received from a transmission using a linear code over the field $GF(5)$: (224332), (412311) and (324410). Correct each of these vectors to the nearest code word if the generator matrix was $G = [I|A]$ and $A = \begin{pmatrix} 1 & 1 & 4 \\ 0 & 2 & 4 \\ 2 & 3 & 4 \end{pmatrix}$.

6. What is the information rate for the binary Hamming code of length 15?

7. What finite field is used to create the Hamming code of length 156?

8. Find the columns of a parity-check matrix for a binary Hamming code of length 15.

9. How many bets need to be placed in a football pool of 4 games in order to be sure of winning at least second prize?

10. If C is a k dimensional code over $GF(q)$, then how many words are in C?

11. If B is the following matrix, then how many words of the form $\mathbf{u}B$ are there? That is, how many codewords does B generate? We are working over $GF(3)$.

$$B = \begin{pmatrix} 1 & 2 & 1 & 1 \\ 2 & 2 & 1 & 2 \\ 0 & 1 & 2 & 0 \end{pmatrix}$$

12. How many codewords would we have expected using B in the previous question? Why is B an ineligible choice for a generator matrix of a linear code?

13. Find a code C equivalent to

$$D = \begin{pmatrix} 2 & 1 & 2 & 1 \\ 1 & 2 & 2 & 1 \end{pmatrix}$$

that is in the form $[I|A]$ over the field $GF(3)$. Then find $C^{\perp}$, the code dual to C.

14. The matrix G below is a generator matrix for a Hamming code of length 7. The vector $\mathbf{y} = (1110101)$ is received. Given that the only possible transmission errors are binary vectors of weight 1, what was the transmitted message?

$$G = \begin{pmatrix} 1 & 0 & 0 & 0 & 0 & 1 & 1 \\ 0 & 1 & 0 & 0 & 1 & 0 & 1 \\ 0 & 0 & 1 & 0 & 1 & 1 & 0 \\ 0 & 0 & 0 & 1 & 1 & 1 & 1 \end{pmatrix}$$

15. Repeat question 14 with $\mathbf{y} = (1001100)$.

16. Use $\mathbf{y} = (0111011)$ to do question 14 again.

17. Consider a binary code C with generator matrix G given by

$$G = \begin{pmatrix} 1 & 0 & 0 & 1 & 1 & 0 \\ 0 & 1 & 0 & 1 & 0 & 1 \\ 0 & 0 & 1 & 0 & 1 & 1 \end{pmatrix}$$

Find the generator matrix for $C^{\perp}$. We are working in binary.

18. For the generator matrix G above, find all of the code words belonging to C.

19. What is the information rate for the code given in Problem 18?

20. Using the parity-check matrix H, of a given binary code with generator matrix G, determine whether a transmission error occurred given that the received message was $(1,0,1,0,1,0)$, where

$$H = \begin{pmatrix} 1 & 1 & 1 & 1 & 0 & 0 \\ 1 & 1 & 0 & 0 & 1 & 0 \\ 1 & 0 & 1 & 0 & 0 & 1 \end{pmatrix}$$

21. Find the Hamming distance between the two vectors $\mathbf{u}, \mathbf{v}$ given by $\mathbf{u} = (1, 1, 0, 0, 1, 0, 1, 0)$, $\mathbf{v} = (0, 1, 0, 1, 0, 1, 0, 1)$

22. If the probability of a bit error is p, what is the probability that a received message obtained from the binary repetition code of length 5 is received correctly?

20.12 Solutions

1. The code words are $(0,0)G = (0,0,0,0)$, $(1,0)G = (1,0,1,2)$, $(2,0)G = (2,0,2,1)$, $(0,1)G = (0,1,2,1)$, $(0,2)G = (0,2,1,2)$, $(1,1)G = (1,1,0,0)$, $(2,2)G = (2,2,0,0)$, $(1,2)G = (1,2,2,1)$ and $(2,1)G = (2,1,1,2)$. The code words in $G^{\perp}$ are those vectors which are orthogonal to every row in the matrix G. They are $(0,0,0,0)$, $(0,0,1,1)$, $(0,0,2,2)$, $(1,2,0,1)$, $(2,1,0,2)$, $(1,2,1,2)$, $(2,1,2,1)$, $(1,2,2,0)$, and $(2,1,1,0)$.

2. We apply row operations to G. First, we interchange the two rows of G. Then we subtract 2 times the new top row from the bottom row. This produces the matrix $G_2 = \begin{pmatrix} 1 & 2 & 0 & 1 \\ 0 & 0 & 2 & 1 \end{pmatrix}$, which is row reduced. Interchanging the second and third columns produces $G' = \begin{pmatrix} 1 & 0 & 2 & 1 \\ 0 & 2 & 0 & 1 \end{pmatrix}$.

3. We set $2t + 1 = 8$. The solution is $t = 3.5$, but t must be an integer. So the correct answer is $t = 3$; the code can correct up to t errors.

4. $\text{rank}(G) = 3$.

5. We construct a parity-check matrix $H = \begin{pmatrix} 4 & 0 & 3 & 1 & 0 & 0 \\ 4 & 3 & 2 & 0 & 1 & 0 \\ 1 & 1 & 1 & 0 & 0 & 1 \end{pmatrix}$. Multiplying each of these vectors by H^t, we obtain $(224332) H^t = (300)$, $(412311) H^t = (043)$ and $(324410) H^t = (324)$. Since (300) is 3 times the fourth column of H, we need to subtract 3 from the fourth component of (224332) to obtain the code word (224032). Since (043) is 3 times the second column of H, we need to subtract 3 from the second component of (412311) to obtain the code word (432311). Since (324) is not a multiple

of any column of H, we know that more than one error has been made. In this case, it is not possible to determine a unique nearest code word to (324410).

6. We know that $2^r - 1 = 15$, where r is the rank of H. Thus, $r = 4$, and it follows that G has $15 - 4$ rows, giving an information rate of $\frac{11}{15}$.

7. We need to find values of q and n such that $q^n + q^{n-1} + \cdots + q + 1 = 156$. Since $156 - 1 = 155$, the value of q must divide 155. We try $q = 5$ and notice that $1 + 5 + 25 + 125 = 156$. So the field is $GF(5)$.

8. We simply take all possible nonzero columns. One such example is

$$
\begin{pmatrix}
1 & 1 & 1 & 1 & 0 & 1 & 1 & 1 & 0 & 0 & 0 & 1 & 0 & 0 & 0 \\
1 & 1 & 1 & 0 & 1 & 1 & 0 & 0 & 1 & 1 & 0 & 0 & 1 & 0 & 0 \\
1 & 1 & 0 & 1 & 1 & 0 & 0 & 1 & 1 & 0 & 1 & 0 & 0 & 1 & 0 \\
1 & 0 & 1 & 1 & 1 & 0 & 1 & 0 & 0 & 1 & 1 & 0 & 0 & 0 & 1
\end{pmatrix}
$$

9. If $m = 4$, then $n = 2$. Hence the required number of bets is $3^{m-n-1} = 3^1 = 3$.

10. For each dimension, we choose an element of $GF(q)$. Since $GF(q)$ has q elements, it follows that C has q^k codewords.

11. As we let $\mathbf{u}$ range over all 3^3 possibilites, we get 9 codewords. They are (0000), (0120), (0210), (2212), (2002), (2122), (1121), (1211), (1001).

12. Since B has 3 rows, and we are working over $GF(3)$, there should have been 3^3 possible codewords. However, we found there were only 9. This is because B has rank 2, that is, the first and second rows of B sum to the third row. This makes B a poor choice because it means that several messages will get encoded to the same codeword, and thus we will not be able to decode uniquely. For example, the word (0000) could have been made by the message (000) or by (112).

To fix this problem, we could simply remove the third row of B to get the matrix

$$
B' = \begin{pmatrix}
1 & 2 & 1 & 1 \\
2 & 2 & 1 & 2
\end{pmatrix}
$$

which could be used to generate a linear code.

13. We take the matrix D and switch the rows to get

$$
\begin{pmatrix}
1 & 2 & 2 & 1 \\
2 & 1 & 2 & 1
\end{pmatrix}
$$

Then adding the rows and storing the result into the second row, we have

$$\begin{pmatrix} 1 & 2 & 2 & 1 \\ 0 & 0 & 1 & 2 \end{pmatrix}.$$

Switching the second and third columns, we get

$$\begin{pmatrix} 1 & 2 & 2 & 1 \\ 0 & 1 & 0 & 2 \end{pmatrix}$$

Finally, adding twice the second row to the first row, we arrive at a matrix for C.

$$[I|A] = \begin{pmatrix} 1 & 0 & 2 & 0 \\ 0 & 1 & 0 & 2 \end{pmatrix}$$

A matrix for $C^{\perp}$ is

$$[-A^t|I] = \begin{pmatrix} 1 & 0 & 1 & 0 \\ 0 & 1 & 0 & 1 \end{pmatrix}$$

14. A parity check matrix is

$$H = \begin{pmatrix} 0 & 1 & 1 & 1 & 1 & 0 & 0 \\ 1 & 0 & 1 & 1 & 0 & 1 & 0 \\ 1 & 1 & 0 & 1 & 0 & 0 & 1 \end{pmatrix}$$

and when we take the product $\mathbf{y}H^t$ we get (101). Since (101) is the second column of H, we conclude that there was an error in the second position. To correct this error, we add 1 to the second position, meaning that (1010101) was sent.

15. We have $\mathbf{y}H^t = (000)$, from which we conclude that there was no error in transmitting $\mathbf{y}$.

16. We find that $\mathbf{y}H^t = (111)$, and as (111) is the fourth column of H, we know that there was an error in the fourth position. Correcting this error, we reason that (0110011) was sent for $\mathbf{y}$.

17. In $GF(2)$, $A = -A$. Therefore, the generator matrix for $C^{\perp}$ is given by

$$H = [-A^t|I] = [A^t|I] = \begin{pmatrix} 1 & 1 & 0 & 1 & 0 & 0 \\ 1 & 0 & 1 & 0 & 1 & 0 \\ 0 & 1 & 1 & 0 & 0 & 1 \end{pmatrix}$$

18. To find the code words for C, we just need to carry out the multiplication $\mathbf{v}G$, where

v is a code word in C. Doing so yields the following vectors:

$$(0,0,0,0,0,0) \quad (0,1,0,1,0,1) \quad (1,0,0,1,1,0) \quad (1,1,0,0,1,1)$$
$$(0,0,1,0,1,1) \quad (0,1,1,1,1,0) \quad (1,0,1,1,0,1) \quad (1,1,1,0,0,0)$$

19. The information rate is $\frac{3}{6} = \frac{1}{2}$.

20. To determine if an error was sent, we need to compute $(1,0,1,0,1,0)H^t$ where H is the parity-check matrix. Since we end up with the vector $(0,0,0)$, we can conclude that no transmission error occurred. An alternative (albeit much longer) solution can be obtained by determining the generator matrix G and identifying all of the code words in C. By doing so, we notice that the code word $(1,0,1,0,1,0)$ is contained in our code list.

21. 6.

22. The binary repetition code of length 5 is given by $(0,0,0,0,0)$, $(1,1,1,1,1)$. Suppose we send the message $(0,0,0,0,0)$. If our received message contains no less than 3 zeros, we can assume that the transmitted message was $(0,0,0,0,0)$. The probability of at least 3 zeros is:

$$(1-p)^5 + 5(1-p)^4 p + 10(1-p)^3 p^2$$

Chapter 21

Linear Cyclic Codes, Shift Registers and CRC

Goals, Discussion We present the basic material concerning linear cyclic codes from a very elementary point of view while covering all the main ideas. We also show that linear cyclic codes and shift registers are equivalent. Lengthy discussions of polynomial rings are avoided.

New, Noteworthy Building on the results in earlier chapters we extend the theory of linear feedback shift registers (LFSRs) by explaining the surprising and elegant correspondence with linear cyclic codes. The work shows that the nature of the tap polynomial governs the length n of the maximum period. The period for **any** input will then divide n.

Here we also present two different methods for calculating the code dual to a given cyclic linear code. This is discussed in detail in Chapter 22 as well.

The algebraic theory of shift registers is closely connected to the protocol of the CRC (cylic redundancy checks), which are widely used on the internet and elsewhere. Here we summarize some of the results.

21.1 Cyclic Linear Codes

As motivation we mention some practical advantages of these codes. They are compactly described and easy to store. Encoding and decoding becomes quite simple. The main codes used in industry, namely the Reed–Solomon codes, are special kinds of cyclic linear codes.

A code C is said to be **cyclic** if the following holds. Whenever $\mathbf{c}$ is in C with $\mathbf{c} = (x_1, x_2, \ldots, x_n)$ then the n-tuple $\mathbf{c}_1 = (x_n, x_1, x_2, \ldots, x_{n-1})$ is also in C. The code-word $\mathbf{c}_1$ denoted by $\sigma(\mathbf{c})$ is called the **cyclic shift** of $\mathbf{c}$. Not all cyclic codes are **linear**.

Example 21.1 *Let* $C = \{(1,1,1,1)\}$. *Then* C *is cyclic but* C *is not linear.*

Example 21.2 *Let* $C = \{(0,0,0,0),(1,1,1,1)\}$. *Then* C *is cyclic and linear.*

Here is a **construction for a linear cyclic code** C. First, pick any finite field F and any positive integer n for the length of C. Next, pick **any** vector $\mathbf{w} = (w_1, w_2, \ldots, w_n)$.

Step 1 Form all vectors S obtained by repeatedly shifting $\mathbf{w}$ i.e. all vectors of the form $\mathbf{w}, \sigma\mathbf{w}, \sigma^2(\mathbf{w}), \ldots$. Here $\mathbf{w} = \sigma^0(\mathbf{w})$ where σ^0 is the null shift.

Step 2 The set $\langle S \rangle$ of all possible linear combinations of vectors in S is then a cyclic linear code C containing $\mathbf{w}$.

Theorem 21.3 C *is linear and cyclic. It is the smallest linear cyclic code containing* $\mathbf{w}$.

Proof. Certainly any cyclic code containing $\mathbf{w}$ must contain all vectors in S. Now if C is linear and C contains S then C must contain $\langle S \rangle$.

It remains to show that C is cyclic. Let $\mathbf{x}$ be any vector in C. We want to show that $\sigma(\mathbf{x})$ is in C. First, suppose that $\mathbf{x}$ is in S. Then $x = \sigma^i(\mathbf{w})$. Then $\sigma(\mathbf{x}) = \sigma\sigma^i(\mathbf{w}) = \sigma^{i+1}(\mathbf{w})$ so $\sigma(\mathbf{x})$ is in S.

Next, let $\mathbf{x}$ be a linear combination of two or more vectors in S. It suffices to consider the case when $\mathbf{x}$ is a linear combination of two such vectors.

So let $x = \alpha\mathbf{u} + \beta\mathbf{v}$ with α, β in F and u, v in S. We want to show that $\sigma(\mathbf{x})$ is in C. Now $\mathbf{u} = \sigma^i(\mathbf{w})$. Then $\sigma(\mathbf{u}) = \sigma^{i+1}(\mathbf{w})$ so $\sigma(\mathbf{u})$ is in S. Similarly, $\sigma(\mathbf{v})$ is in S. We have $\sigma(\mathbf{x}) = \sigma(\alpha\mathbf{u} + \beta\mathbf{v}) = \alpha\sigma(\mathbf{u}) + \beta\sigma(\mathbf{v})$ since σ is a linear mapping. Since $\sigma(\mathbf{u}), \sigma(\mathbf{v})$ are in S we have that $\sigma(\mathbf{x})$ is a linear combination of 2 vectors in S. Thus, $\sigma(\mathbf{x})$ is in C. ∎

Example 21.4 *Take* $n = 4$, $F = \{0,1\}$ *and* $\mathbf{w} = (1,0,1,0)$ *Then* S *consists of the vectors* $\{(1,0,1,0),(0,1,0,1)\}$. *Now* $\langle S \rangle = C$ *consists of* $\{(1,0,1,0),(0,1,0,1),(0,0,0,0),(1,1,1,1)\}$.

Example 21.5 *Let* $n = 3$, $F = \{0,1\}$, $\mathbf{w} = (1,0,1)$. *Then* $S = \{(1,0,1),(1,1,0),(0,1,1)\}$. *Here* $C = \langle S \rangle = S$ *together with* $(0,0,0)$.

The code C in **Theorem 21.3** is said to be **generated by** $\mathbf{w}$ **and we write** $C = \langle \mathbf{w} \rangle$.

Now suppose we start off with some linear cyclic code C. Can we always find a $\mathbf{w}$ in C so that $C = \langle \mathbf{w} \rangle$ as in Theorem 21.3? The answer is yes. In fact there is a unique way of choosing $\mathbf{w}$ in such a way that $\mathbf{w}$ corresponds to a **generator polynomial**.

For the details we have to "go algebraic". First, for any code word $\mathbf{c}$ in a code C of length n with $\mathbf{c} = (c_0, c_1, \ldots, c_{n-1})$ we associate a polynomial $\mathbf{c}(x) = c_0 + c_1 x + c_2 x^2 + \cdots + c_{n-1} x^{n-1}$ of degree at most $n-1$ over the field F. This polynomial is called the **code word polynomial** or the **associated polynomial**. On the other hand, given any polynomial *of degree less than n* say $c_0 + c_1 x + \cdots + c_{n-1} x^{n-1}$ we can associate with it a vector (which is not necessarily a code word) namely the vector $(c_0, c_1, \ldots, c_{n-1})$.

Example 21.6 *In Example 21.4 the code word* (1010) *has the associated polynomial* $1 + x^2$.

Again, we have $\mathbf{c}(x) = c_0 + c_1 x + \cdots + c_{n-1} x^{n-1}$. Then $x\mathbf{c}(x) = c_0 x + c_1 x^2 + c_2 x^3 + \cdots + c_{n-2} x^{n-1} + c_{n-1} x^n$. Now $x\mathbf{c}(x)$ has a degree bigger than $n-1$ so, at first glance, we cannot get a code word of length n from it. However, let's look at $Rem\left[x\mathbf{c}(x), x^n - 1\right]$. (see Chapter 19) The remainder will be some polynomial of degree less than n. Now $c_{n-1} x^n = c_{n-1}(x^n - 1) + c_{n-1}$. Thus the remainder is just the polynomial $c_{n-1} + c_0 x + c_1 x^2 + \cdots + c_{n-2} x^{n-2}$. This polynomial gives the vector $(c_{n-1}, c_0, c_1, \ldots, c_{n-2})$. But this is just $\sigma(\mathbf{c})$, the (cyclic) shift of $\mathbf{c}$!

Principle 1. Cyclically shifting a code word is equivalent to multiplying the corresponding polynomial by x and dividing by $(x^n - 1)$ (i.e. reducing mod$(x^n - 1)$). Multiplication by x^t is equivalent to t successive cyclic shifts.

Example 21.7 *In Example 21.4, take the code word* $\mathbf{c} = (0101)$. *The corresponding polynomial is* $x + x^3$. *Here* $n = 4$. *When we multiply by* x *we get the polynomial* $x^2 + x^4$. *We divide this by* $x^4 + 1$. *The high school long division method is presented as follows.*

$$x^4 + 1 \overline{)\, x^4 + x^2 }$$
$$\underline{x^4 + 1}$$
$$x^2 - 1$$

So the remainder is $x^2 - 1 = x^2 + 1$ since we are working in binary. This gives the code word (1010). We have $\sigma(0101) = (1010)$ as promised by the principle above. Now since $x\mathbf{c}(x)$ corresponds to $\sigma(\mathbf{c})$ we have that $x^2\mathbf{c}(x)$ corresponds to $\sigma^2(\mathbf{c})$ and so on.

Principle 2. Multiplying $\mathbf{c}(x)$ by any polynomial corresponds to a sequence of operations involving multiplying a code word by a scalar, cyclically shifting, and adding.

Principle 3. Let $\mathbf{c}(x)$ be the polynomial associated with a code word $\mathbf{c}$ in a cyclic linear code C of length n. Then, to get the polynomial associated with the cyclic shift of $\mathbf{c}$, multiply $\mathbf{c}(x)$ by x and replace x^n by 1 if x^n appears in the product.

Example 21.8 *In Example 21.7,* $\mathbf{c}(x) = x + x^3$. *Then* $x\mathbf{c}(x) = x(x + x^3) = x^2 + x^4$. *On putting* $x^4 = 1$ *we get* $x^2 + 1$ *as in Example 21.7.*

21.2 Generators for Cyclic Codes

Let C be a cyclic code of length n. Each code word $\mathbf{c} = (c_0, c_1, c_2, \ldots, c_{n-1})$ in C yields a polynomial $\mathbf{c}(x) = c_0 + c_1 x + c_2 x^2 + \cdots c_{n-1} x^{n-1}$ whose degree is at most $n - 1$. Let W be the set of all such polynomials. In W choose the polynomial g_1 of smallest degree, denoted by t, so $g_1(x) = \alpha_0 + \alpha_1 x + \alpha_2 x^2 + \cdots + \alpha_t x^t$, $\alpha_t \neq 0$. Dividing by $\alpha_t \neq 0$ we get a **monic polynomial** of minimal degree, monic meaning that the coefficient of the highest degree is 1. Denote this polynomial by $g(x)$. We have $g(x) = g_0 + g_1 x + g_2 x^2 + \cdots + x^{n-k}$, $t = n - k$.

We refer to $g(x)$ as **the generator polynomial** of C. With the above notation we have the following result.

Theorem 21.9 *There exists a unique polynomial $g(x)$ of smallest degree with leading co-efficient 1 obtained from the polynomials $c(x)$, c in C. Moreover $g(x)$ has the following properties.*

1. *$g(x)$ divides $c(x)$, for every c in C, with $g(x) = g_0 + g_1 x + \cdots + x^{n-k}$.*

2. *$g(x)$ divides $x^n - 1$. Thus x does not divide g and 0 is not a root of $g(x)$.*

3. *The code words corresponding to $g(x), xg(x), x^2 g(x), \ldots, x^{k-1} g(x)$ form a basis for a generator matrix G of C and k is the dimension of C. If $\mathbf{w}$ is the code word corresponding to $g(x)$ then $C = \langle \mathbf{w} \rangle$*

4. *We can write G as follows*

$$
G = \begin{pmatrix} g_0 & g_1 & \cdots & \cdots & g_{n-k-1} & 1 & 0 & 0 & & 0 & 0 \\ 0 & g_0 & g_1 & \cdots & \cdots & g_{n-k-1} & 1 & & \cdots & & \\ & & & & & & & & & & \\ 0 & 0 & 0 & & & g_0 & g_1 & \cdots & g_{n-k-1} & 1 \end{pmatrix}. \tag{21.1}
$$

5. *Conversely, if $g(x)$ is a monic polynomial of degree $n - k$ such that $g(x)$ divides $x^n - 1$ then there exists a linear cyclic code of length n, dimension k with generator polynomial $g(x)$.*

Remark 21.10 *A linear cyclic code C may contain polynomials other than the generator polynomial that also generate C. For instance, the code in Example 21.5 is generated by $\mathbf{w}$ whose associated polynomial is $1 + x^2$. However, the generator polynomial for C is $1 + x$.*

Remark 21.11 *In Theorem 21.3 of Section 21.1 the generator of the code C is the gcd of $(x^n - 1, \mathbf{w}(x))$ where gcd denotes greatest common divisor.*

Remark 21.12 *In Example 21.5 of Section 21.1 $g(x) = 1 + x$. Note that $g(x)$ divides $1 + x^3$ since, in Z_2, $1 + x^3 = (1 + x)(1 + x + x^2)$. The code word $\mathbf{w}$ gives the polynomial $1 + x^2$. Then $\gcd(1 + x^2, 1 + x^3) = \gcd((1 + x)(1 + x), (1 + x)(1 + x + x^2)) = 1 + x$, since we are working in binary.*

21.3 The Dual Code and The Two Methods

Let C be a linear cyclic code of length n, dimension k with generator polynomial $g(x)$ of degree $n-k$. Then $g(x)$ divides x^n-1 so that $x^n-1 = g(x)h(x)$, with $\deg(h) = n-(n-k) = k$. The dual code of C is cyclic of dimension $n-k$, length n. Thus, this dual code will have a generator polynomial of degree k. So "the numbers are right" for $h(x)$ to be the generator of the dual code $C^\perp$. This is almost correct. We have the following result.

Theorem 21.13 *The generator polynomial for $C^\perp$ is the "reverse polynomial" $h_1(x) = x^k h\left(\frac{1}{x}\right)$ which is obtained by reversing the order of the coefficients of $h(x)$, where $h(x)$ has degree k. The roots of $h_1(x)$ are the reciprocals of the roots of $h(x)$.*

Example 21.14 *We work over $GF(5)$ with $n = 4$ and $g(x) = (x-1)(x-2) = x^2 - 3x + 2$. Then $h(x) = (x-3)(x-4) = x^2 + 3x + 2$. Then the reverse polynomial $h_1(x)$ is given by $1 + 3x + 2x^2$. The roots of $h_1(x)$ are always the inverses or reciprocals of the roots of $h(x)$: in this case the roots of $h_1(x)$ are $\left\{\frac{1}{3}, \frac{1}{4}\right\} = \{2, 4\}$.*

Theorem 21.15 *Let $g(x) = g_0 + g_1 x + \cdots + g_{n-k-1} x^{n-k-1} + x^{n-k}$ be the generator polynomial for a linear cyclic code C of length $n = q-1$ over some finite field $F = GF(q)$. Let $\mathbf{u} = (u_0, u_1, \ldots, u_{n-1})$ be an element of the dual code $C^\perp$. Let $\mathbf{u}(x) = u_0 + u_1 x + \cdots u_{n-1} x^{n-1}$ and let $\mathbf{u}_1(x)$ be the reverse polynomial given by $\mathbf{u}_1(x) = u_{n-1} + u_{n-2} x + u_{n-3} x^2 + \cdots + u_0 x^{n-1}$. Then $g(x)\mathbf{u}_1(x)$ is a polynomial multiple of $x^n - 1$, so $g(x)\mathbf{u}_1(x) = f(x)(x^n - 1)$.*

Conversely, if $\mathbf{u}_1(x)$ is a polynomial given by $\mathbf{u}_1(x) = u_{n-1} + u_{n-2} x + \cdots + u_0 x^{n-1}$ with $g(x)\mathbf{u}_1(x) = f(x)(x^n - 1)$ then $(u_0, u_1, \ldots, u_{n-1})$ is in the dual code $C^\perp$. Similar results apply if we reverse the roles of g and h.

Sketch of proof. This is a matter of calculation linking up the dot product of vectors and the product of the corresponding polynomials: see for example Peterson and Weldon [PW72]. ∎

If the roots of $h(x)$ are $\beta_1, \beta_2, \ldots, \beta_i$ with $\beta_1, \beta_2, \ldots, \beta_i$ all non-zero then the roots of $h_1(x)$ are $\frac{1}{\beta_1}, \frac{1}{\beta_2}, \ldots, \frac{1}{\beta_i}$.

If $g(x)$ has all its roots say $\alpha_1, \alpha_2, \ldots, \alpha_t$ in the ground field $GF(q)$ and $n = q-1$ we then have two different methods of calculating $C^\perp$. For, let $g(x) = (x - \alpha_1)(x - \alpha_2) \ldots (x - \alpha_t)$ where all the α_i are non-zero. Then, from Chapter 19 $g(x)$ divides $x^{q-1} - 1 = \prod_\alpha (x - \alpha)$, $\alpha \neq 0$ in $GF(q)$.

We multiply out to get the code word corresponding to g. This code word is of the form $\mathbf{c} = ((-1)^t \alpha_1 \alpha_2 \ldots \alpha_t, \ldots - (\alpha_1 + \alpha_2 + \alpha_t), 1, 0, 0, 0, 0)$. Since $g(\alpha_1) = 0$ we see that the dot product of $\mathbf{c}$ with $(1, \alpha_1, \alpha_1^2, \ldots, \alpha^t, \alpha^{t+1} \ldots \alpha^{q-1})$ is 0. Similar results hold for $\alpha_2, \ldots, \alpha_t$. Also it follows that $\sigma^i(\mathbf{c})$, $0 \leq i \leq n - k - 1$ is orthogonal to this same vector, namely $(1, \alpha_1, \alpha_1^2, \ldots, \alpha_1^{q-1})$. Similar remarks hold if we replace α_1 by $\alpha_2, \alpha_3, \ldots, \alpha_t$. Thus the t rows $(1, \alpha_i, \alpha_i^2, \ldots, \alpha_i^{q-1})$, $1 \leq i \leq t$ give a basis for the dual code $C^\perp$. The fact that these

rows are linearly independent can be shown using the "Vandermonde trick" or other methods
mentioned in Chapter 22. This is **Method 1**.

Method 2 involves the following idea. We have $h(x) = \frac{x^{q-1}-1}{g(x)} = (x - \alpha_{t+1})\ldots(x -$
$\alpha_{q-1-t})$. Then the roots of $h_1(x)$ are $\frac{1}{\alpha_{t+1}}, \ldots \frac{1}{\alpha_{q-1-t}}$ so the generator of $C^\perp$ is $(x - \gamma_1)(x -$
$\gamma_2)\ldots(x - \gamma_{q-1-t})$ with $\gamma_i = \frac{1}{\alpha_i}$, $t+1 \le i \le q-1-t$.

Example 21.16 *Let* $F = GF(5)$ *and let* $g(x) = (x - 1)(x - 2) = x^2 - 3x + 2$. *Then,*
$n = q - 1 = 5 - 1 = 4$, $k = 2$ *and a generator matrix* G *for the cyclic code* C *generated by*
$g(x)$ *is as follows:*

$$G = \begin{pmatrix} 2 & -3 & 1 & 0 \\ 0 & 2 & -3 & 1 \end{pmatrix}. \tag{21.2}$$

From Method 1 we get that $C^\perp$ *has generator matrix given by* $\begin{pmatrix} 1 & 1^2 & 1^3 & 1^4 \\ 2 & 2^2 & 2^3 & 2^4 \end{pmatrix} =$

$\begin{pmatrix} 1 & 1 & 1 & 1 \\ 2 & 4 & 3 & 1 \end{pmatrix}.$

Method 2 gives that $h(x) = (x - 3)(x - 4)$. *Then* $h_1(x)$ *has roots* $\{\frac{1}{3}, \frac{1}{4}\} = \{2, -1\}$.
Therefore $C^\perp$ *has generator polynomial* $(x - 2)(x + 1) = x^2 - x - 2$ *and generator ma-*
trix $\begin{pmatrix} -2 & -1 & 1 & 0 \\ 0 & -2 & -1 & 1 \end{pmatrix}$. *Of course this generator matrix will give the same code as the*
previous generator matrix.

21.4   Linear Feedback Shift Registers and Codes

In Chapter 10 we discussed recurrence relations of the type $\sum_{j=0}^{k} c_j x_{i+j} = 0$ over the binary
field with $c_0 = c_k = 1$. For $i = 0$, this then gives $x_k = c_0 x_0 + c_1 x_1 + \cdots + c_{k-1} x_{k-1}$. We want
to generalize to arbitrary fields. Also, it will be convenient to change the notation from c
to h.

So we consider the recurrence given by

$$\sum_{j=0}^{k} h_j x_{i+j} = 0, \quad i = 0, 1, 2, \ldots \tag{21.3}$$

with $h_0 \ne 0$, $h_k = 1$ and with each coefficient h_j in some finite field $F = GF(q)$. From
equation 21.3 we get the following

$$x_{i+k} = -\sum_{j=0}^{k-1} h_j x_{i+j}, \quad i = 0, 1, 2, \tag{21.4}$$

(If we put $i = 0$ we get $x_k = -\sum_{j=0}^{k-1} h_j x_j$ which is what we had in the binary case in Chapter

10). The solutions of the recurrence relation are described in the following theorem.

Theorem 21.17 *Let $h(X) = \sum_{j=0}^{k} h_j x^j$ with $h_0 \neq 0$, $h_k = 1$, so that h is the tap polynomial of the associated LFSR. Let n be the smallest positive integer for which $X^n - 1$ is divisible by $h(X)$. Set $g(X) = \frac{X^n - 1}{h(X)}$. Then*

(a) *The solutions of equation 21.3 are periodic, with fundamental period dividing n. These solutions are the output of the linear feedback shift register with recurrence generated by $x_k = -\sum_{j=0}^{k-1} h_j x_j$.*

(b) *Let $x_0, x_1, \ldots, x_{n-1}$ be a solution of the recurrence and form the polynomial $u(X) = x_0 + x_1 X + \cdots + x_{n-2} X^{n-2} + x_{n-1} X^{n-1}$ corresponding to the fundamental period (= "first period" in [PW72]) of that solution. Then the set of polynomials u is the set of polynomials that are associated with the code words in C. Here C is the cyclic code of length n generated by $g_1(X)$, the reverse of the polynomial $g(X)$, over the field F.*

(c) *The period of the sequence obtained from $g_1(X)$ is equal to n.*

Sketch of proof. It can be shown that $g_1(x)$ yields a solution to the recurrence having period n. In general, let $(x_0, x_1, \ldots, x_{k-1}, x_k, x_{k+1}, \ldots, x_n)$ be any solution of period n. The dot product

$$(h_0, h_1, \ldots, h_{k-1}, h_k, 0, 0, 0, \ldots)(x_0, x_1, \ldots, x_{k-1}, x_k, x_{k+1}, \ldots, x_{n-1})$$

is 0, and that also holds if we dot product the $n - k$ cyclic shifts - including the null shift - of the vector on the left with the vector on the right. This says that $(x_0, x_1, \ldots, x_{k-1}, x_k, \ldots, x_{n-1})$ is in the dual code of the cyclic code D of length n generated by $h(X)$. In fact, this holds for any solution of the recurrence, whether or not it has period n. Conversely, any vector in the dual code yeilds a solution of the recurrence. Since $g(X)h(X) = X^n - 1$, the dual code $D^{\perp} = C$ is generated by the reverse polynomial $g_1(X)$ of $g(X)$.

For a complete proof we refer the reader to Peterson and Weldon [PW72]. ∎

Example 21.18 *(See [PW72].) Let $h(X) = X^4 + X^3 + X + 1$ over the binary field $F = \{0, 1\}$. Then $h(X) = (X+1)^2(X^2+X+1)$. The smallest n such that $h(X)$ divides $X^n - 1 = X^n + 1$ is 6. Thus $g(X) = \frac{X^6 + 1}{h(X)} = X^2 + X + 1$. Now $g_1(X) = X^2 + X + 1$. The generator matrix for C is*

$$G = \begin{pmatrix} 1 & 1 & 1 & 0 & 0 & 0 \\ 0 & 1 & 1 & 1 & 0 & 0 \\ 0 & 0 & 1 & 1 & 1 & 0 \\ 0 & 0 & 0 & 1 & 1 & 1 \end{pmatrix}. \tag{21.5}$$

The solution (111000) has period 6. The solution obtained by adding the first 2 rows is (100100) and has period 3. From Theorem 21.17 the period must divide 6. The corresponding

LFSR here has length 4 with recurrence generated by the equation $x_4 = x_3 + x_1 + x_0$, since the tap polynomial is $x^4 + x^3 + x + 1$. For further examples we refer to the problems.

21.5   Finding the Period of an LFSR

To keep things simple here we work in binary. Referring to the notation in Chapter 10, let Γ be an LFSR of length k generated by the recurrence given by $x_k = c_{k-1}x_{k-1} + c_{k-2}x_{k-2} + \cdots + c_1 x_1 + 1$. This gives $c(x)$, the tap polynomial, given by $x^k + c_{k-1}x^{k-1} + c_{k-2}x^{k-2} + \cdots + c_1 x + 1$. We now find the smallest value of n such that $c(x)$ divides $x^n - 1 = x^n + 1$. Then $c(x)g(x) = x^n + 1$. The polynomial g has degree $n - k$. Then, from Theorem 21.9 in Section 21.2 the cyclic code C generated by the reverse of g has dimension k since the reverse polynomial also has degree k. The code words in C form a vector space of dimension k and correspond exactly to the k dimensional solution space of the LFSR since any vector of length k can be the initial state. Then, **no matter what the input is, the period divides n**.

Remark 21.19 *We can see that n exists from the combinatorial argument in Chapter 10: in fact, $n \leq 2^k - 1$. Or we can think algebraically about an extension field containing the roots of $c(x)$: This also gives that $n \leq 2^k - 1$.*

In Table 21.1 we give factorizations of $x^n + 1$ over the binary field.

n	factorization
1	$1 + x$
2	$(1 + x)^2$
3	$(1 + x)(1 + x + x^2)$
4	$(1 + x)^4$
5	$(1 + x)(1 + x + x^2 + x^3 + x^4)$
6	$(1 + x)^2(1 + x + x^2)^2$
7	$(1 + x)(1 + x + x^3)(1 + x^2 + x^3)$
8	$(1 + x)^8$
9	$(1 + x)(1 + x + x^2)(1 + x^3 + x^6)$
10	$(1 + x)^2(1 + x + x^2 + x^3 + x^r)^2$
11	$(1 + x)^2(1 + x + \cdots + x^{10})$
12	$(1 + 4)^4(1 + x + x^2)^4$
13	$(1 + x)(1 + x + \cdots + x^{12})$
14	$(1 + x)^2(1 + x + x^3)^2(1 + x^2 + x^3)^2$
15	$(1 + x)(1 + x + x^2)(1 + x + x^2 + x^3 + x^4)(1 + x + x^4)(1 + x^3 + x^4)$
16	$(1 + x)^{16}$

Table 21.1: Factorization of $x^n + 1$ over $GF(2)$

21.6 Cyclic Redundancy Check (CRC)

In the internet the CRC (**cyclic redundancy check**) and the Internet checksum have different functions. The CRC is used to detect link-level transmission errors while the Internet checksum, used by most internet protocols, is designed to detect higher-level transmission errors. For interesting details on this we refer to the preprent (as of 2000) entitled "When the CRC and TCP Checksum Disagree" by Johnathan Stone and Craig Partridge.

CRCs use polynomial arithmetic. As mentioned in chapter 19 the polynomials over any field form a ring analogous to the ring of integers modulo n. We will use the binary field.

Associated with each binary string, say 11101, there is a binary polynomial: in this case the polynomial is $1x^0 + 1x^1 + 1x^2 + 0x^3 + 1x^4$, which yields the polynomial $1 + x + x^2 + x^4$. Also, given a polynomial in binary, we obtain a corresponding binary string. As we have seen in this chapter it is sometimes convenient to reverse the polynomial associated with a string, and vice versa.

How do polynomials relate to error detection? Lets start with an example, corresponding to the binary parity check. Suppose we transmit a binary message M which has *even parity*, i.e. the string has an even number of ones in it. Form the corresponding polynomial $M(x)$. Now, since the message has an even number of ones in it, $M(x)$ is divisible by $x + 1$.

Example 21.20 $M = (10111101)$. *Then*

$$
\begin{aligned}
M(x) &= 1 + x^2 + x^3 + x^4 + x^5 + x^7 \\
&= (1 + x^2) + (x^3 + x^4) + (x^5 + x^7) \\
&= (1 + x)(1 + x) + x^3(1 + x) + x^5(1 + x)(1 + x)
\end{aligned}
$$

In summary, $M(x)$ **is divisible by** $1 + x$ **since** M **has even parity.**

Now suppose that M is transmitted and that R is received. In polynomial terms

$$R(x) = M(x) + E(x)$$

where $E(x)$ corresponds to the error term.

Example 21.21 *In the above suppose that there is a bit error in the received string in, say, the third position. Then the received string is* $R = (10011101)$, *giving the polynomial* $R(x) = 1 + x^3 + x^4 + x^5 + x^7$. *Now* $R(x)$ *is not divisible by* $x + 1$. *One way to see this is to argue that if* $x + 1$ *divides a polynomial that polynomial must have an even number of terms. But* $R(x)$ *has 5 (non-zero) terms. We conclude that there has been an error and ask for a re-transmission. Alternatively, we have*

$$M(x) - R(x) = -E(x).$$

Or, since we are working in binary,

$$M(x) + R(x) = E(x).$$

Now, $x+1$ divides $M(x)$. If $x+1$ also divides $E(x)$ then $x+1$ divides $R(x)$. This the error would go undetected if $x+1$ divides $E(x)$. In our case, $E = (00100000)$, $E(x) = x^2$ and $E(x)$ is not divisible by $x+1$, so the transmission error is detected.

The above is the basic idea. The most commonly used CRC is CRC-32. It uses the following polynomial.

$$f(x) = 1 + x + x^2 + x^4 + x^5 + x^7 + x^8 + x^{10} + x^{11} + x^{12} + x^{16} + x^{22} + x^{23} + x^{26} + x^{32}$$

Using the above notation, suppose $M(x)$ is divisible by $f(x)$. The binary string for f (it is usually reversed) is a string of length 33 - don't forget the constant term. *So, if there are 32 or fewer errors - but at least one error - in the first 32 bits, then this will be detected.* To see this, $E(x) \neq 0$ has degree less than 32. Then $f(x)$ cannot divide $E(x)$ so a transmission error will be detected.

What about detecting a pair of errors? Suppose $E(x) = x^a + x^b$ with $a < b$.

Then $E(x) = x^a(x^{b-a} + 1) = x^a(x^n + 1)$. Then $f(x)$ can only divide $E(x)$ if $f(x)$ divides $x^n + 1$. But this is precisely the reason why we chose the particular polynomial $f(x)$! For $f(x)$ to divide $x^n + 1$, n must be very big. In fact, $f(x)$ is **primitive**! Thus n is at least $2^{32} - 1$. (We have met primitive polynomials before when discussing shift registers)

We conclude that the CRC of degree 32 detects all two-bit errors in the first $2^{32} - 1$ positions.

For further mathematical discussions on CRC we refer to Paul Garrett [Gar04].

21.7   Problems

1. In each case decide if the given code is (i) cyclic, and (ii) linear.

   (a) The binary code $\{(0000), (1111), (1010), (0101)\}$.

   (b) The ternary code $\{(0000), (1111), (2222), (1100), (2200), (0011), (0022), (1122), (2211)\}$, where ternary means over $Z_3 = \{0, 1, 2\}$.

2. Which of the following codes are (i) cyclic? (ii) linear?

   (a) The ternary code $\{(000), (121), (212), (111), (222), (202), (010), (101), (020)\}$.

   (b) The binary code $\{(101), (110), (011)\}$.

3. Repeat the Problem 2 for the following codes.

(a) The ternary code $\{(000), (211), (121), (112)\}$.

(b) The code of length n over $GF(q)$ such that each nonzero codeword has weight r.

4. We have a notion of **equivalent codes**, which is discussed in Chapter 20. It turns out that when we permute the columns of a code C to get another code D, then we have an equivalent pair of codes. For example, if we take C and switch the fourth column with the second column in all codewords, then we have a new, equivalent code. For those codes in the previous three questions that are linear but not cyclic, find an equivalent code that is linear and cyclic by switching the order of the columns. For those codes that are not linear, find a code that is linear by adjoining codewords to the existing nonlinear code.

5. Find all of the codewords v of length 6 over Z_3 such that $x^2 v(x) = v(x)$, when we reduce modulo $(x^6 - 1)$.

6. Form the code $\langle S \rangle$ from the vector $\mathbf{w} = (021)$ over Z_3 as detailed in the text.

7. Repeat Problem 6 with $\mathbf{w} = (00011)$, working over the binary field. Is the generator polynomial of the resulting code equal to $\mathbf{w}(x)$?

8. How could we alter the linear cyclic code construction procedure to create a linear cyclic code containing the codewords $\mathbf{w}_1$, $\mathbf{w}_2$, of equal length, instead of just the codeword $\mathbf{w}$?

9. Working over Z_5, express the codewords in $A = \{(1243), (1414)\}$ as linear combinations of the codewords in $B = \{(1130), (0113)\}$, and vice versa.

10. Starting with the polynomial $g(x) = 3 + 4x + x^2$, which generates a linear cyclic code over Z_5 of length 4, use the two methods described in the text to find generator matrices for the code dual to $\langle g(x) \rangle$. How can we see that these two codes are the same in this case?

11. Working in binary suppose we have a shift register Γ generated by the recurrence $x_2 = x_1 + x_0$.

(a) Find the corresponding tap polynomial, $h(x)$.

(b) Find the smallest n such that $h(x)$ divides $x^n - 1$.

(c) Find the maximum period of Γ over various input.

(d) Give an input which generates an output having maximum period.

(e) Given an arbitrary input for Γ, what can you say about the period of the output of the resulting LFSR?

12. Repeat Problem 11, using instead the recurrence $x_3 = x_2 + x_0$.

13. Working over Z_3, repeat the previous question with the recurrence $x_3 = 2x_0$.

21.8   Solutions

1. (a) Linear and cyclic.

   (b) Linear and not cyclic; $\sigma(1100) = (0110)$ is not in the code.

2. (a) Linear but not cyclic; $\sigma(121) = (112)$ is not a codeword.

   (b) Cyclic but not linear; (000) is not a codeword.

3. (a) Cyclic and not linear; $(211) + (121) = (002)$ is not a codeword.

   (b) This code is always cyclic; if we shift a codeword of weight r, we still have a codeword of weight r. However, the code is not guaranteed to be linear. Consider the code of weight 0 with length 3 over Z_3, where $(111) + (221) = (001)$, but (001) has weight 2 and is therefore not a codeword.

4. (a) No change

   (b) We switch the second and third columns to arrive at the code

   $$\{(0000), (1111), (2222), (1010), (2020), (0101), (0202), (1212), (2121)\}. \quad (21.6)$$

   (c) No such cyclic code exists. To see this, one can take all possible permutations of the columns without finding such a code.

   (d) We can make the code linear by adjoining (000) to the code.

   (e) We adjoin the codewords $(002), (020)$ and (200). Note that after this is complete, the code is still cyclic.

   (f) To make this code linear, we follow the method as described in the text and take the set of all linear combinations of these codewords. In fact, this process is exactly what we did for the previous two codes.

5. The condition implies that if we cyclically shift the six components of a codeword two places to the right, we get the same word. That is, $\sigma^2(v) = v$, so if $v = (v_0, v_1, v_2, v_3, v_4, v_5)$, we have $v = (v_0, v_1, v_2, v_3, v_4, v_5) = (v_4, v_5, v_0, v_1, v_2, v_3) = \sigma^2(v)$. Thus, $v_0 = v_2 = v_4$ and $v_1 = v_3 = v_5$; Therefore, v must be of the form $(ababab)$, where a, b are any elements in $Z_3 = GF(3)$. The 9 possibilities are: $(000000), (111111), (222222), (010101), (101010), (121212), (212121), (020202)$ and (202020).

6. $S = \{(021), (102), (201)\}$, so that

$$\langle S \rangle = \left\{ \begin{array}{ccc} (000) & (012) & (021) \\ (102) & (111) & (120) \\ (201) & (210) & (222) \end{array} \right\}$$

7. $S = \{(11000), (01100), (00110), (00011), (10001)\}$ and taking all linear combinations, we get

$$\langle S \rangle = \left\{ \begin{array}{cccc} (00000) & (00011) & (00101) & (00110) \\ (01001) & (01010) & (01100) & (01111) \\ (10001) & (10010) & (10100) & (10111) \\ (11000) & (11011) & (11101) & (11110) \end{array} \right\} \tag{21.7}$$

with generator $g = (11000)$, which in this case is not $\mathbf{w}$.

8. When constructing S, we would simply take it to be the set of all cyclic shifts of both $\mathbf{w}_1$ and $\mathbf{w}_2$. Construcing $\langle S \rangle$ from S in the same manner, we would find that $\langle S \rangle$ is a linear cyclic code containing both $\mathbf{w}_1$ and $\mathbf{w}_2$. To prove that it all works, use the fact that the shift operator σ is a linear mapping.

9. We get that

$$(1, 1, 3, 0) = 4(1, 2, 4, 3) + 2(1, 4, 1, 4)$$
$$(0, 1, 1, 3) = 2(1, 2, 4, 3) + 3(1, 4, 1, 4)$$

and

$$(1, 2, 4, 3) = 1(1, 1, 3, 0) + 1(0, 1, 1, 3)$$
$$(1, 4, 1, 4) = 1(1, 1, 3, 0) + 3(0, 1, 1, 3)$$

10. Using method 1, we get

$$C_1^{\perp} = \left(\begin{array}{cccc} 2^0 & 2^1 & 2^2 & 2^3 \\ 4^0 & 4^1 & 4^2 & 4^3 \end{array} \right) = \left(\begin{array}{cccc} 1 & 2 & 4 & 3 \\ 1 & 4 & 1 & 4 \end{array} \right) \tag{21.8}$$

and using method 2, we get that

$$C_2^{\perp} = \left(\begin{array}{cccc} 1 & 1 & 3 & 0 \\ 0 & 1 & 1 & 3 \end{array} \right). \tag{21.9}$$

These can be seen to be equivalent by the work of Problem 9; since each row in $C_1^{\perp}$ can be expressed as a linear combination of the rows in $C_2^{\perp}$, and vice versa, it follows that the set of all linear combinations of the rows of $C_1^{\perp}$ is equivalent to the set of all linear combinations of the rows of $C_2^{\perp}$. That is, the codes are the same.

11. (a) The tap polynomial is $h(x) = x^2 + x + 1$.

   (b) $n = 3$.

   (c) The maximum period is three.

   (d) Any nonzero input generates output with maximum period. The output sequence could be $\{110110110\ldots\}$.

   (e) If the input is nonzero, then the output has period three, as noted above. If the input is zero, then the output is $\{00000\ldots\}$, which has period 1. In either case, the period divides 3.

12. (a) The tap polynomial is $h(x) = x^3 + x^2 + 1$.

   (b) $n = 7$.

   (c) The maximum period is seven.

   (d) Any nonzero input generates output with maximum period. The output sequence could be $\{10011101001110\ldots\}$.

   (e) If the input is nonzero, then the output has period seven, as noted above. If the input is zero, then the output is $\{00000\ldots\}$, which has period 1. In either case, the period divides 7.

13. (a) The tap polynomial is $h(x) = x^3 + 1$.

   (b) $n = 6$.

   (c) The maximum period is six.

   (d) An input that achieves maximum period is $(x_0 x_1 x_2) = (120)$. The output is then $\{120210120210\ldots\}$

   (e) Regardless of what the input is, the output period will divide 6.

Chapter 22

Reed–Solomon and MDS Codes, and Bruen-Thas-Blokhuis

Goals, Discussion We present the main features of Reed–Solomon (RS) codes – both encoding and decoding – which are central for many industrial applications. These codes generalize to MDS codes and arcs. We discuss the main results due to one of authors and others on the fifty-year old problem of finding the longest arc and therefore the longest MDS code, of a given dimension over a given field. This question can also be phrased as a version of "The main coding theory problem" discussed in Chapter 18. The longer the code the bigger the minimum distance so the question is of considerable interest in coding theory. Our result is that at least when q is large the longest MDS code possible is equivalent to an extended Reed–Solomon code.

New, Noteworthy We avoid general BCH codes which are usually covered as an introduction to the RS codes. Instead we get down to business quickly. We prefer to present RS codes in the context of general MDS codes. We dispel the (erroneous) notion that in order to obtain the MDS property the generator polynomial $g(x)$ must have successive powers of a field generator as its roots.

In [McE78] the author states the following: "It is fairly clear that the deepest and most impressive theoretical result in coding theory (block or convolutional) is the algebraic decoding of BCH-RS codes".

We want to comment on this. Certainly the work is deep and impressive although the author is light on historical background. In a fascinating paper (Alexander Barg, At the Dawn of the Theory of Codes, Mathematical Intelligence **15**, 1993, 20-26) the author points out that in fact the decoding problem was solved in 1795 by Gaspard-Clair-François-Marie baron Riche de Prony. Later it was solved by the legendary Srinivasa Ramanujan in a two-page paper in 1912. Barg also refers to related work of several others, including work of Peterson and Gorenstein-Zierler.

However, it is our opinion that the geometrical work on MDS codes and arcs developed by the late Beniamino Segre must be put on an equally high pedestal. Segre was the pioneer in exploiting the famous Hasse-Weil theorem in algebraic geometry in the study of arcs and MDS codes. We give a brief discussion of this in the last section.

22.1   Cyclic Linear Codes and Vandermonde

Let us work over some finite field $F = GF(q)$: then, necessarily, $q = p^m$, where p is a prime and m is a positive integer. We put $n = q - 1$.

Let $\alpha_1, \cdots, \alpha_{n-k}$ be any set of $n - k$ distinct elements of F. Put

$$g(x) = (x - \alpha_1) \cdots (x - \alpha_{n-k}).$$

Since

$$\prod_{\alpha \neq 0 \text{ in } F} (x - \alpha) = x^{q-1} - 1 \text{ (see Chapter 19)},$$

it follows that $g(x)$ divides $x^n - 1 = x^{q-1} - 1$ and (see Chapter 21) we can construct the cyclic code C of length $q - 1$ with generator polynomial $g(x)$. Then, as in Chapter 21, the code words formed from the polynomials

$$\{g(x), xg(x), \cdots, x^{n-k-1}g(x)\}$$

form a basis for C, which has dimension k. Thus C is a linear code of length n and dimension k, i.e., C is an (n, k) **linear code**.

Let $g(x) = g_0 + g_1 x + \cdots + g_{n-k} x^{n-k}$. Then the corresponding code word for g of length n is $(g_0, g_1, \cdots, g_{n-k}, 0, \cdots, 0)$. Since $g(\alpha_i) = 0$, we have that

$$g_0 + g_1 \alpha_i + \cdots + g_{n-k} \alpha_i^{n-k} = 0, \quad 1 \leq i \leq n - k.$$

It follows that the dot product of the code word $(g_0, g_1, \cdots, g_{n-k}, 0, \cdots, 0)$ and each of its k cyclic shifts with the vector

$$(1, \alpha_i, \alpha_i^2, \cdots, \alpha_i^{n-k}, \alpha_i^{n-k+1}, \cdots, \alpha_i^{n-1})$$

is zero, $1 \leq i \leq n - k$.

We have shown the following result.

Theorem 22.1 *Each row of the matrix*

$$H = \begin{pmatrix} 1 & \alpha_1 & \alpha_1^2 & \cdots & \alpha_1^{n-1} \\ 1 & \alpha_2 & \alpha_2^2 & \cdots & \alpha_2^{n-1} \\ \vdots & \vdots & \vdots & \vdots & \vdots \\ 1 & \alpha_{n-k} & \alpha_{n-k}^2 & \cdots & \alpha_{n-k}^{n-1} \end{pmatrix}$$

lies in the dual space of the code C generated by $g(x) = (x - \alpha_1) \cdots (x - \alpha_{n-k})$.

Now the matrix H is of size $(n - k) \times n$. The transpose, H^t, is then of size $n \times (n - k)$ and is of the following form:

$$H^t = \begin{pmatrix} 1 & 1 & \cdots & 1 \\ \alpha_1 & \alpha_2 & \cdots & \alpha_{n-k} \\ \alpha_1^2 & \alpha_2^2 & \cdots & \alpha_{n-k}^2 \\ \vdots & \vdots & \vdots & \vdots \\ \alpha_1^{n-1} & \alpha_2^{n-1} & \cdots & \alpha_{n-k}^{n-1} \end{pmatrix}.$$

The first $n - k$ rows of H^t give a Vandermonde matrix of the type:

$$V = \begin{pmatrix} 1 & 1 & \cdots & 1 \\ \alpha_1 & \alpha_2 & \cdots & \alpha_{n-k} \\ \alpha_1^2 & \alpha_2^2 & \cdots & \alpha_{n-k}^2 \\ \vdots & \vdots & \vdots & \vdots \\ \alpha_1^{n-k-1} & \alpha_2^{n-k-1} & \cdots & \alpha_{n-k}^{n-k-1} \end{pmatrix}.$$

Now, as mentioned in Chapter 18 we have a Vandermonde determinant, so

$$\det V = \prod (\alpha_i - \alpha_j),$$

where the product is over all pairs (i, j) with $1 \leq i < j \leq n - k$. Since $\alpha_i \neq \alpha_j$ if $i \neq j$, we see that the determinant of V is non-zero. So the first $n - k$ rows of H^t are linearly independent. In fact, by manipulating determinants, we see that any set of $n - k$ rows of H^t is a linearly independent set. Thus the first $n - k$ columns (and in fact any $n - k$ columns) of H are linearly independent. Since H is of size $(n - k) \times (n - 1)$, we have that, a priori, $\text{rank}(H) \leq n - k$. Putting these together shows that $\text{rank}(H) = n - k$.

We have now shown the following result.

Theorem 22.2 *(a) A parity check matrix of C, corresponding to the dual code $C^\perp$, is given by the matrix H of Theorem 22.1.*

(b) Any set of $n - k$ columns of H is linearly independent.

Proof. The dimension of C is k. Thus, the dimension of $C^{\perp}$ is $n - k$. Each row of H is in $C^{\perp}$. H is of size $(n - k) \times n$ and the rank of H is $n - k$. This proves Part (a). Part (b) has already been shown. ∎

22.2   The Singleton Bound

This calculates the maximum possible distance that an (n, k) linear code can have.

Theorem 22.3 *Let C be any linear (n, k) code over any finite field of length n and dimension k with minimum distance d. Then $d \leq n - k + 1$. Moreover, $d = n - k + 1$ if and only if no set of $n - k$ or fewer columns of the parity-check matrix H is linearly dependent.*

Proof. There exists a code word $\mathbf{c}$ of weight m in the code C if and only if $\mathbf{c}H^t = \mathbf{0}$ where H^t is the transpose of H. This is equivalent to saying that some m columns of H are linearly dependent. [This is easier than it looks. For example, $\mathbf{c} = (1, 1, -1, 0, \cdots, 0)$ is in C if and only if the sum of the first two columns in H equals the third column.]

Now, since C has dimension k, the rank of H is $n - k$. Thus the maximum number of linearly independent columns of H is $n - k$. So some set of $n - k + 1$ columns of H are linearly dependent and C then has a word of length $n - k + 1$. ∎

If C is a linear (n, k) code with $d = n - k + 1$ then C is called a **Maximum Distance Separable Code** (= **MDS code**). Later on in a subsequent chapter we will discuss **nonlinear MDS codes**.

Theorem 22.4 *If C is MDS then so is $C^{\perp}$, the dual code of C.*

Proof. Let C have length n and dimension k. From the above, no set of $n - k$ (or fewer) columns of H is linearly dependent. H is of size $(n - k) \times n$. Now suppose a code word in $C^{\perp}$ is zero on as many as $n - k$ coordinates. Then the corresponding square submatrix of H of size $(n - k) \times (n - k)$ has rank at most $n - k - 1$. It follows that the corresponding $n - k$ columns of H would be linearly dependent which is impossible since C is MDS. Thus no code word in $C^{\perp}$ is zero on as many as $n - k$ coordinates. Therefore the minimum distance d of $C^{\perp}$ is at least $n - (n - k - 1) = k + 1$. From the Singleton bound $d \leq k + 1 = n - (n - k) + 1$ so that $C^{\perp}$ is an MDS code. ∎

From the discussion above and Theorem 22.2 part (b) in Section 22.1 we have the following result.

Theorem 22.5 *The linear code C generated by $g(x) = (x - \alpha_1) \cdots (x - \alpha_{n-k})$ is an MDS code where $n = q - 1$ and $\alpha_1, \cdots, \alpha_{n-k}$ are any distinct elements in $F = GF(q)$.*

Alternative proof of the first part of Theorem 22.3 In Chapter 18 we showed that for any code of length n, minimum distance d over an alphabet of size q and having exactly M code words then $M \leq q^{n-d+1}$. Here $M = q^k$, so $k \leq n - d + 1$, giving $d \leq n - k + 1$.

We want to develop several other properties of linear MDS codes. Some of these results can actually be shown using the work in Chapter 18.

Theorem 22.6 *Let G be a generator matrix for a linear (n, k) code C with $k \le n$. Then*

(1) *If C is MDS, then any set of k columns of G is linearly independent.*

(2) *If any set of k columns of G is linearly independent, then C is MDS.*

(3) *If C is MDS, then the linear code generated by deleting any set of t columns is MDS, $t \le n - k$.*

(4) *C is MDS if and only if every $k \times k$ submatrix of G is non singular.*

Proof. Let C be MDS. Then $C^{\perp}$ is MDS from Theorem 22.4. Then no set of $n - (n - k)$ or fewer columns of $(C^{\perp})^{\perp}$ is linearly independent from Theorem 22.3. But $(C^{\perp})^{\perp} = C$ and Part (1) follows.

For Part (2), note that the condition implies that $C^{\perp}$ is MDS from Theorem 22.3. But then, from Theorem 22.4, $(C^{\perp})^{\perp} = C$ is MDS.

Part (3) is proved as follows. Any subset of a linearly independent set is linearly independent. From Part (1) the columns of C are linearly independent. The result now follows from Part (2).

Part (4) follows by noting that since G has rank k, the given condition is equivalent to the statement that any set of k columns of G is linearly independent. ∎

22.3 Reed–Solomon Codes

These are the main codes used in industry. The applications are everywhere, ranging from computer based disk drives to CD players, satellite communications and space probes. According to [MS78] they first appeared *as codes* in the paper by Reed and Solomon in 1960. However, they had already been explicitly constructed by K.A. Bush [Bus52] in 1952 using the language of statistics and orthogonal arrays. Geometrically (see [BTB88]) they have been known for a very long time as **normal rational curves**.

Let F be any finite field with $|F| = q = p^m$. Let α be a generator of the multiplicative group of F. Put $n = q - 1$. Define a **Reed–Solomon** code over F to be the linear cyclic code C generated by $g(x) = (x - \alpha^b)(x - \alpha^{b+1}) \cdots (x - \alpha^{b+n-k-1})$. For example, if $b = 1$, then

$$g(x) = (x - \alpha)(x - \alpha^2) \cdots (x - \alpha^{n-k})$$

and C is the linear cyclic code generated by $g(x)$, with $0 \le k < n$, $n = q - 1$.

The integer b can have any value. For instance, if $b = 0$ we get $x - 1$ as the first term.

Theorem 22.7 C *is a cyclic linear code of length n and dimension k. Moreover, C is an MDS code of minimum distance $d = n - k + 1 = \deg(g) + 1$, where $\deg(g)$ is the degree of g. A parity check matrix for C is given by the matrix H in Theorem 22.2 of Section 22.1, where $\alpha_i = \alpha^{b+i-1}, 1 \le i \le n - k$.*

Proof. The fact that C has dimension k follows from Chapter 21. The rest follows using also Theorem 22.5. ∎

22.4   Reed–Solomon Codes and the Fourier Transform Approach

As in the previous section, let C be the linear (n, k) code generated by $g(x)$ with

$$g(x) = (x - \alpha)(x - \alpha^2) \cdots (x - \alpha^{n-k}).$$

In what follows, we describe these codes in a completely different way using, implicitly, the Mattson–Solomon polynomial (which we cannot discuss here.). This method is also called the **Discrete Fourier Transform approach**.

Let $f(x) = f_0 + f_1 x + \cdots + f_{k-1} x^{k-1}$ denote a polynomial of degree at most $k - 1$. We have the the following result (see Van Lint [vL98] and MacWilliams, Sloane [MS78]).

Theorem 22.8 $C = (c_0, c_1, \cdots, c_{n-1})$ *with $c_i = f(\alpha^i)$, $0 \le i \le n - 1$, and with f being any polynomial of degree at most $k - 1$.*

Remark.   Note that we have moved from a representation involving the original $n - k$ field elements to *a representation involving* **all non-zero elements** *of the field F.*

We examine the set of all polynomials over $F = GF(q)$ of degree at most $k - 1$. They form a vector space F^k of dimension k. A basis for this space is given by the polynomials $1, x, \cdots, x^{k-1}$. From this and Theorem 22.8 we get the following result, where β_i denotes $\alpha^i, 0 \le i \le q - 2$.

Theorem 22.9 *A generator matrix for C is given by the matrix G below. Given k information bits $f = (f_0, f_1, \cdots, f_{k-1})$ the encoding of f is just the code word $(f(\beta_0), f(\beta_1), \cdots, f(\beta_{q-2}))$, where $f(x) = f_0 + f_1 x + \cdots + f_{k-1} x^{k-1}$.*

$$G = \begin{pmatrix} 1 & 1 & \cdots & 1 \\ \beta_0 & \beta_1 & \cdots & \beta_{q-2} \\ \beta_0^2 & \beta_1^2 & \cdots & \beta_{q-2}^2 \\ \vdots & \vdots & \vdots & \vdots \\ \beta_0^{k-1} & \beta_1^{k-1} & \cdots & \beta_{q-2}^{k-1} \end{pmatrix}.$$

Remark. The elements $\beta_0, \beta_1, \cdots \beta_{q-2}$ give a complete listing of all the non-zero elements of F. The matrix G is a generator for a linear cyclic code of length $n = q - 1$ and dimension k. G has size $k \times (q - 1)$. The code C is MDS with minimum distance $d = n - k + 1$. Also, $C^{\perp}$ is MDS. Thus any set of k columns of G are linearly independent, $0 \le k < q - 1$.

Importance of Reed–Solomon Codes.

1. If we need a code of length less than q they can be used. Since they are MDS the minimum distance is as big as possible.

2. Then can be combined and concatenated with other codes to build strong error correcting codes.

3. The encoding of f is very easy from Theorem 22.9.

4. They are very useful for correcting bursts of errors either individually or when interleaved with other codes.

5. They have a highly developed decoding theory, discussed later.

6. If $q = 2^r$, we can represent them as binary codes since each element of $F = GF(2^r)$ can be represented as a binary string of length r.

22.5 Correcting Burst Errors, Interleaving

Very often, errors are not random but occur in clusters or bursts. This is sometimes due to physical constraints. For example, for compact disk players, laser tracking makes it desirable that in the binary representation of each code word, there must be at least two and at most ten zeros between any two ones.

A **burst of length** b is a vector whose only non-zero elements are among b successive coordinate positions the first and last of which are non-zero.

Binary codes obtained from Reed–Solomon codes (RS codes) are particularly useful for burst error correction because of their big minimum distance d. Such a code, (over the field $GF(q)$ with $q = 2^r$), can correct up to $\lfloor \frac{d-1}{2} \rfloor$ errors, where $\lfloor x \rfloor$ denotes the integer part of x: for example, $\lfloor \frac{5}{2} \rfloor = 2$.

Theorem 22.10 (see [MS78]) *A binary burst of length b can affect at most m adjacent symbols from $GF(2^r)$, where $m \le b \le (m - 1)r + 1$.*

To see this, think of the situation where the left-most of the b bits is the right-most element of a symbol from $GF(2^r)$.

Thus, if d (or e) is much bigger than m long bursts can be corrected.

Interleaving is a technique that is used for improving the burst error correcting capability of a code.

Lets take an example. Suppose we have a code C of length n, and we take 3 code words in C, namely, the 3 code words:

$$\begin{aligned} \mathbf{x} &= X_1, \cdots, X_n, \\ \mathbf{y} &= Y_1, \cdots, Y_n, \\ \mathbf{z} &= Z_1, \cdots, Z_n. \end{aligned}$$

We could send the 3 code words $\mathbf{x}, \mathbf{y}, \mathbf{z}$ one after the other. Another option is to transmit column by column. Then our transmitted code-word $\mathbf{w}$ would look like this:

$$\mathbf{w} = X_1 Y_1 Z_1 \cdots X_n Y_n Z_n.$$

Now suppose that each code word in C can correct a burst of length 1. Then each code word $\mathbf{w}$ corrects all bursts of length 3. We have **interleaved the code** C **to depth 3**. In general, we have the following result (see [HLL+92]).

Theorem 22.11 *Let C be a b burst error correcting code. If C is interleaved to depth w then all bursts of length at most bw will be corrected provided that each interleaved code word will be subjected to at most one burst of errors.*

The proof depends on the fact that any burst of errors of length at most bw in an interleaved code word will cause a burst error pattern of length at most b in a code word $\mathbf{c}$ of C.

22.6   Decoding Reed–Solomon Codes, Ramanujan and Berlekamp–Massey

Let C be a Reed–Solomon code as described in Section 22.3 with $b = 0$ and generator

$$g(x) = (x - 1)(x - \alpha) \cdots (x - \alpha)^{n-k-1}.$$

To simply the treatment (and to tie in with shift registers) we assume that F is a field of characteristic 2, so $-1 = 1$.

Recall that the parity check matrix is

$$H = \begin{pmatrix} 1 & \beta_0 & \cdots & \beta_0^{q-1} \\ 1 & \beta_1 & \cdots & \beta_1^{q-1} \\ \vdots & \vdots & \vdots & \vdots \\ 1 & \beta_{n-k-1} & \cdots & \beta_{n-k-1}^{q-1} \end{pmatrix},$$

where $\beta_i = \alpha^i$ and α is a generator of the field $F = GF(2^r)$.

Suppose that the code word $\mathbf{c}$ with corresponding polynomial $\mathbf{c}(x)$ is transmitted and that the string $\mathbf{w}$ with corresponding polynomial $\mathbf{w}(x)$ is received. Then we may calculate the $2t$ **syndromes**

$$s_i = \mathbf{w}(\alpha^i) \text{ for } 0 \le i \le n - k - 1, \ n = q - 1.$$

Write

$$\mathbf{w}(x) = \mathbf{c}(x) + \mathbf{E}(x),$$

where $\mathbf{E}(x)$ is the **error polynomial**.

The code C has dimension k and minimum distance $d = n - k + 1$.

Let $t = \frac{d-1}{2} = \lfloor \frac{n-k}{2} \rfloor$. Then the code C can correct up to t errors. Here x denotes the integer part of x. So when decoding we assume that at most t transmission errors occur.

To obtain as much error-correction power as possible *we assume that d is odd* so $\frac{d-1}{2} = t$ is an integer. Since C is an MDS code, $d = n - k + 1$. Then the dimension of the dual code $C^\perp$ is $n - k = d - 1 = 2t$.

Since $\mathbf{c}$ is in the code C the polynomial $\mathbf{c}(x)$ is a multiple of $g(x)$. Since $g(\alpha^i) = 0$, we have then that $\mathbf{c}(\alpha^i) = 0$. Thus

$$\mathbf{w}(\alpha^i) = \mathbf{c}(\alpha^i) + \mathbf{E}(\alpha^i) = \mathbf{E}(\alpha^i),$$

for $0 \le i \le n - k - 1$. Now $\mathbf{E}(x)$ is some polynomial of degree at most t. Suppose there are exactly e errors, $e \le t$. Let

$$E(x) = b_1 x^{i_1} + b_2 x^{i_2} + \cdots + b_e x^{i_e}.$$

Then, for any j with $0 \le j \le n - k - 1$ we have

$$\begin{aligned} s_j &= E(\alpha^j) \\ &= b_1(\alpha^j)^{i_1} + b_2(\alpha^j)^{i_2} + \cdots + b_e(\alpha^j)^{i_e} \\ &= b_1(\alpha^{i_1})^j + b_2(\alpha^{i_2})^j + \cdots + b_e(\alpha^{i_e})^j \\ &= b_1 a_1^j + b_2 a_2^j + \cdots + b_e a_e^j, \end{aligned}$$

where $a_1 = \alpha^{i_1}, a_2 = \alpha^{i_2}, \cdots, a_e = \alpha^{i_e}$ are the so-called "**error locations**" and $b_1, b_2, \cdots, b_e$ are the **error magnitudes**.

If $e < t$ it is convenient to define $a_i = 0$ for $e + 1 \leq i \leq t$ even though the corresponding error locations do not exist.

So the **decoding problem** comes down to solving the nonlinear system of $2t$ equations given by

$$\sum_{i=1}^{t} b_i a_i^j = s_j \tag{22.1}$$

for the $2e$ unknowns $a_1, a_2, \cdots, a_e$ and $b_1, b_2, \cdots, b_e$.

However, an astonishing fact, pointed out by Hill [Hil86] is that the system was solved over 90 years ago! The solution was provided in a little known two-page paper in 1912 by the legendary Indian mathematician Srinivasa Ramanujan [Ram12]. Ramanujan's elegant method used partial fractions and power series. (In fact, as mentioned above and pointed out by Barg, the solution goes back to de Prony in 1795.)

A more modern method involves the **error-locator polynomial**

$$\sigma_A(x) = (x - a_1) \cdots (x - a_e). \tag{22.2}$$

Note that the roots of σ_A are $a_1, \cdots, a_e$. Recall that we are in characteristic 2. Expanding this polynomial we get

$$\sigma_A(x) = \sigma_0 + \sigma_1 x + \cdots \sigma_{e-1} x^{e-1} + x^e, \tag{22.3}$$

where the σ_i are the elementary symmetric functions, so that we also have $\sigma_A(a_i) = 0$, $1 \leq i \leq e$. If we multiply both sides of equation (22.3) by $b_i a_i^j$, substitute $x = a_i$ and sum both sides from $i = 1$ to $i = t$, we get the equation

$$s_{j+e} = s_j \sigma_0 + s_{j+1} \sigma_1 + \cdots + s_{j+e-1} \sigma_{e-1} \tag{22.4}$$

as in [HLL+92]. This equation certainly has the look of something we have seen in this book more than once: *shift-registers!*

In fact, putting $j = 0$ we get a shift register with e registers and recurrence coefficients $\sigma_0, \sigma_1, \cdots, \sigma_{e-1}$ as in Chapters 13 and 21. The problem of finding the error locator polynomial is solved if we can find the shortest LFSR that generates the syndromes.

This is where the **Berlekamp–Massey algorithm** discussed in Chapter 16 comes in to play, leading to an iterative solution of the decoding problem. For details and many examples, we refer the reader to [PW72] and to [HLL+92].

22.7 An Algorithm for Decoding and an Example

We have from the above that.

$$s_{j+e} = s_j \sigma_0 + s_{j+1}\sigma_1 + \cdots + s_{j+e-1}\sigma_{e-1}.$$

We know the values $s_0, s_1, \cdots, s_{e-1}$ and so we get a linear system of the following form:

$$\begin{pmatrix} s_0 & s_1 & \cdots & s_{e-1} \\ s_1 & s_2 & \cdots & s_e \\ \vdots & \vdots & \vdots & \vdots \\ s_{e-1} & s_e & \cdots & s_{2e-2} \end{pmatrix} \begin{pmatrix} \sigma_0 \\ \sigma_1 \\ \vdots \\ \sigma_{e-1} \end{pmatrix} = \begin{pmatrix} s_e \\ s_{e+1} \\ \vdots \\ s_{2e-1} \end{pmatrix}. \tag{22.5}$$

Denote the coefficient matrix in the above by P. On the assumption that $a_1, \cdots, a_e$ are distinct and $a_1, a_2, \cdots, a_e, b_1, b_2, \cdots, b_e$ are all non-zero, we can solve equation (22.5) for $\sigma_0, \sigma_1, \cdots, \sigma_{e-1}$ thereby obtaining the error locator polynomial $\sigma_A(x)$. Since the roots of $\sigma_A(x)$ are $a_1, a_2, \cdots, a_e$, we can then find these. Finally, we can find $b_1, b_2, \cdots, b_e$ from the linear system in equation (22.1) above. These can be represented in matrix form as follows.

$$\begin{pmatrix} a_1^0 & a_2^0 & \cdots & a_e^0 \\ a_1^1 & a_2^1 & \cdots & a_e^1 \\ \vdots & \vdots & \vdots & \vdots \\ a_1^{e-1} & a_2^{e-1} & \cdots & a_e^{e-1} \end{pmatrix} \begin{pmatrix} b_1 \\ b_2 \\ \vdots \\ b_e \end{pmatrix} = \begin{pmatrix} s_0 \\ s_1 \\ \vdots \\ s_{e-1} \end{pmatrix}. \tag{22.6}$$

Now we have found the locations of the errors (corresponding to the a_i) and the "magnitude" of the errors corresponding to the b_i, $1 \le i \le e$, completing the decoding.

To summarize, we have our Reed–Solomon decoding algorithm.

Reed–Solomon Decoding Algorithm.

1. Calculate the $2t$ syndromes $s_0, s_1, \cdots, s_{2t-1}$, $s_i = E(\alpha^i)$, where E is the error polynomial.

2. The rank of the matrix P in equation (22.5) yields the number e.

3. Having found e solve the system (22.5) for $\sigma_0, \sigma_1, \cdots, \sigma_{e-1}$.

4. Find the roots of the error locator polynomial. These give the *error locations* $a_1, a_2, \cdots, a_e$.

5. Solve the linear system in (22.6) for $b_1, b_2, \cdots, b_e$. We now have found the errors and their locations.

A Worked Example. Let C be the linear code generated by $g(x)$, where

$$g(x) = (x + \alpha^0)(x + \alpha)(x + \alpha^2)(x + \alpha^3) = (x + 1)(x + \alpha)(x + \alpha^2)(x + \alpha^3).$$

Here, α is a generator for a field F of order $2^3 = 8$ obtained from the irreducible polynomial $u(x) = x^3 + x + 1$. (See Chapter 19.) Thus $n = q - 1 = 7$. The degree of the polynomial $g(x)$ (which is 4) is $n - k = 7 - k$. So $7 - k = 4$, giving $k = 3$. Thus the code C has dimension 3. Since C is MDS, the minimum distance is $n - k + 1 = 7 - 3 + 1 = 5$. It follows that C can correct as many as t errors where $t = \frac{5-1}{2} = 2$.

Now let us suppose that the received vector is the vector $\mathbf{w} = (1, \alpha, 0, \alpha, \alpha^6, 1, 0)$. On the assumption that at most 2 errors have been made, we want to find the transmitted code word from our algorithm.

First we calculate the syndromes. The syndrome polynomial $\mathbf{w}(x)$ is $1 + \alpha x + \alpha x^3 + \alpha^6 x^4 + x^5$. We calculate now $\mathbf{w}(1) = \mathbf{w}(\alpha^0), \mathbf{w}(\alpha), \mathbf{w}(\alpha^2), \mathbf{w}(\alpha^3)$. We have $\mathbf{w}(1) = 1 + \alpha + \alpha + \alpha^6 + 1$. Since the field F has characteristic 2, we have $1 + 1 = 0$, $\alpha + \alpha = 0$. Thus

$$\mathbf{w}(1) = \mathbf{w}(\alpha^0) = s_0 = \alpha^6.$$

Then $\boxed{s_0 = \alpha^6}$.

Now $\mathbf{w}(\alpha)$ can be written as the dot product of $\mathbf{w}$ with the vector $(1, \alpha, \alpha^2, \alpha^3, \alpha^4, \alpha^5, \alpha^6)$. So we have

$$(1, \alpha, 0, \alpha, \alpha^6, 1, 0) \cdot (1, \alpha, \alpha^2, \alpha^3, \alpha^4, \alpha^5, \alpha^6) = 1 + \alpha^2 + \alpha^4 + \alpha^3 + \alpha^5.$$

(Since $\alpha^7 = 1$). Now using the addition tables (or remembering that $\alpha^3 = \alpha + 1$ so $\alpha^4 = \alpha^2 + \alpha$) we get $\mathbf{w}(\alpha) = \alpha^5$. Therefore $\boxed{s_1 = \alpha^5}$.

To get s_2, we dot product $\mathbf{w}$ with $(1, \alpha^2, (\alpha^2)^2, \cdots, (\alpha^2)^6)$. Then $\boxed{s_2 = 1}$. Similarly, $\boxed{s_3 = 0}$.

Our system (22.5) now gives the following

$$\begin{pmatrix} \alpha^6 & \alpha^5 \\ \alpha^5 & 1 \end{pmatrix} \begin{pmatrix} \sigma_0 \\ \sigma_1 \end{pmatrix} = \begin{pmatrix} s_2 \\ s_3 \end{pmatrix} = \begin{pmatrix} 1 \\ 0 \end{pmatrix}.$$

So we have the system

$$\begin{aligned} \alpha^6 \sigma_0 + \alpha^5 \sigma_1 &= 1, \\ \alpha^5 \sigma_0 + \sigma_1 &= 0. \end{aligned}$$

The rank of the system is 2 so $e = 2$. Solving gives $\sigma_0 = \alpha^3$ and $\sigma_1 = \alpha^5 \sigma_0 = \alpha^8 = \alpha$.

The error locator polynomial is given by the quadratic $\alpha^3 + \alpha x + x^2$. The roots are $x = 1$ and $x = \alpha^3$. Thus $a_1 = 1$ and $a_2 = \alpha^3$. To solve for b_1, b_2, we have from equation (22.6) that

$$\begin{pmatrix} a_1^0 & a_2^0 \\ a_1^1 & a_2^1 \end{pmatrix} \begin{pmatrix} b_1 \\ b_2 \end{pmatrix} = \begin{pmatrix} \alpha^6 \\ \alpha^5 \end{pmatrix},$$

i.e.,

$$\begin{pmatrix} 1 & 1 \\ 1 & \alpha^3 \end{pmatrix} \begin{pmatrix} b_1 \\ b_2 \end{pmatrix} = \begin{pmatrix} \alpha^6 \\ \alpha^5 \end{pmatrix}.$$

Solving gives $b_2 = 1$ and $b_1 = \alpha^2$. Summing up then, on the assumption of at most two errors, we get that the two errors occurred in the α^0 and α^3 locations with magnitudes α^2 and 1. The received vector was $\mathbf{w} = (1, \alpha, 0, \alpha, \alpha^6, 1, 0)$. The (likely) error pattern is $(\alpha^2, 0, 0, 1, 0, 0, 0)$. Thus the (likely) transmitted code word is the sum of these, namely,

$$(1 + \alpha^2, \alpha, 0, \alpha + 1, \alpha^6, 1, 0) = (\alpha^6, \alpha, 0, \alpha^3, \alpha^6, 1, 0).$$

This is indeed a code word $\mathbf{c}$ in C (see Problem 1)

22.8 Long MDS Codes and a Solution of a Fifty Year-Old Problem

In Section 22.3, we discussed a Reed–Solomon code C with generator matrix

$$G = \begin{pmatrix} 1 & 1 & \cdots & 1 \\ \beta_0 & \beta_1 & \cdots & \beta_{q-2} \\ \beta_0^2 & \beta_1^2 & \cdots & \beta_{q-2}^2 \\ \vdots & \vdots & \vdots & \vdots \\ \beta_0^{k-1} & \beta_1^{k-1} & \cdots & \beta_{q-2}^{k-1} \end{pmatrix},$$

where the non-zero elements of a given finite field $F = GF(q)$ are denoted by $\beta_0, \beta_1, \cdots, \beta_{q-2}$. The matrix G is of size $k \times n = k \times (q-1)$ with $k \leq n-1$. The code C is MDS so that (see Section 22.2) any $k \times k$ submatrix of G is non singular. Thus, any set of k columns are linearly independent. The rows of G are linearly independent: however, from the previous remark, the **truncated rows** of G i.e. the rows corresponding to any k column positions are also linearly independent. Let us record some properties of MDS codes, some of which were discussed also in Chapter 18.

We have that no two code words in C can agree on as many as $k = n - d + 1$ positions. For suppose two distinct code words in C agreed in k positions. Then their difference which is a non-zero vector in C has at most $n - k$ non-zero entries. But, since C is MDS, any non-zero vector in C has at least $d = n - k + 1$ non-zero entries.

Take any k-tuple of non-zero entries from F, say $\gamma_1, \cdots, \gamma_k$. Pick any set of coordinate (column) positions, say the first k positions. No two code words in C have $\gamma_1, \cdots, \gamma_k$ in their first k coordinate positions. So, from this, the total number of code words in C is at most $(q) \cdots (q) = q^k$. But $|C| = q^k$ since C has dimension k. It follows that there is a *unique code word which has* $\gamma_1, \cdots, \gamma_k$ *in its first* k *coordinate positions*. The same holds true no matter which k positions we choose.

An interesting property of MDS codes which follows from the above is this. Even if we lose the data in any $d-1$ positions of a code word we can still recover the entire word since we are left with $n - (d-1) = k$ data pieces.

(A related idea involves the use of MDS error correction in the RAID (Redundant Array of Independent Disks) system which is a collection of drives collectively acting as a storage system which can tolerate the failure of an individual drive without losing data and which can act independently of each other.)

It is possible to extend G by two more columns, namely

$$
\begin{pmatrix} 1 \\ 0 \\ \vdots \\ 0 \\ 0 \end{pmatrix} \text{ and } \begin{pmatrix} 0 \\ 0 \\ \vdots \\ 0 \\ 1 \end{pmatrix}
$$

while still preserving the MDS property, and get a doubly extended RS code D. (When $k = 3$ and $F = GF(q) = GF(2^m)$ we can get a further extension. In this case, we can take the matrix

$$
G = \begin{pmatrix} 1 & \cdots & 1 & 1 & 0 & 0 \\ \alpha_1 & \cdots & \alpha_{q-1} & 0 & 0 & 1 \\ \alpha_1^2 & \cdots & \alpha_{q-1}^2 & 0 & 1 & 0 \end{pmatrix}
$$

so the length of the MDS code C is now $q - 1 + 3 = q + 2$. Similarly, for C here the code $C^\perp$ also has length $q + 2$.)

We may now pose the following question.

Question. For given k, q what is the length $m(k, q)$ of the longest linear MDS code of dimension k over $GF(q)$?

Conjecture.

$$
m(k, q) = \begin{cases} q + 1 & \text{for } 2 \leq k \leq q, \\ k + 1 & \text{for } q < k, \end{cases}
$$

apart from the cases, $m(3, q) = m(q - 1, q) = q + 2$ if $q = 2^r$, as detailed above.

Let us return to the double-extended RS code D of length $q+1$ with generator matrix G, obtained from G by adjoining 2 extra columns. A typical column of G looks like $\begin{pmatrix} 1 \\ t \\ \vdots \\ t^{k-1} \end{pmatrix}$

with t in $GF(q)$. (Allowing t to be infinite gives $\begin{pmatrix} 0 \\ 0 \\ \vdots \\ 1 \end{pmatrix}$.) We can regard this column

either as a vector in a k-dimensional vector space over $GF(q)$ **or** as a point in a projective space $\Sigma = PG(k-1, q)$ of dimension $k-1$ over $GF(q)$.

Lets take $k - 1 = 2$. Then our point looks like $\begin{pmatrix} 1 \\ t \\ t^2 \end{pmatrix}$ which we can identify with the

Euclidean point $\begin{pmatrix} t \\ t^2 \end{pmatrix}$ – all of which lie on the curve $y = x^2$, which is a conic.

For $t = 0$, we get the Euclidean point $\begin{pmatrix} 0 \\ 0 \end{pmatrix}$, the origin.

For $t = \infty$, we get a single "infinite point" $\begin{pmatrix} 0 \\ 0 \\ 1 \end{pmatrix}$ or point at infinity which we met in

Chapter 6.

Lets take $k - 1 = 3$. Then our point set is a "twisted cubic" in $PG(3, q)$. In general, we get a **normal rational curve** in $PG(k - 1, q)$.

Any MDS code of dimension k will yield an **arc** S in $\Sigma = PG(k - 1, q)$, i.e., a set of points S such that no k points lie in a hyperplane $PG(k - 2, q)$ where $|S| \geq k \geq 3$, and conversely.

We can now extend the MDS conjecture to arcs. Here is a version.

Conjecture. Any arc in $PG(k - 1, q)$, $2 \leq k \leq q$, has at most $q + 1$ points unless $k = 3$ or $k = q - 1$ and $q = 2^r$ in which case the arc has at most $q + 2$ points.

In Wicker [Wic95] the author refers to the problem related to this conjecture as "one of the more interesting problems in projective geometry over Galois fields".

In fact this famous problem had indeed been open for about 50 years. The arc conjecture was asymptotically proven in the 1960s for q odd following pioneering work by the late B. Segre. A similar result for q even was finally obtained in 1988 by the senior author of this book along with two co-authors (see [BTB88]). Here are two typical results from that paper.

Theorem 22.12 Let K be an arc of $\Sigma = PG(k - 1, q)$, $q = 2^s$, $k \geq 5$ and $|K| \geq q + k - 1 - \sqrt[3]{q}$. Then K lies in a unique normal rational curve.

Theorem 22.13 *In* $PG(k-1,q)$, $q = 2^s$, *let* K *be an arc. Then*

(a) $|K| \leq q+1$ *if* $k = 4$.

(b) $|K| \leq q+1$ *if* $k \geq 5$ *and* $q \geq (k-3)^3$.

(c) *For* $k \geq 6$, *if* $|K| > q+1$ *then* $|K| < q+k-1-\sqrt[3]{q}$.

In the projective plane, it is known that, for an arc K, $|K| \leq q+1$ (q odd) and $|K| \leq q+2$ (q even).

Results similar to Theorem 22.13 for q odd have also been shown. The method of the proof for Theorem 22.13 involved the celebrated Hasse–Weil estimates for algebraic curves along with the geometry of quadrics and some combinatorics. For further background, we refer the reader to [BTB88].

22.9   Problems

1. Why is the vector $\mathbf{c} = (\alpha^6, \alpha, 0, \alpha^3, \alpha^6, 1, 0)$ in C?

2. Using the code C with generator polynomial $g(x)$ over the field of order 8 as in the text, suppose the received vector is the vector $(\alpha^6, \alpha^6, \alpha^4, \alpha^3, \alpha^4, \alpha^2, 1)$. What is the (most likely) transmitted code word?

3. Construct an MDS code of minimum distance 4.

4. Show that the dual of the linear code C with generator matrix given by

$$g(x) = (x-\alpha)(x-\alpha^2)\cdots(x-\alpha^t)$$

is also a Reed–Solomon code.

22.10   Solutions

1. The first two rows of a generator matrix G for C are

$$\begin{pmatrix} \alpha^6 & \alpha^5 & \alpha^5 & \alpha^2 & 1 & 0 & 0 \\ 0 & \alpha^6 & \alpha^5 & \alpha^5 & \alpha^2 & 1 & 0 \end{pmatrix}.$$

The sum of thes is $\mathbf{c}$.

2. $(\alpha^6, \alpha^5, \alpha, \alpha^3, \alpha^4, \alpha^2, 1)$.

3. There are many options. Since the minimum distance is 4, the length of each code word is at least 4. So if we work over a field $F = GF(q)$ of order q, the length, namely $q - 1$, must be at least 4. Thus, $q - 1 \geq 4$ so $q \geq 5$. We work with $GF(5)$ and pick a generator $\alpha = 2$ of the multiplicative group.

Now choose the generator polynomial to construct a code C with

$$g(x) = (x - 2^0)(x - 2^1)(x - 2^2) = (x - 1)(x - 2)(x - 4).$$

4. From the previous chapter, the code $C^\perp$ has a generator polynomial $h_1(x)$ whose roots are the inverses of the elements $\alpha^{t+1}, \alpha^{t+2}, \cdots, \alpha^{q-1} = 1$. The roots of h_1 are then $(\frac{1}{\alpha})^{t+1}, (\frac{1}{\alpha})^{t+2}, \cdots$. Thus $C^\perp$ is a Reed–Solomon code since $\frac{1}{\alpha}$ is also a generator: we take $b = t + 1$ in the definition, with $\frac{1}{\alpha}$ being the generator.

Chapter 23

MDS Codes, Secret Sharing, Invariant Theory

Goals, Discussion We want to give a quick summary of nonlinear MDS codes and their applications to secret sharing schemes and combinatorics. The MacWilliams identities are presented and used in connection with ideas from invariant theory applied to linear codes related to projective planes. In particular, the "computer algebra theorem of the twentieth century" – namely the nonexistence of a projective plane of order ten – is discussed. We also mention Euler's famous "36 officers" problem.

New, Noteworthy Most of this material has only appeared in research papers, so, from a textbook point of view, it is new.

23.1   General MDS Codes

In Chapter 22 we discussed **linear** MDS codes, using Reed-Solomon codes as examples. Basically, what we need for a linear MDS code C over a finite field $F = GF(q)$ is a $k \times n$ generator matrix G of rank k over F, with $n \geq k$, such that **every set of** k columns of G is linearly independent (We know from linear algebra that **some set** of k columns is linearly independent since the row rank equals the column rank).

The code C will have the property that, given any k entries $\beta_1, \beta_2, \ldots, \beta_k$ from F in columns $y_1, y_2, \ldots, y_k$ say, there is a unique code word $\mathbf{c}$ in C such that $\mathbf{c}$ has entry β_1 in column y_1, entry β_2 in column $y_2, \ldots$, and entry β_k in column y_k. We have that two code words in C cannot agree in as many as k positions, i.e. $d(\mathbf{c}_1, \mathbf{c}_2) \geq n - (k-1) = n + 1 - k$ for any two code words $\mathbf{c}_1, \mathbf{c}_2$ in C, where d denotes the Hamming distance. Since G has rank k, the code C can be regarded as a k-dimensional vector space with basis the rows of

G so C has exactly q^k code words.

In general an (n, k, q) MDS code, linear or not, is defined as follows. *It is a set C of q^k code words of length n over an alphabet A of size q with $k \le n$, where q is some positive integer, satisfying the following property P.*

Property P   *No two code words in C agree in as many as k positions.*

Explanation of property P.   Choose any set $y_1, y_2, \ldots, y_k$ of k column positions. For example, we might choose the first k column positions of C. Now map each code word $\mathbf{c}$ to its entries in these column positions. So we have a projection mapping $f : C \to A^k$, the set of k-tuples over A. From property P, this projection mapping f is one-to-one. Since $|C| = q^k = |A^k|$ we have that f is an onto mapping. It follows that the following condition holds.

Condition Q   *Given any k elements $\beta_1, \beta_2, \ldots, \beta_k$ from the alphabet A and any set of k column positions $y_1, y_2, \ldots, y_k$ there is a unique code word $\mathbf{c}$ having entry β_i in column position y_i, for $1 \le i \le k$.*

We also discussed this condition in Chapter 18.

23.2   The Case $k = 2$, Bruck Nets

Suppose $k = 2$. Then C is a set of q^2 code words of length n, where $2 \le n$, over some alphabet A of size q satisfying the following property P (or the analogous condition Q).

Property P   *No two code words in C agree in as many as 2 positions.*

Example 23.1 *Suppose A is the binary alphabet of size 2, so $q = 2$. Let $n = 3$. So $|C| = 4$, and each word in C is a binary triple. Such a code is given as follows.*

$$C = \{(000), (011), (101), (110)\}$$

It is convenient to discuss C from a geometric standpoint. We define a structure N of points and lines as follows.

Points of N: All q^2 code words of C

Lines of N: All code words in C that have a given symbol in a given position

It follows that there are qn lines of N altogether. Each line has q points.

To see that there are qn lines altogether, let's suppose that we are looking at all code words $\mathbf{c}$ having a symbol, α_1 say, in the first position. Fix any other position, say the second position and pick any α in A. There is exactly one code word c having α_1 in the first position and α in the second position. There are q choices for α. So, altogether, there are exactly q code words c having α_1 in the first position. So each line will have q points. We are using property P here.

Letting α_1 vary we get q distinct lines corresponding to the different symbols that a code word can have in the first position. These lines form a **parallel class** of q lines. This means that

1. No two of these lines meet, and

2. Each point of N lies on exactly one line of this parallel class.

Examining the code words having a given symbol from A in the ith position, $1 \le i \le n$, we get altogether a set of qn lines. Each line consists of all the code words that have a given symbol in a given position. The lines are therefore arranged into n parallel classes of lines satisfying the following axioms.

1. Two lines from the same parallel class are parallel, i.e. they have no points in common.

2. Two lines from distinct parallel classes meet in a unique point.

To see Axiom 2, we can use condition Q Section 23.1. Now we make the following definition.

A **Bruck net** N of order q and degree n is a set of q^2 points and qn lines, $n \ge 3$, satisfying the following axioms.

1. Any 2 points lie on at most one line.

2. Given any line l and any point P of l there is a unique line on P failing to meet l. This line is said to be **parallel** to l.

From the above it can be seen that any MDS code with $k = 2$ yields a Bruck net of order q and degree n.

For the converse, let N be such a Bruck net. There will then be n parallel classes of lines, as defined earlier. Label the n parallel classes as $1, 2, \ldots, n$ in some fashion, and label the q lines as $1, 2, \ldots, q$ within each parallel class. Then, with each point P associate a code word $(x_1, x_2, \ldots, x_n)$, where x_i is the number of the line in the ith parallel class that goes through P.

In this way we get an MDS code with q^2 code words of length n with $k = 2$. *We have thus shown that MDS codes with $k = 2$ are equivalent to Bruck nets.*

Example 23.2 *In Example 23.1, we have $q^2 = 2^2 = 4$ points and $qn = (2)(3) = 6$ lines. Let's draw a Bruck net of order 2 and degree 3.*

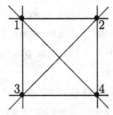

Figure 23.1: Bruck net of order 2 and degree 3

We have 4 points and 6 lines and 3 parallel classes of lines, namely horizontal, vertical and diagonal (Note that the line 14 is parallel to the line 23).

Then, labeling the lines within each parallel class as 0 or 1, we get, for each point, a code word as follows.

$$
\begin{array}{ccc}
3 & \longrightarrow & (000) \\
4 & \longrightarrow & (011) \\
1 & \longrightarrow & (101) \\
2 & \longrightarrow & (110)
\end{array}
$$

This is just the code described in Example 23.1 above!

23.3   Upper Bounds on MDS Codes, Bruck-Ryser

For $k = 2$, one can show that $n \le q + 1$. If $n = q + 1$ the corresponding Bruck net N is in fact an **affine plane** of order q. This means that **any 2 points of N are joined by a unique line**.

An example is obtained by using a field $F = GF(q)$ as alphabet when q is a prime power with $q = p^n$.. The affine plane $\Pi_A = AG(2, q) = AG(2, F)$ over F is easily described. The points of Π_A are all possible ordered pairs (a, b) with a, b in F. The lines are sets of points (x, y) satisfying linear equations of the form $y = mx + b$ or $x = c$, where m, b, c are fixed elements of F (This is exactly the construction of the Euclidean plane when $F = R$, the field of real numbers, following Descartes). These particular affine planes satisfy **Desargues axiom**, but there also exist non-Desarguesian planes of order q. (The Desargues axiom states that if two triangles are in perspective from a point in a projective plane then the intersections of corresponding lines are collinear.) The MDS code obtained from $AG(2, F)$ is equivalent to a linear code, but the code obtained from a non-Desarguesian plane will **not** be equivalent to a linear code.

Given any affine plane of order q we can adjoin $q + 1$ "points of infinity" to end up with a projective plane of order q. These were discussed in Chapter 22. Conversely, given a projective plane of order q we obtain an affine plane of order q by deleting a line. So the two concepts are, in a sense, equivalent. The affine plane of order q has q^2 points and $q^2 + q$ lines. The projective plane of order q has $q^2 + q + 1$ points and lines.

It is not known for which values of q affine (or projective) planes exist. If q is a prime power they certainly exist, as we have seen, by using a field of order q. This existence question has been studied intensively but there has been no general result since the **Bruck-Ryser Theorem** [BR49] in 1949 - generalized to the **Bruck-Ryser-Chowla Theorem** for symmetric designs - which rules out all of those values of q with q mod $4 = 1$ or q mod $4 = 2$, whenever q is not the sum of 2 integer squares. This theorem rules out infinitely many values of q, for example, $q = 6, 14, \ldots$, but leaves open the possibility of a plane of order q for infinitely many values of q such as $q = 10, 18, \ldots$ (since $10 = 3^2 + 1^2$ and $18 = 3^2 + 3^2$) which are not prime powers. We shall have more to say about the case $q = 10$ shortly. *There is no known plane whose order is not a prime power.*

Even when there is no plane of order q the question remains as to how big n can be in a Bruck net for a given q. A famous case occurs when $q = 6$, dating back to Euler and the "36 officers problem" in 1782. The 36 officer problem is as follows. "Officers can have 6 distinct ranks. The problem is to choose 36 officers from 6 regiments, no two officeres from the same regiment having the same rank, such that when they are on parade in a 6 by 6 square, no colomn or row shall contian 2 officers of the same regiment or of the same rank". It was shown by G. Tarry in 1901 that for $q = 6$, n is less than or equal to 3. Tarry's result is equivalent to showing that the 36 officer's problem has no solution. Tarry's method of proof was exhaustive trial and error. A shorter proof was constructed by D. Stinson [Sti84]. Hill [Hil86] discusses this in the context of "the main coding theory problem." (The usual version of that problem is to find the largest code of a given length and given minimum distance.) A famous theorem of Bruck, which also relates to work on **derivation** by T.G. Ostrom, is that if n is close to $q+1$ then the corresponding Bruck net can be embedded in a unique plane. Bruen [Bru71b] showed the existence of large unimbeddable nets close to the Bruck bound using a centuries-old geometric result known as **Galluci's Theorem** or the **Theorem of Dandelin**. Further references are given in Beutelspacher-Rosenbaum [Beu98] and Alderson [Ald02].

For $k \geq 3$ very little is known. It is known that $n \leq q + k - 1$. If $k \geq 4$, $q > 2$ and there exists an MDS code with $n = q + k - 1$ then 36 divides q so that q cannot be a prime power. This result is proved in [BS83] and [BS88], where related results are given. For $k = 3$ and $n = q + 2$, the only known examples are related to Desarguesian planes. We refer the reader to [BS88]. A specific example due to Alderson [Ald02] is described in Figure 23.2, with $n = 6$, $k = 3$ and $q = 4$.

$$
\begin{array}{llll}
(1,1,1,1,1,1) & (1,1,2,2,2,2) & (1,1,3,3,3,3) & (1,1,4,4,4,4) \\
(1,2,4,3,2,1) & (1,2,3,4,1,2) & (1,2,2,1,4,3) & (1,2,1,2,3,4) \\
(1,3,2,4,3,1) & (1,3,1,3,4,2) & (1,3,4,2,1,3) & (1,3,3,1,2,4) \\
(1,4,3,2,4,1) & (1,4,4,1,3,2) & (1,4,1,4,2,3) & (1,4,2,3,1,4) \\
(2,1,4,2,3,1) & (2,1,3,1,4,2) & (2,1,2,4,1,3) & (2,1,1,3,2,4) \\
(2,2,1,4,4,1) & (2,2,2,3,3,2) & (2,2,3,2,2,3) & (2,2,4,1,1,4) \\
(2,3,3,3,1,1) & (2,3,4,4,2,2) & (2,3,1,1,3,3) & (2,3,2,2,4,4) \\
(2,4,2,1,2,1) & (2,4,1,2,1,2) & (2,4,4,3,4,3) & (2,4,3,4,3,4) \\
(3,1,2,3,4,1) & (3,1,1,4,3,2) & (3,1,4,1,2,3) & (3,1,3,2,1,4) \\
(3,2,3,1,3,1) & (3,2,4,2,4,2) & (3,2,1,3,1,3) & (3,2,2,4,2,4) \\
(3,3,1,2,2,1) & (3,3,2,1,1,2) & (3,3,3,4,4,3) & (3,3,4,3,3,4) \\
(3,4,4,4,1,1) & (3,4,3,3,2,2) & (3,4,2,2,3,3) & (3,4,1,1,4,4) \\
(4,1,3,4,2,1) & (4,1,4,3,1,2) & (4,1,1,2,4,3) & (4,1,2,1,3,4) \\
(4,2,2,2,1,1) & (4,2,1,1,2,2) & (4,2,4,4,3,3) & (4,2,3,3,4,4) \\
(4,3,4,1,4,1) & (4,3,3,2,3,2) & (4,3,2,3,2,3) & (4,3,1,4,1,4) \\
(4,4,1,3,3,1) & (4,4,2,4,4,2) & (4,4,3,1,1,3) & (4,4,4,2,2,4)
\end{array}
$$

Figure 23.2: An MDS code with $k = 3, n = 6, q = 4$ with $4^3 = 64$ code work

23.4   MDS Codes and Secret Sharing Schemes

Suppose a combination lock with 100 dial positions has combination 13-32-93. The classic way of sharing this secret combination would be to first dissect the secret into three shares, each share consisting of a position and a two-digit number. Each share is then distributed to different users, 13 to Alice, 32 to Bob, and 93 to Charles, and each user is told the position of his or her number within the secret combination.

There are two difficulties with this secret sharing protocol. Firstly, knowledge of any given share reduces the security of the secret. In the above example, before being given her share, Alice can determine the secret with a probability of $1/100^3$. After she is given her share, this probability is increased to $1/100^2$. If Alice were privy to Bob's share, the probability would be further increased to $1/100$.

Secondly, all shares are required to determine the secret. Indeed, if Alice were to lose her share then the remaining share holders would be left with only experimentation to determine the secret.

These two shortcomings led to the formulation of (S, T) **threshold schemes** for secret sharing due to Shamir, Blakely and others (see Trappe and Washington [TW01] and also Beutelspacher and Rosenbaum [BE71]) in which a secret is dissected into an *ordered list of S shares*. As above, a share consists of its value and position within the list. We have two requirements or postulates.

(P1) Knowledge of T shares reveals the secret

(P2) Knowledge of $T - 1$ or fewer shares reveals no information regarding the secret.

Now, to construct such a scheme, assume that the given secret is a member of some finite alphabet A. Consider an (n, k) MDS code C over A, linear or not. So the code has length n and any k entries from A in k given positions determine a unique code word.

We may construct an $(S = n - 1, T = k)$ threshold scheme as follows. The code C is made known to all. The secret is the final coordinate a_n, where $a = (a_1, a_2, \ldots, a_n)$ is a given code word. The share list is $\{a_1, a_2, \ldots, a_{n-1}\}$. Since any k coordinate positions in C form an information set, property (P1) is satisfied for $T = k$. Moreover, given t shares where $t < k$, no information regarding the secret is divulged. For example, suppose the first t coordinates are known. There are $|A|^{k-t}$ words in C having $a_1, a_2, \ldots, a_t$ as the first t coordinates.

Let B be this set of code words. If α and β are elements of A, then, among the words of B, just as many, namely $|A|^{k-t-1}$, have final coordinate α as have β. So property (P2) is satisfied.

This scheme will work for any MDS code, linear or not. It is more general than previous schemes such as the Shamir scheme, which is closely related to *linear* MDS codes.

23.5 MacWilliams Identities, Invariant Theory

The celebrated MacWilliams identities relate the weight distribution of a linear code C to that of the dual code $C^{\perp}$. Let C have length n and denote by A_i the number of code words in C that have weight equal to i. Then the **weight enumerator** of C is defined as follows.

$$W_C(X, Y) = \sum_{i=0}^{n} A_i X^i Y^{n-i}$$

Let B_i denote the number of vectors of weight i in the dual code $C^{\perp}$.

In the binary case we have the following result, which shows that the weight enumerator polynomial of the dual code $C^{\perp}$ is entirely determined by W_C and is in fact a linear transformation of $W_C(X, Y)$. We have **the MacWilliams identities**

$$W_{C^{\perp}}(X, Y) = \frac{1}{|C|} W_C(X - Y, X + Y) \tag{23.1}$$

where $|C|$ denotes the number of code words in C.

Now suppose that $C = C^{\perp}$. Then $n = 2k$. Define

$$f(X, Y) = 2^{n-k} \sum_{i=0}^{n} A_i X^i Y^{n-i} \tag{23.2}$$

Multiplying across by $|C| = 2^k$ in equation 23.1 we obtain

$$f(X,Y) = f(2^{-\frac{1}{2}}(Y - X), 2^{-\frac{1}{2}}(Y + X)) \tag{23.3}$$

In other words, we have

$$f(X,Y) = f(\frac{1}{\sqrt{2}}(Y - X), \frac{1}{\sqrt{2}}(Y + X)) \tag{23.4}$$

23.6   Codes, Planes, Blocking Sets

Figure 23.3: Bruce L. Rothschild, U.C.L.A, 2004 and Chris Fisher, U of Regina

Let Π be a finite projective plane of order n. In this section we assume that $n \bmod 4 = 2$, that is, the remainder when n is divided by 4 equals 2. Thus 2 divides n but 4 does not divide n. .An example (the only known example) is when $n = 2$ (see Chapter 20).

Let A be any incidence matrix for Π. Thus the $n^2 + n + 1$ points of Π give a $(0,1)$ matrix A of size $(n^2 + n + 1) \times (n^2 + n + 1)$. If we denote the points and lines of Π by $P_1, P_2, \ldots, P_{q^2+q+1}$ and $L_1, L_2, \ldots, L_{q^2+q+1}$ then $A = (a_{ij})$ where

$$a_{ij} = \begin{cases} 1 \text{ if } L_i \text{ contains } P_j \\ 0 \text{ otherwise.} \end{cases}$$

Now the rows of A generate a linear code D, of length $n^2 + n + 1$ working over the binary field. Using a parity check for D, we arrive at the code C. One of the show-pieces of the theory, which uses the Smith canonical form for matrices in the proof, was mentioned in Chapter 18. Applied to our special case it gives the following result.

Theorem 23.3 *The dimension of C is $\frac{1}{2}(n^2 + n + 2)$.*

Example Let Π denote the projective plane of order 2 (the Fano plane) as diagrammed on Page 367. It has 7 points, namely $P_1, P_2, \cdots, P_7$ and seven lines $l_1, l_2, \cdots, l_7$ labelled as follows.

$$
\begin{array}{lll}
l_1 = \{P_1, P_6, P_5\}, & l_2 = \{P_1, P_7, P_4\}, & l_3 = \{P_1, P_2, P_3\}, \\
l_4 = \{P_5, P_4, P_3\}, & l_5 = \{P_5, P_7, P_2\}, & l_6 = \{P_3, P_7, P_6\}, \\
l_7 = \{P_6, P_4, P_2\}. & &
\end{array}
$$

Then the incidence matrix A is as follows.

$$
A = \begin{pmatrix}
1 & 0 & 0 & 0 & 1 & 1 & 0 \\
1 & 0 & 0 & 1 & 0 & 0 & 1 \\
1 & 1 & 1 & 0 & 0 & 0 & 0 \\
0 & 0 & 1 & 1 & 1 & 0 & 0 \\
0 & 1 & 0 & 0 & 1 & 0 & 1 \\
0 & 0 & 1 & 0 & 0 & 1 & 1 \\
0 & 1 & 0 & 1 & 0 & 1 & 0
\end{pmatrix}.
$$

A suitable set of 4 rows of A serves as a generator matrix for the code D. To obtain a generator matrix for C, simply adjoin a column of 1's to a generator matrix for A. A generator matrix for a code equivalent to C is given in Chapter 20.

The minimum weight of D is 3, corresponding to the lines. In the general case, the minimum weight of a code word in D is $n + 1$.

The minimum weight of C is $n + 2$ corresponding to the "extended lines" and the hyperovals which are sets of $n + 2$ points with no 3 collinear. For $n = 2$ these are just quadrangles of which there are exactly 168.

Note that $n^2 + n + 2$ is the length of the code C. Each row of a generator matrix of C corresponds to a line of Π. This line has $n + 1$ points, which is odd, since n is even. Thus, when the parity check is applied, the corresponding line will have a one at the "infinite coordinate" i.e. in column number $n^2 + n + 2$.

Any two distinct lines meet in a unique finite point. But they also each have. 1 in column number $n^2 + n + 2$. So two distinct lines intersect in two coordinate positions when regarded as code words in C. Now the lines generate C. It follows that C is contained in $C^{\perp}$. Therefore, $\dim(C) \leq \dim(C^{\perp})$, where dim denotes the dimension. But $\dim(C^{\perp}) = n^2 + n + 2 - \dim(C) = \frac{1}{2}(n^2 + n + 2) = \dim(C)$. We now have the following result.

Theorem 23.4 C is self-dual, i.e. $C = C^{\perp}$.

From Equation 23.4 of the previous section $f(X, Y)$ is invariant under the matrix

$$\begin{pmatrix} a & b \\ c & d \end{pmatrix} = \frac{1}{\sqrt{2}} \begin{pmatrix} -1 & 1 \\ 1 & 1 \end{pmatrix} = M_1 \qquad (23.5)$$

(Here we say that f is invariant under the matrix $M = \begin{pmatrix} a & b \\ c & d \end{pmatrix}$ if $f(X, Y) = f(aX + bY, cX + dY)$.)

Each row $\mathbf{u}, \mathbf{v}$ in the generator matrix for C has weight $(n+1) + 1 = n + 2$. So each row has weight divisible by 4. Now, for binary, $\text{wt}(\mathbf{u} + \mathbf{v}) = \text{wt}(\mathbf{u}) + \text{wt}(\mathbf{v}) - 2\text{wt}(\mathbf{u} \cap \mathbf{v})$, where $\text{wt}(\mathbf{u} \cap \mathbf{v})$ means the number of ones that $\mathbf{u}, \mathbf{v}$ have in common. Since $C \subset C^{\perp}$, $\text{wt}(\mathbf{u} \cap \mathbf{v})$ is even. In summary, each word in C has weight divisible by 4. Thus f is invariant under the matrix

$$\begin{pmatrix} \omega & 0 \\ 0 & 1 \end{pmatrix} = M_2, \qquad (23.6)$$

where $\omega^2 = -1$. Then (See Broué and Euguehard [BE71], Assmus and Mattson [AM74])

$$(M_1 M_2)^3 = \frac{1 + \omega}{\sqrt{2}} \begin{pmatrix} 1 & 0 \\ 0 & 1 \end{pmatrix}$$

which is a **primitive 8th root of unity** and acts to multiply X and Y by this primitive 8th root. We have shown the following result.

Theorem 23.5 *The length $n^2 + n + 2$ is divisible by 8.*

This is a nice example of the use of codes in combinatorial questions. As an immediate consequence we have the following result.

Theorem 23.6 *There are no projective planes of order n with $n \mod 8 = 6$.*

Proof. We have that $n^2 + n + 2$ would have to be divisible by 8 from Theorem 23.5. This contradicts the fact that $n \mod 8 = 6$ since in that case $n^2 + n + 2$ leaves a remainder of 4 (not 0) upon division by 8. ∎

Let us return to the linear code D for which C is the parity check code. We know that each word in C has weight divisible by 4. Thus each code word in D has a weight w such that $w \mod 4 = 0$ or 1.

The code words of smallest weight in D are the lines, which have weight $n + 1$. The next question is this.

Question What is the smallest weight of a code word $\mathbf{c}$ in D having odd weight bigger than $n + 1$?

Given such a code word **c** we associate a set S of points of Π corresponding to those columns of **c** with 1 in them. Assume that $|S| < 2n+3$. Such a set S then has the following properties.

1. Each line of Π contains at least one point of S.

2. Each line of Π contains at least one point not in S.

(Bruen studied such sets for his doctoral dissertation at Toronto. His friend and classmate J. C. Fisher urged that they be called "amiable sets." Bruen went with the name **blocking sets**. This name was vindicated by existing papers in the literature due to di Paola and others [dP69] where the terminology "blocking coalitions" was used, stemming from the connections to game theory).

The following result is shown in [Bru70] and [Bru71a].

Theorem 23.7 *If S is a blocking set then $|S| \geq n + \sqrt{n} + 1$. If equality occurs, then S is a projective subplane of order $\sqrt{n}$ i.e. a Baer subplane.*

It should be mentioned that, working independently in Hungary, Jan Pelikán [Pel70] had a result that came close to Theorem 23.7. The case of multiple blocking sets was not handled for another 20 years [Bru92].

The case of equality in Theorem 23.7 is of particular interest when $n = 10$. For a quick history, we refer the reader to Cameron and van Lint [CvL80].

The issue was whether or not there was a plane of order 10. This was the first case that was left open by the Bruck-Ryser theorem.

With invariant theory it was known that the entire weight distribution was known if the values A_{12}, A_{15}, A_{16} were found, where A_{12}, A_{15}, A_{16} denote the number of code words in D of weight $12, 15, 16$, respectively. A computer search showed that A_{12} was zero. Bruen had shown in [Bru71a] that A_{15} was zero unless Π had a subplane of order 2. The possible configurations were also classified there by hand. A computer search by Denniston [Den69] and independently by MacWilliams, Sloane, and Thompson [MST73] showed that A_{15} was zero.

We now come to A_{16}. The configurations corresponding to words of weight 15 and weight 16 are closely connected. Using Bruen's classification of code words of weight 15, Larry Carter, in his dissertation at Berkeley, showed that A_{16} was zero.

Finally Lam et al. [LTS89] showed that there were no code words of weight 19 by a lengthy computer analysis, leading to the conclusion that *there is no projective plane of order 10*. The result garnered much publicity and was discussed in the New York Times. A shorter proof of Theorem 23.7 appears, jointly with B.L. Rothschild [Lower bounds on blocking sets, Pacific J. Math., **118**, 1985,303–311].

23.7   Binary Linear Codes of Minimum Distance 4

The determination of codes in the title seems quite specialized. But, in fact, an amazing amount of discrete mathematics is involved in determining the "long" binary codes of minimum distance 4. Related topics include sum-free sets in groups and Tutte's tangential two-block conjecture for matroids, which implies the four-colour theorem in graph theory. For details we refer the reader to Bruen-Wehlau [BW99].

Chapter 24

Key Reconciliation, Linear Codes, New Algorithms

The authors are very grateful to Professor David Wehlau of the Royal Military College (RMC) in Kingston, Ontario and the CSE for his help with this work. Work related to this chapter will appear in a joint paper co-authored by Dr. Wehlau and the authors of this book.

Goals, Discussion In traditional symmetric key cryptography, the communicating parties **A**, **B** must both be in possession of a common secret key to communicate in secret. Here we give a very broad generalization of this. Namely, we assume that **A**, **B** are each in possession of keys **U**, **V** of a suitable common length N such that the mutual information $I(X:Y)$ of the corresponding random variables X,Y is non-zero. Under this assumption we show that a common secret key can be constructed by "public discussion." This new algorithm is already being used commercially.

New, Noteworthy Most of the material in this chapter is new. We show how to effectively use hash functions derived from linear codes in connection with the new algorithms. Our methods can be modified to apply to error correction but for reasons of space, we focus only on the cryptographic applications. This chapter brings together several ideas from previous chapters.

24.1 Symmetirc and Public Key Cryptography

The ancient difficulty for establishing a common cryptographic secret key between two communicating parties Alice and Bob is nicely summarized by the catch-22 dictum of Lomonaco

[Lom99], to wit: "in order to communicate in secret one must communicate in secret". In other words, to establish a common secret key Alice and Bob must already have a "shared secret".

In this chapter we show how to establish such a common secret key by public discussion under the modest requirement that Alice and Bob are initially in possession of keys A and B respectively of a common length N whcih are not necessarily equal but are such that the mutual information $I(A, B)$ is non-zero. This assumption is tantamount to assuming only that the corresponding statistical variables are correlated. The common secret key generated will enjoy the property of "perfect secrecy" in the sense of Shannon. Our method then provides a profound generalization of symmetric key cryptography. Such initial keys arise in quantum cryptography using the laws of quantum physics. Other methods, using classical physics (such as measurements of a bit stream from a common satellite), have also been proposed for generating the correlated keys A and B.

Finally, we note that if Alice and Bob are in possession of a "shared secret" such as an error-correcting code then a common secret key can also be generated with a variation of the well-known McEleice cryptosystem. However, we point out that this key will no longer enjoy perfect secrecy but only computational security. Some such systems are in commercial use.

24.2   General Background

In his foundational paper [Sha48], Shannon pointed out that for cryptographic systems, two different kinds of security methodologies can be considered as follows.

1. **Computational Security**. Here, although the system can be broken given sufficient time and computational resources, it is assumed that this cannot be done in a sufficiently short time frame as to pose a real threat.

2. **Perfect Shannon Security (Perfect Secrecy)**. Here the system can mathematically never be broken even assuming that an adversary has unlimited time and computational resources.

The main algorithm for (1) is the well-known RSA public key algorithm. The main algorithm in connection with (2) is the so-called *Vernam cipher* or *one-time pad*, which was used extensively during the Cold War in the 1950's and 1960's and still remains in use for military purposes. A secret message encoded with a one-time pad will remain secret forever. As Schneier [Sch96] puts it, "Even after the aliens from Andromeda land with their massive spaceships and undreamed-of computing power, they will not be able to read the Soviet spy messages encrypted with one-time pads (unless they can go back in time and get the one-time pads). "

The difficulty with (1) is that no system has yet been devised that is provably computationally secure. In particular, the security, even computational security, of the RSA system rests on mathematical assumptions about the difficulty of factoring integers which remain unproven and may turn out to be false.

The problem with (2), as exemplified by the one-time pad, is the difficulty of providing the two communicating parties Alice and Bob with a secret key to enable encryption. During the Cold War, and throughout history, the two parties were provided the secret key by using trusted couriers[1].

One of the goals of this chapter is to provide some new results relating to the problem of key establishment. We aim to show how a secret key exchange can be designed by **public discussion**. Public discussion here means that Alice and Bob can exchange information in the open, (e.g., over the Internet, so that anyone can listen in) and yet end up with a common secret key concerning which any eavesdropper will have only a vanishingly small amount of information as measured in Shannon bits.

Such a procedure has, of course, been sought for a considerable amount of time.

We must point out that there are two aspects to our procedure, namely

(i) Physical

(ii) Mathematical

For (i) *we take it as given that Alice and Bob are each in possession of a binary string of length N*. The physical process that generates these two strings is assumed to have the property that the two strings are not completely random with respect to each other, i.e., their bit correlation (which is their total number of bit agreements divided by N) is bounded away from $1/2$.

Such physical systems arise in various applications. We mention the well-known quantum encryption procedure developed by Bennett, Brassard, and others (see, for example, [BBR88]). Other physical examples have also been patented and are currently in industrial use. It appears that quantum encryption is progressing so quickly that it will soon become a reality (see, for example, [Eco03]). We will not dwell on the details here other than to exploit them mathematically.

Assuming the existence of the two sufficiently long strings A and B, in the respective possession of Alice and Bob, *we show how to construct a common secret key by public discussion using information theory and other mathematical techniques*. In the case where an eavesdropper, Eve, has no initial information about the strings A and B, we prove that Eve will have no information about the common secret key distilled from A and B. The situation where Eve has some initial information about either A or B (or both) is

---

[1] As Professor Lomonaco so eloquently put it, we are confronted by a catch-22. *Catch-22 of modern cryptography*: In order to communicate in secret we must first communicate in secret.

more complicated but can also be handled using our methods. We also point out that our algorithm for constructing a common secret key has an **intruder detection system** built into it.

Wireless devices cannot support long keys required by popular public key cryptosystems such as RSA. It is hoped that the methods in this chapter may lead to a key exchange procedure in which the keys are reasonably short and therefore suitable for wireless systems.

Unfortunately, in previously published discussions of these techniques, the question of the length of the final secret key Alice and Bob obtain has not been specified. Of course, this is crucial in practical applications. In practical situations, Alice and Bob know ahead of time the desired length n of their secret key. They need to know what length N their initial keys should be. Indeed, if N is chosen too small they may end up with the empty string as their final key. Conversely, if N is too large they will waste time and computing resources generating a final key whose length significantly exceeds n. We provide here a method for computing N from n. Our analysis may also be used by Alice and Bob to predict how much time and effort will be required to generate their common secret key.

The basic method requires the choice of a block length, ℓ. We show here that for information theoretic and experimental reasons, choosing ℓ to be an integer power of 2 gives especially good results.

To verify that the secret key possessed by Alice is indeed exactly equal to the secret key possessed by Bob, we design a new kind of hash function customized for this purpose, using a new approach based on the theory of error-correcting codes.

24.3   The Secret Key and the Reconciliation Algorithm

We consider two communicating parties (cryptographic stations), Alice and Bob, who are each in possession of a binary string $A = (a_1, a_2, \ldots, a_N)$, $B = (b_1, b_2, \ldots, b_N)$ respectively. We define the bit correlation between A and B, denoted by $\text{bitcorr}(A, B)$ as follows

$$\text{bitcorr}(A, B) = \frac{\#\{i \mid a_i = b_i, 1 \leq i \leq N\}}{N}.$$

Thus we have $\text{bitcorr}(A, B) = 1 - d(A, B)/N$ where $d(A, B)$ denotes the usual Hamming distance between A and B, i.e., the number of indices i for which $a_i \neq b_i$.

Let us explain our philosophy. For the key agreement protocol denoted by KAP we assume for the moment that X and Y are two strings of equal length with $1/2 < \text{bitcorr}(X, Y) < 1$. Then, if U and V are substrings of X, Y and if f is a suitable (hash) function such that $f(U) = f(V)$, it is likely that $\text{bitcorr}(U, V) > \text{bitcorr}(X, Y)$. Assume further that we partition X and Y into corresponding substrings $U_1, U_2, \ldots, U_t$ and $V_1, V_2, \ldots, V_t$ and that we modify, by reducing or even deleting, those substrings U_i, V_i for which $f(U_i) \neq f(V_i)$.

Then, for the remnant strings X_1 and Y_1 it is reasonable to expect that bitcorr$(X_1, Y_1) >$ bitcorr(X, Y). By pursuing this iteratively, after a suitable number of iterations, say s, the two initial strings "converge" to two equal substrings $X_s = Y_s$.

This is the essence of the method. Of course, the procedure is complicated by the fact that care must be taken to account for any information that might be revealed to the eavesdropper, Eve, during the entire process.

Special cases of the above procedure were suggested as an algorithm for quantum encryption in the paper [BBB+90]. A general, but abstract, approach is outlined by Maurer ([Mau93]). Of course, there is nothing new in this general philosophy. It dates to the basic methodology used in error-correcting codes going back to Hamming and Shannon in the 1940's

Let us begin. In the first phase of the KAP, called the *convergence phase*, the following procedure is applied repeatedly.

1. A random permutation σ (better still a *shuffle* of a certain type, as described in Section 24.8), is applied equally to each of the two keys A and B.

2. A block size $\ell \geq 2$ is chosen.

3. The two keys are divided into blocks of length ℓ. If the length of the keys is not evenly divisible by ℓ, the excess bits are discarded.

4. For each block, Alice and Bob publicly compare the total parity of that block.

5. If the parities agree then the two blocks are tentatively accepted as agreeing. In this case the remnant string from that block consists of the entire block, save the last bit. Alice and Bob each delete this last bit from their block to offset the information revealed when announcing their two parities.

   If the parities do not agree, Alice and Bob perform a binary search to locate an error.

6. Having started with two keys A and B and applied the previous 5 steps, Alice and Bob are now in possession of two remnant keys A_1 and B_1. They now apply steps 1–5 using A_1 and B_1 in the roles of A and B. They continue to iterate steps 1–5 until the estimated correlation of the remnant keys satisfies a certain halting condition described in Section 24.4.

Each time steps 1-5 of the convergence phase is performed we say that one *round* of the convergence phase has been performed.

To elaborate on how Alice and Bob carry out the binary search in step 5 when the parities disagree, we first define functions $P_1, P_2, P_3, \ldots$ inductively as follows. The function $P_i(S_A, S_B)$ for $i \geq 1$ is defined on ordered pairs of strings, S_A and S_B of common length i. The function value of P_i is an ordered pair of strings of a common length.

For $i = 1$ we define $P_i(S_A, S_B) = (\emptyset, \emptyset)$ where $\emptyset$ is the null string of length 0. For $i = 2$, write $S_A = (\alpha_1, \alpha_2)$ and $S_B = (\beta_1, \beta_2)$. If $\alpha_1 \neq \beta_1$, we put $P_2(S_A, S_B) = ((\alpha_2), (\beta_2))$. Otherwise, we put $P_2(S_A, S_B) = (\emptyset, \emptyset)$.

For $i \geq 3$, define procedure P_i as follows. Put $t := \lceil i/2 \rceil$ and write $S_A = (\alpha_1, \ldots, \alpha_t, \alpha_{t+1}, \ldots, \alpha_i)$ and $S_B = (\beta_1, \ldots, \beta_t, \beta_{t+1}, \ldots, \beta_i)$. Thus S_A is the concatenation of M_A and N_A where $M_A := (\alpha_1, \ldots, \alpha_t)$ and $N_A := (\alpha_{t+1}, \ldots, \alpha_i)$. Define M_B and N_B similarly from S_B. Also put $M'_A := (\alpha_1, \ldots, \alpha_{t-1})$ and $M'_B := (\beta_1, \ldots, \beta_{t-1})$.

If the parities of M_A and M_B agree, then $P_i(S_A, S_B) := (R_A, R_B)$. Here R_A is the concatenation of the string M'_A with that string which is the first component of the ordered pair $P_{i-t}(N_A, N_B)$. Similarly R_B is the concatenation of the string M'_B with the second component of the ordered pair $P_{i-t}(N_A, N_B)$.

On the other hand, suppose the parities of M_A and M_B disagree. Then $P_i(S_A, S_B) := (Q_A, Q_B)$. Here Q_A is the concatenation of the first component of $P_{t-1}(M'_A, M'_B)$ with the string N_A. Similarly, Q_B is the concatenation of the second component of $P_t(M'_A, M'_B)$ with the string N_B.

Having thus defined $P_i(S_A, S_B)$ we now easily obtain recursively Alice's and Bob's remnant strings for each block as follows. Denote the strings in their blocks of length ℓ by S_A and S_B, respectively. Also, let S'_A be the string obtained from S_A by deleting the last bit and define S'_B similarly. Alice and Bob publicly announce the parities of S_A and S_B. If the parities of S_A and S_B agree, then the remnant strings are S'_A and S'_B as specified in step 5. Should the parities disagree, then the remnant strings are obtained recursively as $P_\ell(S'_A, S'_B)$, with the relevant parities being announced at each stage by Alice and Bob. Having computed a remnant string for each block, both Alice and Bob concatenate all of their remnants from each block to obtain their whole remnant string.

To illustrate, let us describe the procedure for blocks of size 2 and blocks of size 3.

Blocks of size 2   Let $N = 6$ and $\ell = 2$. Then $A = (a_1, \ldots, a_6)$ and $B = (b_1, \ldots, b_6)$. We first consider the blocks $S_A = (a_1, a_2)$ and $S_B = (b_1, b_2)$. Alice announces the parity $(a_1 + a_2) \pmod 2$, and Bob announces $(b_1 + b_2) \pmod 2$. If the two announced parities are equal, the remnant string from this block for Alice is (a_1) and the remnant string from this block for Bob is the string (b_1). If the announced values are different, the remnant string for both Alice and Bob is the null string. Alice then proceeds to her next block comprised of the string (a_3, a_4), and Bob proceeds to his next block (b_3, b_4). They then proceed to their third and final blocks (a_5, a_6) and (b_5, b_6), respectively. To end this round, they concatenate their three remnant strings, obtaining an overall remnant string.

To give a concrete example, suppose $A = (1, 1, 0, 0, 0, 1)$ and $B = (1, 1, 0, 1, 1, 0)$. For the first block the remnant string for both Alice and for Bob is (1). For the second block the remnant string for both Alice and Bob is the null string. For the third block the remnant string for Alice is (0) and for Bob it is (1). Thus the overall remnant string for Alice is $A_1 = (1, 0)$, and for Bob it is $B_1 = (1, 1)$.

Blocks of size 3 Again let $N = 6$. Write $A = (a_1, \ldots, a_6)$ and $B = (b_1, \ldots, b_6)$. Alice's first block is the string $S_A = (a_1, a_2, a_3)$, and Bob's first block is $S_B = (b_1, b_2, b_3)$. Alice announces $(a_1 + a_2 + a_3) \pmod 2$, and Bob announces $(b_1 + b_2 + b_3) \pmod 2$. If these announced parities agree then Alice keeps the remnant string (a_1, a_2) and Bob keeps the remnant string (b_1, b_2).

Assume on the contrary that the announced parities differ. Then Alice announces (a_1) and Bob announces (b_1). If these two bits agree, then the remnant string for this block of length 3 for Alice and for Bob is the null string. If the two bits disagree, then the remnant string is (a_2) for Alice and (b_2) for Bob. Alice and Bob then proceed to calculate their remnant strings from the second and final block. Concatenating their remnant strings from each of their two blocks gives them their overall remnant string.

As an example, let $A = (1, 1, 0, 0, 0, 1)$ and $B = (1, 1, 0, 1, 1, 0)$ as above. Then the remnant string for both Alice and for Bob from their first block is $(1, 1)$. The remnant string from the second block of length 3 is (0) for Alice and (1) for Bob. Thus the overall remnant string is $A_1 = (1, 1, 0)$ for Alice and $B_1 = (1, 1, 1)$ for Bob.

Remark 24.1 *In the above two concrete examples, the initial bit correlation, $\mathrm{bitcorr}(A, B) = 3/6$. When $\ell = 2$ the bit correlation of the remnant strings is still $1/2$, whereas for $\ell = 3$ the bit correlation has risen to $2/3$. We will show that, in the general case, the bit correlation increases to 1 with arbitrarily high probability. We then employ a suitable hash function to ensure that the two remnant strings are actually equal.*

Remark 24.2 *Note that the "binary search procedure" differs from the usual binary search in that we never "get lucky" early on. If their first comparison, on blocks of length $k + 1$, indicates that the blocks disagree, Alice and Bob have to each discard 1 bit and then perform their binary search using P_k. This binary search consists of either $\lceil \log_2(k+1) \rceil$ or $\lfloor \log_2(k+1) \rfloor$ comparisons, and, for each of these comparisons, Alice and Bob must both discard a corresponding bit from their original block.*

Remark 24.3 *As indicated above, the motivation for deleting a bit in a sub-string in various steps of the algorithm is to offset information revealed when the parities are publicly announced.*

24.4 Equality of Remnant Keys: the Halting Criterion

Let us recap the situation. Alice and Bob initially have generated, by physical means or otherwise, two strings or keys A, B of common length N such that the corresponding random variables are not independent. In fact, without loss of generality, it can be assumed that $1/2 < x = \text{bitcorr}(A, B) \leq 1$. The case where $x < 1/2$ can be reduced to the above case; this will be clarified at the end of this section.

Alice and Bob carry out one round of the convergence phase, using block size ℓ_1, obtaining remnant keys A_1 and B_1. It will transpire (see Section 24.6) that, on average, $\text{bitcorr}(A_1, B_1) > x = \text{bitcorr}(A, B)$. The expected correlation of A_1 and B_1 is given by $x_1 := \phi_{\ell_1}(x)$.

Performing the algorithm, after a suitable number of rounds, say i rounds, Alice and Bob end up with two keys A_i, B_i of common length $n := N_i$ and whose expected correlation, $y := x_i = \phi_{\ell_i}(x_{i-1}) = \phi_{\ell_i}(\phi_{\ell_{i-1}}(\cdots(\phi_{\ell_1}(x))\cdots))$, satisfies the following condition:

$$\textbf{Halting Criterion} \qquad\qquad\qquad n(1 - y) \leq \theta$$

Here $\theta < 1$ is a suitable pre-determined positive constant.

Write $p = 1 - y$. Then the expected number of disagreements between the keys A_i and B_i is np. The number of disagreements is a *Bernoulli process* which in turn is approximated by a normal distribution with mean np and variance $np(1 - p)$. For a given positive number t one can estimate the probability that $np > t$. In particular, this holds for $t = 1$. By a suitable choice of θ we can make this probability arbitrarily small. Thus the probability that A_i and B_i differ in even a single bit can be made arbitrarily small.

To further ensure that A_i and B_i are indeed identical we employ a hash function f based on an error-correcting code of a suitable minimum distance, say 4.

If $f(A_i) = f(B_i)$, we can then be assured that $A_i = B_i$. In the unlikely case that $f(A_i) \neq f(B_i)$ Alice and Bob repeat the convergence algorithm until the hash function agrees on the two remnant keys.

Let us clarify our earlier statement that we can assume $\text{bitcorr}(A, B) > 1/2$ provided that $\text{bitcorr}(A, B) \neq 1/2$.

Theorem 24.4 *If* $\text{bitcorr}(A, B) \neq 1/2$ *we may assume that* $\text{bitcorr}(A, B) > 1/2$.

Proof. If Alice and Bob know in advance that $x = \text{bitcorr}(A, B) < 1/2$ then $\text{bitcorr}(A, \overline{B}) = 1 - x > 1/2$ where $\overline{B}$ is the Boolean complement of B obtained by interchanging 0's and 1's in B. Bob can then replace B by $\overline{B}$.

Assume that nothing about $\text{bitcorr}(A, B)$ is known in advance save that $\text{bitcorr}(A, B) \neq 1/2$. If Alice and Bob begin with block size ℓ then they can estimate $x = \text{bitcorr}(A, B)$ from from the length of the remnant strings A_1 and B_1. (See Theorem 24.14). Since $L_\ell(x)$ is not a one-to-one function, its value indicates a few possible values for x. Applying another round of the convergence procedure (with another block length ℓ') gives a new remnant length $L_{\ell'}(L_\ell(x))$. This gives further information about the original value x. After a few rounds Alice and Bob may estimate x and the current correlation x'. Then if the value of

x' is less than $1/2$ Bob simply replaces B by $\overline{B}$. ∎

Remark 24.5 *There is, remarkably, another approach that Alice and Bob may use in which they always choose block lengths which are integer powers of 2 as follows. Suppose $x = $ bitcorr$(A, B) \neq 1/2$ and the block lengths chosen are all powers of 2. Then either the two strings will converge as usual if $x > 1/2$ or else they will diverge if $x < 1/2$. In the latter case, if the initial keys are sufficiently long, $A_i = \overline{B_i}$ with high probability (at least $1 - \theta$). Thus the two strings A_i and B_i output from the convergence phase will very probably satisfy either $A_i = B_i$ or else $A_i = \overline{B_i}$, depending on whether bitcorr$(A, B) > 1/2$ or bitcorr$(A, B) < 1/2$. Publicly testing a few bits of A_i and B_i will then settle the question one way or the other.*

Remark 24.6 *Choosing the block length ℓ to be an integer power of 2 has an additional pair of advantages. Experiments show (see Table 24.4) that for many values of x the optimal value of ℓ is a power of 2. Furthermore, for ℓ a power of 2, $\phi_\ell(x) = \phi_\ell(1 - x)$.*

Remark 24.7 *In practice, Alice and Bob will have decided in advance on the desired length of the final common secret key (as well as on the value of the constant θ.) On the basis of these two values together the value of the correlation y can be calculated. For example, we have worked with $n = 160$ and $\theta = 0.25$ where $y > \frac{639}{640}$. Knowing the values of n, θ, and bitcorr(A, B), the required length N of the initial keys can be estimated with high precision. We give some details of this in Section 24.6.*

24.5 Linear Codes: the Checking Hash Function

Traditionally, in error correction, linear codes are used as follows. A message m is embedded in a code word $\widetilde{m}$ belonging to some linear code C where $\widetilde{m}$ is obtained from m by adjoining to m certain "parity" bits. The vector $\widetilde{m}$ is transmitted to a receiver. Classical approaches, on the assumption of few errors, attempt to decode $\widetilde{m}$ and thus to recover m.

Here we provide a new approach, as follows. Recall that in cryptographic and other applications a hash function f is constructed to help decide with high probability whether two binary vectors u and v are equal. Consider the special situation where it is known (with high probability) that the Hamming distance between u and v is less than some small integer t. In other words, it is known that the number of bits where u and v differ is very likely less than t.

Consider next an $r \times n$ matrix H which is the parity check matrix of a code, C, of minimum distance at least t. This implies that the subspace, C, of vectors perpendicular to each row of H contains only one vector of Hamming weight less than t, namely the zero vector.

For each row w of H, define a function f_w, by taking for any binary vector z of length n the value $f_w(z)$ to be the dot product of the row w with z. Then, given vectors u and v as above such that $f_w(u) = f_w(v)$ for all r rows w of H, it follows that $u + v$ is an element of the code, C, which has minimum distance t. Therefore, either $u = v$ or else the Hamming distance between u and v is at least t.

Let us summarize the above remarks.

Theorem 24.8 *Let u, v be binary vectors of length n whose Hamming distance is less than t. Let f be the syndrome, i.e., let f be the hash function given by $f(z) := (f_{w_1}(z), f_{w_2}(z), \ldots, f_{w_r}(z))$ where $w_1, w_2, \ldots, w_r$ are the r rows of the $r \times n$ parity check matrix H of a code of minimum distance at least t. Here $f_{w_i}(z)$ is the dot product of the row w with the binary vector z of length n. Then, if $f(u) = f(v)$, it follows that $u = v$.*

Motivated by this, we define the hash function $f(z) := (f_{w_1}(z), f_{w_2}(z), \ldots, f_{w_r}(z))$ where $w_1, w_2, \ldots, w_r$ are the r rows of the check matrix H.

This technique of constructing hash functions applies in particular to the cryptographic situation in Section 24.4 where, in the notation there, $u = A_i$ and $v = B_i$.

Using the notation of the previous section, we have now the following result.

Theorem 24.9 *Fix any number ϵ, no matter how small, with $0 < \epsilon < 1$. Let A_i and B_i be Alice and Bob's remnant keys. Assume $f(A_i) = f(B_i)$ where f is the hash function of Theorem 24.8. Then with probability at least $1 - \epsilon$, $A_i = B_i$ and Alice and Bob are in possession of a common secret key.*

Remark 24.10 *In Section 24.4, compliance with the stopping condition ensured that $A_i = B_i$ with high probability. If, subsequently, it is also verified that $f(A_i) = f(B_i)$, then the probability of equality is even higher.*

A formula for the combined probability, although complicated, can be determined but we do not pursue this here, in order to keep our discussion to a reasonable length.

Example 24.11 *Suppose that n is some integer with $64 < n \leq 128$ and that A and B are binary vectors of length n. We will construct an 8×128 parity check matrix H. First, we construct a 7×128 matrix $\overline{H}$ where the 128 columns of $\overline{H}$ are all distinct. Take the first eight columns of $\overline{H}$ to be*

$$\begin{pmatrix} 0 & 1 & 0 & 0 & 0 & 0 & 0 & 0 \\ 0 & 0 & 1 & 0 & 0 & 0 & 0 & 0 \\ 0 & 0 & 0 & 1 & 0 & 0 & 0 & 0 \\ 0 & 0 & 0 & 0 & 1 & 0 & 0 & 0 \\ 0 & 0 & 0 & 0 & 0 & 1 & 0 & 0 \\ 0 & 0 & 0 & 0 & 0 & 0 & 1 & 0 \\ 0 & 0 & 0 & 0 & 0 & 0 & 0 & 1 \end{pmatrix}$$ *The remaining 120 distinct columns*

of $\overline{H}$ may be arranged in any order, say in lexicographic order. Now H is obtained from $\overline{H}$ by adding a row of length 128 consisting entirely of 1's to the top of $\overline{H}$.

Then, as can be seen from projective geometry, H is the parity check matrix for a code of minimum distance 4. We form 8 functions $f_1, f_2, \ldots, f_8$ by defining $f_i(z)$ to be the dot *product of the ith row of H with z where z is a binary vector of length 128. Now if $n < 128$, we extend A and B to to new binary strings A' and B' of length 128 by adding 0's to the right end of A and B. Note that the Hamming distance between A' and B' is the same as the Hamming distance between A and B. If $f_i(A') = f_i(B')$ then either $A' = B'$ or else the Hamming distance from A' to B' is at least 4. Thus, clearly, either $A = B$ or else the Hamming distance from A to B is at least 4.*

Security Finally, consider the extra condition that it is desired to conceal information about the values of A and B from some eavesdropper, Eve, who has learned say the eight values $f_1(A'), f_2(A'), \ldots, f_8(A')$. In this case the first eight bits may be deleted from A (and B) leaving shortened strings $\overline{A}$ and $\overline{B}$ of length $n - 8$. Although eight bits have been lost from A and B, this is compensated for by the fact that Eve's knowledge of the eight values $f_j(A)$ provides her with no information about A (or B).

24.6 Convergence and Length of Keys

As above, Alice and Bob are in possession of two keys A and B of common length N with $x = \text{bitcorr}(A, B)$, where $1/2 < x \leq 1$. Let $\phi_\ell(x)$ denote the expected correlation of the two keys after applying one round of the convergence procedure of Section 24.3 where ℓ denotes the block length. Our goal now is to calculate $\phi_\ell(x)$ and also the expected length of various remnant keys.

We begin by considering the case $\ell = 2$.

Theorem 24.12

$$\phi_2(x) = \frac{x^2}{2x^2 - 2x + 1}$$

Thus, $\phi_2(x)$ is a strictly increasing function on the interval (1/2,1). In particular, for every $x \in (1/2, 1)$ and for all $\epsilon > 0$, there exists $i \in N$ such that the ith iteration $\phi_2^i(x)$ is greater than $1 - \epsilon$.

Proof. We divide A and B into sub-blocks of length 2, discarding the last bit of A and B if N is odd. Consider a typical pair of sub-blocks (α_1, α_2) and (β_1, β_2) belonging to Alice and Bob, respectively. If the parities $\alpha_1 + \alpha_2 \pmod 2$ and $\beta_1 + \beta_2 \pmod 2$ disagree both sub-blocks are deleted and there are no surviving (remnant) bits from these sub-blocks. If the two parities agree, then the surviving (remnant) bits are α_1 and β_1. The two parities only agree if either $\alpha_1 = \beta_1$ and $\alpha_2 = \beta_2$ or else if $\alpha_1 \neq \beta_1$ and $\alpha_2 \neq \beta_2$. Thus the conditional probability that $\alpha_1 = \beta_1$ given that the parities are equal is $\frac{x^2}{x^2 + (1-x)^2}$. Averaging over the disjoint blocks of length 2, we conclude that the expected value for the bit correlation of the two remnant strings A_1 and B_1 is $\phi_2(x) = \frac{x^2}{x^2 + (1-x)^2} = \frac{x^2}{2x^2 - 2x + 1}$. Thus $\phi_2(x) > x$ if

	$\alpha + \beta$	Probability	Remnant Length	Remnant Agreements
1	(0,0,0)	x^3	2	2
2	(0,0,1)	$x^2(1-x)$	0	0
3	(0,1,0)	$x^2(1-x)$	0	0
4	(1,0,0)	$x^2(1-x)$	1	1
5	(0,1,1)	$x(1-x)^2$	2	1
6	(1,0,1)	$x(1-x)^2$	2	1
7	(1,1,0)	$x(1-x)^2$	2	0
8	(1,1,1)	$(1-x)^3$	1	0

Table 24.1:

and only if $x > 2x^2 - 2x + 1$ if and only if $x \in (1/2, 1)$. The final assertion of the theorem then follow by a calculus argument. ∎

We proceed to the case $\ell = 3$.

Theorem 24.13

$$\phi_3(x) = \frac{3x^3 - 3x^2 + 2x}{6x^3 - 8x^2 + 3x + 1}$$

Thus, $\phi_3(x)$ is a strictly increasing function on the interval (1/2,1). In particular, for every $x \in (1/2, 1)$ and for all $\epsilon > 0$, there exists $i \in N$ such that the ith iteration $\phi_3^i(x)$ is greater than $1 - \epsilon$.

Proof. As before, we divide A and B into sub-blocks of length 3, discarding leftover bits if any. Again we consider a typical pair of sub-blocks $\alpha := (\alpha_1, \alpha_2, \alpha_3)$ and $\beta := (\beta_1, \beta_2, \beta_3)$.

An analysis of the KAP algorithm reveals the following.

1. There are remnant keys of length 2, namely (α_1, α_2) for Alice and (β_1, β_2) for Bob if $\alpha_1 + \alpha_2 + \alpha_3$ (mod 2) equals $\beta_1 + \beta_2 + \beta_3$ (mod 2).

2. There are remnant keys of length 1, namely (α_2) and (β_2) if $\alpha_1 + \alpha_2 + \alpha_3$ (mod 2) differs from $\beta_1 + \beta_2 + \beta_3$ (mod 2) and also α_1 (mod 2) differs from β_1 (mod 2).

3. In the remaining case, the remnant keys are null.

Note that the parity of α is equal to the parity of β if and only if the bitwise sum (exclusive OR), $\alpha + \beta$, has an even number of 1's.

We construct a table as follows. The eight entries in the first column enumerate the eight possibilities for $\alpha + \beta$. The second column expresses the probability of occurrence of the triple in its row. The third column is the length of the two corresponding remnant strings from the sub-blocks. The fourth column is the number of bit agreements between these two remnant strings. See Table 24.1.

Case 1 corresponds to rows 1, 5, 6 and 7. Case 2 corresponds to rows 4 and 8. The remaining rows correspond to Case 3.

From the entries in Table 24.1 we can easily calculate $\phi_3(x)$ by dividing the average number of remnant bit agreements by the average length of the remnants.

Thus the numerator is $2x^3 + x^2(1-x) + 2 \times (x(1-x)^2)$ and the denominator is $2x^3 + x^2(1-x) + 3 \times (2x(1-x)^2) + (1-x)^3$ giving the desired formula for $\phi_3(x)$. Thus $\phi_3(x) > x$ if and only if $\frac{3x^3-3x^2+2x}{6x^3-8x^2+3x+1} > x$ if and only if $3x^2 - 3x + 2 > 6x^3 - 8x^2 + 3x + 1$ if and only if $6x^3 - 11x^2 + 6x - 1 < 0$. Since $6x^3 - 11x^2 + 6x - 1 = 6(x-1/3)(x-1/2)(x-1)$ we see that $\phi_3(x) > x$ for all $x \in (1/2, 1)$. The final assertion of the theorem again follows easily using calculus. ∎

Let $L_\ell(x)$ be the fraction of the initial key that is expected to remain after one round of the convergence algorithm. We also consider $M_\ell(x) = \phi_\ell(x) \cdot L_\ell(x)$ which when multiplied by N gives the predicted number of bit agreements in the remnant keys after one round of the convergence algorithm. A remarkable fact is that, after one round of the KAP algorithm, Alice and Bob can estimate the initial correlation by comparing the length of their remnant keys with that predicted by the function L_ℓ. After this round they can also monitor any distortion of signals from an intruder ("intruder detection") by means of knowledge of $L_\ell(x)$.

Suppose we begin with a pair of keys A and B of length n and correlation x. Applying one round of the convergence algorithm we obtain the pair A_1 and B_1. Then $L_\ell(x) \cdot n$ is the expected length of the keys A_1 and B_1 and $M_\ell(x) \cdot n = \phi_\ell(x) \cdot L_\ell(x) \cdot n$ is the expected number of bits which agree in the two new keys A_1 and B_1.

We summarize the resulting formulae for $M_\ell(x)$ and $L_\ell(x)$ for $\ell \le 24$.

Theorem 24.14 *Let A and B denote binary strings of length N and let $0 \le x \le 1$. Then for $2 \le \ell \le 24$ the values for the quantities $M_\ell(x)$ and $L_\ell(x)$ are as in Tables 24.2 and 24.3. The quantity $\phi_\ell(x)$ is then calculated from the formula $\phi_\ell(x) = M_\ell(x)/L_\ell(x)$.*

To be certain that the convergence phase terminates one needs to prove that if $x > 1/2$ then $\phi_\ell(x) > x$ for each block size ℓ used. If each of the block sizes used is at most 24, then by Theorem 24.14 we know that $\phi_\ell(x) > x$. Furthermore, using the explicit formulae for $L_\ell(x)$ we can accurately predict the length of the final common secret key from the initial length N and the correlation x of the initial keys. Explicitly, the expected length of the keys A_i and B_i, produced by the convergence phase, is given by the expression $L_{\ell_i}(x_{i-1})N_{i-1} = L_{\ell_i}(L_{\ell_{i-1}}(\cdots(L_{\ell_1}(x))\cdots))N$. Taking into account an estimate for the number of bits required to be discarded while verifying equality with the hash function allows Alice and Bob to estimate the length, n, of their final secret key. Similarly, they may begin with n and use this desired length to determine the length, N, of the initial key required to produce a common final key of length n.

We emphasize again that **the initial bit correlation**, even if unknown, **can be estimated with high precision** based on the length of the remnant key after one round of

ℓ	$\mathbf{M}_\ell(x)$
2	$(x^2)/2$
3	$(3x^3 - 3x^2 + 2x)/3$
4	$(10x^4 - 18x^3 + 11x^2)/4$
5	$(18x^5 - 38x^4 + 30x^3 - 10x^2 + 4x)/5$
6	$(40x^6 - 100x^5 + 94x^4 - 38x^3 + 5x^2 + 4x)/6$
7	$(88x^7 - 272x^6 + 348x^5 - 236x^4 + 90x^3 - 18x^2 + 6x)/7$
8	$(192x^8 - 704x^7 + 1120x^6 - 1016x^5 + 580x^4 - 214x^3 + 49x^2)/8$
9	$(400x^9 - 1648x^8 + 2984x^7 - 3112x^6 + 2050x^5 - 872x^4 + 231x^3 - 33x^2 + 8x)/9$
10	$(832x^{10} - 3808x^9 + 7744x^8 - 9168x^7 + 6916x^6 - 3388x^5 + 1036x^4 - 172x^3 + 9x^2 + 8x)/10$
11	$(1728x^{11} - 8704x^{10} + 19632x^9 - 26048x^8 + 22420x^7 - 12996x^6 + 5096x^5 - 1328x^4 + 225x^3 - 25x^2 + 10x)/11$

12. $(3584x^{12} - 19712x^{11} + 48896x^{10} - 72000x^9 + 69696x^8 - 46448x^7 + 21784x^6 - 7288x^5 + 1764x^4 - 306x^3 + 33x^2$
$+8x)/12$

13. $(7424x^{13} - 44800x^{12} + 123456x^{11} - 205248x^{10} + 229072x^9 - 180704x^8 + 103320x^7 - 43176x^6 + 13098x^5 - 2812x^4$
$+407x^3 - 37x^2 + 12x)/13$

14. $(15360x^{14} - 100864x^{13} + 305408x^{12} - 564608x^{11} + 711104x^{10} - 644384x^9 + 432048x^8 - 216528x^7 + 80568x^6$
$-21624x^5 + 3916x^4 - 408x^3 + 13x^2 + 12x)/14$

15. $(31744x^{15} - 225280x^{14} + 742912x^{13} - 1509888x^{12} + 2114688x^{11} - 2161408x^{10} + 1664576x^9 - 982400x^8 + 446904x^7$
$-156048x^6 + 41184x^5 - 7956x^4 + 1053x^3 - 81x^2 + 14x)/15$

16. $(65536x^{16} - 499712x^{15} + 1781760x^{14} - 3944448x^{13} + 6070272x^{12} - 6888448x^{11} + 5964288x^{10} - 4020096x^9$
$+2132160x^8 - 893152x^7 + 294800x^6 - 75992x^5 + 15028x^4 - 2206x^3 + 225x^2)/16$

17. $(133120x^{17} - 1079296x^{16} + 4107264x^{15} - 9743360x^{14} + 16136192x^{13} - 19795712x^{12} + 18622336x^{11} - 13713792x^{10}$
$+7996656x^9 - 3708272x^8 + 1364616x^7 - 394768x^6 + 88062x^5 - 14614x^4 + 1680x^3 - 112x^2 + 16x)/17$

18. $(270336x^{18} - 2322432x^{17} + 9396224x^{16} - 23783424x^{15} + 42190848x^{14} - 55675904x^{13} + 56599552x^{12} - 45273856x^{11}$
$+28840864x^{10} - 14704928x^9 + 5990864x^8 - 1931984x^7 + 483236x^6 - 90176x^5 + 11630x^4 - 866x^3 + 17x^2 + 16x)/18$

19. $(548864x^{19} - 4980736x^{18} + 21350400x^{17} - 57444352x^{16} + 108703744x^{15} - 153607680x^{14} + 167923200x^{13}$
$-145129216x^{12} + 100441824x^{11} - 56014624x^{10} + 25187360x^9 - 9086784x^8 + 2600832x^7 - 579464x^6 + 97620x^5$
$-11948x^4 + 1020x^3 - 60x^2 + 18x)/19$

20. $(1114112x^{20} - 10649600x^{19} + 48218112x^{18} - 137437184x^{17} + 276410368x^{16} - 416587776x^{15} + 487615488x^{14}$
$-453193728x^{13} + 338967296x^{12} - 205485184x^{11} + 101154944x^{10} - 40326016x^9 + 12925328x^8 - 3293216x^7 + 6575$
$-101464x^5 + 11968x^4 - 1046x^3 + 57x^2 + 16x)/20$

21. $(2260992x^{21} - 22708224x^{20} + 108281856x^{19} - 325902336x^{18} + 694210560x^{17} - 1112119296x^{16} + 1389661696x^{15}$
$-1386181632x^{14} + 1120237696x^{13} - 740054912x^{12} + 401422560x^{11} - 178885184x^{10} + 65294632x^9 - 19391816x^8$
$+4635488x^7 - 877440x^6 + 128282x^5 - 13930x^4 + 1064x^3 - 56x^2 + 20x)/21$

22. $(4587520x^{22} - 48300032x^{21} + 241958912x^{20} - 766902272x^{19} + 1725038592x^{18} - 2927468544x^{17} + 3889661952x^{16}$
$-4144225280x^{15} + 3597054464x^{14} - 2569691392x^{13} + 1520272128x^{12} - 746976192x^{11} + 304787392x^{10} - 1029557$
$+28598440x^8 - 6454040x^7 + 1158856x^6 - 159596x^5 + 15770x^4 - 970x^3 + 21x^2 + 20x)/22$

23. $(9306112x^{23} - 102498304x^{22} + 538214400x^{21} - 1792081920x^{20} + 4245225472x^{19} - 7608737792x^{18} + 10712315904$
$-12140961792x^{16} + 11261769728x^{15} - 8645957120x^{14} + 5534317440x^{13} - 2966481920x^{12} + 1333880288x^{11}$
$-502825312x^{10} + 158404672x^9 - 41446528x^8 + 8916772x^7 - 1553252x^6 + 214016x^5 - 22464x^4 + 1680x^3 - 80x^2$
$+22x)/23$

24. $(18874368x^{24} - 217055232x^{23} + 1192230912x^{22} - 4161011712x^{21} + 10355343360x^{20} - 19548471296x^{19}$
$+29073178624x^{18} - 34925461504x^{17} + 34474221568x^{16} - 28296159232x^{15} + 19472275456x^{14} - 11295878144x^{13}$
$+5541159936x^{12} - 2301116160x^{11} + 808267648x^{10} - 239427968x^9 + 59503808x^8 - 12311968x^7 + 2098512x^6$
$-290376x^5 + 31948x^4 - 2702x^3 + 161x^2 + 16x)/24$

Table 24.2: Expressions for $M_\ell(x)$ for $2 \le \ell \le 24$

ℓ	$\mathbf{L}_\ell(x)$
2	$(2x^2 - 2x + 1)/2$
3	$(6x^3 - 8x^2 + 3x + 1)/3$
4	$(16x^4 - 32x^3 + 24x^2 - 8x + 3)/4$
5	$(36x^5 - 88x^4 + 86x^3 - 42x^2 + 10x + 2)/5$
6	$(80x^6 - 232x^5 + 276x^4 - 172x^3 + 60x^2 - 12x + 5)/6$
7	$(176x^7 - 608x^6 + 896x^5 - 728x^4 + 350x^3 - 98x^2 + 14x + 4)/7$
8	$(384x^8 - 1536x^7 + 2688x^6 - 2688x^5 + 1680x^4 - 672x^3 + 168x^2 - 24x + 7)/8$
9	$(800x^9 - 3584x^8 + 7136x^7 - 8288x^6 + 6188x^5 - 3080x^4 + 1022x^3 - 218x^2 + 27x + 5)/9$
10	$(1664x^{10} - 8256x^9 + 18432x^8 - 24384x^7 + 21160x^6 - 12576x^5 + 5180x^4 - 1460x^3 + 270x^2 - 30x + 9)/10$
11	$(3456x^{11} - 18816x^{10} + 46496x^9 - 68832x^8 + 67832x^7 - 46736x^6 + 22988x^5 - 8084x^4 + 1998x^3 - 332x^2 + 33x + 7)/11$
12	$(7100x^{12} - 42490x^{11} + 115200x^{10} - 188800x^9 + 208384x^8 - 163296x^7 + 93296x^6 - 39248x^5 + 12096x^4 - 2664x^3$ $+ 396x^2 - 36x + 11)/12$
13	$(14848x^{13} - 95744x^{12} + 284544x^{11} - 515968x^{10} + 636832x^9 - 564928x^8 + 370576x^7 - 182032x^6$ $+ 66984x^5 - 18252x^4 + 3586x^3 - 482x^2 + 39x + 9)/13$
14	$(30720x^{14} - 214016x^{13} + 691712x^{12} - 1374464x^{11} + 1875328x^{10} - 1857856x^9 + 1377376x^8 - 775648x^7$ $+ 332976x^6 - 108248x^5 + 26180x^4 - 4564x^3 + 546x^2 - 42x + 13)/14$
15	$(63488x^{15} - 475136x^{14} + 1658880x^{13} - 3584000x^{12} + 5358080x^{11} - 5870592x^{10} + 4868864x^9 - 3111680x^8 +$ $1544400x^7 - 594880x^6 + 176176x^5 - 39312x^4 + 6370x^3 - 700x^2 + 45x + 11)/15$
16	$(131072x^{16} - 1048576x^{15} + 3932160x^{14} - 9175040x^{13} + 14909440x^{12} - 17891328x^{11} + 16400384x^{10} - 11714560x^9$ $+ 6589440x^8 - 2928640x^7 + 1025024x^6 - 279552x^5 + 58240x^4 - 8960x^3 + 960x^2 - 64x + 15)/16$
17	$(266240x^{17} - 2260992x^{16} + 9035776x^{15} - 22568960x^{14} + 39459840x^{13} - 51251200x^{12} + 51204608x^{11} - 40195584x^{10}$ $+ 25099360x^9 - 12538240x^8 + 5010720x^7 - 1592864x^6 + 397852x^5 - 76440x^4 + 10910x^3 - 1090x^2 + 68x + 12)/17$
18	$(540672x^{18} - 4857856x^{17} + 20611072x^{16} - 54870016x^{15} + 102707200x^{14} - 143546368x^{13} + 155244544x^{12}$ $- 132840448x^{11} + 91172160x^{10} - 50564480x^9 + 22714560x^8 - 8245056x^7 + 2400216x^6 - 552720x^5 + 98460x^4 - 13092x^3$ $+ 1224x^2 - 72x + 17)/18$
19	$(1097728x^{19} - 10403840x^{18} + 46706688x^{17} - 132022272x^{16} + 263415808x^{15} - 394171392x^{14} + 458734592x^{13}$ $- 424885760x^{12} + 317820608x^{11} - 193683456x^{10} + 96557472x^9 - 39378496x^8 + 13083616x^7 - 3510760x^6 + 749812x^5$ $- 124588x^4 + 15538x^3 - 1370x^2 + 76x + 14)/19$
20	$(2228224x^{20} - 22216704x^{19} + 105218048x^{18} - 314720256x^{17} + 666787840x^{16} - 1063632896x^{15} + 1325410304x^{14}$ $- 1321125888x^{13} + 1069758976x^{12} - 710588160x^{11} + 389309184x^{10} - 176227712x^9 + 65798304x^8 - 20155808x^7$ $+ 5017280x^6 - 999616x^5 + 155720x^4 - 18280x^3 + 1520x^2 - 80x + 19)/20$
21	$(4521984x^{21} - 47316992x^{20} + 235732992x^{19} - 743686144x^{18} + 1666752512x^{17} - 2821943296x^{16} + 3746638848x^{15}$ $- 3996544000x^{14} + 3480948480x^{13} - 2502096896x^{12} + 1493944896x^{11} - 743277248x^{10} + 308170096x^9 - 106161952x^8$ $+ 30195096x^7 - 7016936x^6 + 1311108x^5 - 192264x^4 + 21318x^3 - 1682x^2 + 84x + 16)/21$
22	$(9175040x^{22} - 100532224x^{21} + 525598720x^{20} - 1744240640x^{19} + 4123131904x^{18} - 7384875008x^{17} + 10407481344x^{16}$ $- 11829676032x^{15} + 11027932160x^{14} - 8527770112x^{13} + 5510624768x^{12} - 2988261504x^{11} + 1361817600x^{10} - 520937\!8$ $+ 166608112x^8 - 44231440x^7 + 9638728x^6 - 1695584x^5 + 234916x^4 - 24684x^3 + 1848x^2 - 88x + 21)/22$
23	$(18612224x^{23} - 213123072x^{22} + 1166737408x^{21} - 4063166464x^{20} + 10103521280x^{19} - 19087310848x^{18} + 2845874176$ $- 34339606528x^{16} + 34116040704x^{15} - 28240956416x^{14} + 19636367104x^{13} - 11526520064x^{12} + 5726300736x^{11}$ $- 2407951232x^{10} + 855159584x^9 - 255255264x^8 + 63533352x^7 - 13031904x^6 + 2165468x^5 - 284268x^4 + 28378x^3$ $- 2026x^2 + 92x + 18)/23$
24	$(37748736x^{24} - 450887680x^{23} + 2579496960x^{22} - 9406251008x^{21} + 24545329152x^{20} - 48780017664x^{19} + 7671703142$ $- 97942405120x^{17} + 103305461760x^{16} - 91143421952x^{15} + 67845869568x^{14} - 42855499776x^{13} + 23047305216x^{12}$ $- 10565132800x^{11} + 4124555520x^{10} - 1367061248x^9 + 382566912x^8 - 89630784x^7 + 17366304x^6 - 2734240x^5 + 3410\!4$ $- 32432x^3 + 2208x^2 - 96x + 23)/24$

Table 24.3: Expressions for $L_\ell(x)$ for $2 \leq \ell \leq 24$

the algorithm.

24.7   Main Results

Let us summarize. Alice and Bob start off equipped only with two strings A and B of a common length N with non-zero mutual information $I(A, B) \neq 0$, i.e, with bitcorr$(A, B) \neq 1/2$. These keys may be thought of as having been produced by some physical means, as is the case for quantum encryption using the laws of quantum physics. (Other physical means using classical physics have recently been advanced). We have shown that Alice and Bob can then proceed by *public discussion* to obtain a final common secret key. We summarize as follows.

Theorem 24.15 *Assume that two cryptographic stations Alice and Bob are in possession of keys A and B of a common length N such that the corresponding random variables are not independent, i.e., such that $x = $ bitcorr$(A, B) \neq 1/2$. If N is sufficiently long then Alice and Bob can generate, by public discussion, a common secret key of length n using the convergence algorithm. If x is known, the expected value of N required to produce a secret key of desired length n can also be calculated as a function of N. If x is not known in advance it can be estimated from the length of the remnant keys produced by a few rounds of the convergence algorithm. Despite the public discussion, an eavesdropper who has no initial information concerning A and B will have no information whatsoever, measured in Shannon bits, about the final common secret key.*

Remark 24.16 Perfect Secrecy *If an eavesdropper, Eve, is in fact in possession of initial information concerning A or B the procedure can be modified so that, once again, Eve's average information about the final secret key, measured in Shannon bits, can be made arbitrarily small.*

Remark 24.17 Optimal Key Length *In any given industrial application, users will make a choice between, on the one hand, a slow convergence (many rounds of the algorithm) with a long final key and, on the other hand, rapid convergence with a shorter final secret key. Much depends on the initial value of x as indicated in Theorem 24.14. This choice manifests itself in the choice of the value of the block length ℓ used in each successive round of the algorithm. We have in fact devised a procedure for calculating optimal values for ℓ for maximizing the length of the final key. This procedure involves considerations from information theory and we do not reproduce it here, in order to keep our discussion to a reasonable length.*

Remark 24.18 Theory and Practice *In actual experiments, there is strong agreement between theory and practice. We include some experimental data in Tables 24.4 and 24.5. In Table 24.5, the column labeled x gives the initial correlation of two $N = 400$ bit strings.*

The column n_f denotes the expected bit length of the remnant keys, and x_f denotes the expected value of the correlation between the two remnant keys after the indicated number of rounds of the algorithm using optimal choices for the block length. The final column shows the expected number of errors in the remnant strings.

Remark 24.19 Intruder Detection *One classical method of attack on a cryptosystem is the so-called "man in the middle attack" where an active intruder impersonates Alice to Bob and Bob to Alice. The usual suggested remedy for this is a separate authentication channel. For the present algorithm an intruder might attempt to disrupt communications by changing the announced parities en route between Alice and Bob. After the first round, Alice and Bob can estimate the initial correlation from the remnant key. Based on this, Alice and Bob can calculate the correlation at the end of the second round using the function L_ℓ. On the other hand, Alice and Bob can also estimate the correlation at the end of the second round on the basis of the length of the remnant key. If these estimates differ significantly one can conclude that the parity signals have been altered by an intruder. Similarly, Alice and Bob can carry out this procedure on any round after the first round. Thus the algorithm has built-in intruder detection.*

Remark 24.20 *In [BBB+90] the authors correctly suggest that the original correlation $\text{bitcorr}(A, B)$ can be estimated at the end from the "errors corrected during reconciliation". However, as pointed out above $\text{bitcorr}(A, B)$ can be estimated after the first round of the convergence algorithm.*

In [BBB+90] the authors give an example of two strings A and B generated using a quantum channel. These keys were of length 640 and had $x = \text{bitcorr}(A, B) = 0.95625$. There, two rounds of the convergence algorithm were performed using $\ell_1 = 10$ and $\ell_2 = 20$ and yielding strings (A_1, B_1) and (A_2, B_2) with lengths $N_1 = 509$ and $N_2 = 457$ and correlations $x_1 = \text{bitcorr}(A_1, B_1) = 501/509 \approx 0.98428$ and $x_2 = \text{bitcorr}(A_2, B_2) = 455/457 \approx 0.99562$.

The expected values are $N_1 = 640$, $L_{10}(x) = 510.76$, $x_1 = \phi_{10}(0.95625) \approx 0.98473$, $N_2 = 509L_{20}(x_1) \approx 457.12$ or $N_2 = 640L_{20}(\phi_{10}(x)) \approx 459.25$ and $x_2 = \phi_{20}(\phi_{10}(x)) \approx 0.99610$. These numbers predict $(1 - 0.99610)(459.25) \approx 1.79$ errors.

Instead, for example, using $\ell_1 = 14$, $\ell_2 = 32$ and $\ell_3 = 64$ gives the expected values of $N_1 = 640$, $L_{14}(x) \approx 530.56$, $x_1 = \phi_{14}(x) \approx 0.98025$, $N_2 \approx 484.00$, $x_2 \approx 0.99111$, $N_3 \approx 460.45$ and $x_3 \approx 0.99637$.

24.8 Some Details on the Random Permutation

To ensure convergence it is important that in step 1 of the convergence phase Alice and Bob first agree on a random permutation of their two strings to better randomize the location

Correlation Range	Best ℓ
0.5 $< x \leq 0.81260286$	2
$0.81260286 < x \leq 0.90549162$	4
$0.90549162 < x \leq 0.90879524$	5
$0.90879524 < x \leq 0.90967405$	6
$0.90967405 < x \leq 0.91255608$	7
$0.91255608 < x \leq 0.95135877$	8
$0.95135877 < x \leq 0.95264488$	9
$0.95264488 < x \leq 0.95345573$	10
$0.95345573 < x \leq 0.95463845$	11
$0.95463845 < x \leq 0.95512870$	12
$0.95512870 < x \leq 0.95623924$	13
$0.95623924 < x \leq 0.95687848$	14
$0.95687848 < x \leq 0.95795073$	15
$0.95795073 < x \leq 0.97523500$	16
$0.97523500 < x \leq 0.97561411$	17
$0.97561411 < x \leq 0.97588534$	18
$0.97588534 < x \leq 0.97624747$	19
$0.97624747 < x \leq 0.97645235$	20
$0.97645235 < x \leq 0.97680092$	21
$0.97680092 < x \leq 0.97704305$	22
$0.97704305 < x \leq 0.97737559$	23
$0.97737559 < x \leq 0.97752890$	24

Table 24.4: Optimal Values for ℓ.

x	Rounds	n_f	x_f	Expected Errors
0.60	5	10.02	0.9800	0.20
0.61	5	11.29	0.9912	0.10
0.62	4	14.25	0.9847	0.21
0.63	4	16.74	0.9845	0.25
0.64	4	19.64	0.9822	0.34
0.65	4	22.22	0.9844	0.34
0.66	4	23.82	0.9908	0.21
0.67	3	29.09	0.9803	0.57
0.68	4	31.35	0.9866	0.42
0.69	4	37.37	0.9802	0.74
0.70	4	40.57	0.9854	0.59
0.71	4	43.10	0.9913	0.37
0.72	3	52.16	0.9808	1.00
0.73	3	56.78	0.9818	1.03
0.74	3	61.78	0.9827	1.07
0.75	3	64.46	0.9876	0.80
0.76	3	72.95	0.9824	1.28
0.77	3	79.97	0.9815	1.47
0.78	3	83.04	0.9867	1.10
0.79	3	86.35	0.9907	0.79
0.80	2	102.88	0.9781	2.24
0.81	2	105.93	0.9825	1.84
0.82	3	111.56	0.9872	1.42
0.83	3	125.83	0.9826	2.19
0.84	3	137.52	0.9825	2.41
0.85	3	142.61	0.9882	1.68
0.86	3	149.37	0.9917	1.24
0.87	2	175.48	0.9815	3.24
0.88	2	187.59	0.9820	3.37
0.89	2	200.77	0.9826	3.49
0.90	2	206.28	0.9876	2.56

Table 24.5: Reconciling two 400-bit keys

of errors. In practice, we have found it preferable to perform a specific kind of permutation that has some of the properties of a shuffle.

During the first round it is useful to apply a random permutation to randomize the location of the bits where the two keys disagree. However, in subsequent rounds, what is desired is that two or more errors lying in the same block should not be permuted into the same (new) block. Rather than relying on a random permutation to do this, we have found it more useful, experimentally, to use a specific permutation designed to prevent this. For example, an effective procedure is given by applying the permutation that changes the order from $(1, 2, 3, \ldots, n)$ into the order $(1, \ell + 1, 2\ell + 1, \ldots, n - \ell + 1, 2, \ell + 2, 2\ell + 2, \ldots, n - \ell + 2, \ldots, \ell - 1, 2\ell - 1, 3\ell - 1, \ldots, n - \ell - 1, \ell, 2\ell, 3\ell, \ldots, n)$. For example if $\ell = 4$ and $n = 20$, this shuffling produces the order $(1,5,9,13,17,2,6,10,14,18,3,7,11,15,19,4,8,12,16,20)$.

The advantage of this permutation is that it minimizes the occurrences of pairs of bits occurring in the same block (of length ℓ) twice in a row. Furthermore, since it is pre-determined, Alice and Bob do not have to spend time choosing and communicating a random permutation.

Experiments suggest that using such permutations performs better than choosing random permutations and is faster to perform.

24.9 The Case Where Eve Has Non-zero Initial Information

In practice, it may happen that an eavesdropper, Eve, has some (but not all) initial information about the keys A and/or B. Regardless of this, Alice and Bob can perform the convergence algorithm as described in Section 24.3 to achieve a common secret key. If Eve begins with no information, the algorithm is such that she cannot acquire any information from Alice's and Bob's public discussions. However, if Eve's initial information about A or about B is non-zero this initial information, in concert with the public discussion, may generate further information for Eve. The amount of information about the final secret key that Eve possesses, measured in Shannon bits, can, however, be estimated. In [BBB+90] this estimate is made based on physical considerations. In the mathematical setting, additional techniques from information theory can be used to obtain bounds on the amount of Eve's possible information.

The procedure for generating the common secret key, X, by public discussion must now take account of Eve's information.

Let us say that Alice and Bob are in possession of a common key of length n and that the amount of Eve's information about X is at most k Shannon bits. Let the integer $s > 0$ denote a security parameter that Alice and Bob may adjust as desired. A well-known and natural result of Bennett, Brassard, and Robert ([BBR88]) then shows that, by using a suitable hash function, Alice and Bob can construct from X a new common key, Y, of length $n - k - s$ such that Eve's average knowledge about Y is less than $2^{-s}/\ln(2)$ Shannon bits. In effect, then, Alice and Bob are now in possession of a common secret key of length $n - k - s$. The precise details concerning a suitable hash function as described above are given in Section 24.10.

24.10 Hash Functions Using Block Designs

In connection with the hash function mentioned above, it is pointed out in [BBB+90] that a suitable hash function is given by $n - (k - s) = n - t$ independent random subset parities where $t = k - s$.

The details of this are as follows. Let $v = (v_1, v_2, \ldots, v_n)$ be a binary vector of length n. We construct a set of $n - t$ functions $f_1, f_2, \ldots, f_{n-t}$ where $t > 0$ as follows. Choose first a family $F = F_1, F_2, \ldots, F_{n-t}$ of $n - t$ subsets of $\Omega := \{1, 2, 3, \ldots, n\}$. We identify each set F_i with its characteristic function. Thus each F_i is a binary vector of length n. We require that these corresponding $n - t$ binary vectors are linearly independent over the field with two elements.

Define $f_j(v)$ as the dot product of v with the binary vector corresponding to F_j. Finally, put $f(v) = (f_1(v), f_2(v), \ldots, f_{n-t}(v))$. This gives the desired hash function.

In a preferred embodiment, when f is utilized as a hash function to maximize the difficulty of eavesdropping, we want f to be constructed in such a way that it has regularity properties. That is, it is desirable that the subsets in the family F be "well spread out". Ideally, the family F would also have the property that any two elements of Ω lie in a constant number of subsets in F. Further, it is desirable that each subset in F have the same cardinality and that every two different subsets in F intersect in a constant number of elements. Indeed, these are precisely the criteria that motivated the design of experiments in statistics leading to the combinatorial study of designs (see [BJL86]).

In cryptography a condition known as the *Avalanche Criterion* (AC) is used in the analysis of S-boxes or substitution boxes (see, for example, [Mol00, Nic99]). S-boxes take a string as input and produce an encoded string as output. The avalanche criterion requires that if any one bit of the input to an S-box is changed, about half of the bits that are output by the S-box should change their values.

Here we wish to adapt this criterion to hash functions. Given a set of hash functions with values in $\{0, 1\}$, if one bit of the input string is changed then the avalanche criterion requires that about half of the hash functions should change their output values.

Here we show how to use design theory to construct a large set of hash functions which satisfy all of these criteria, as follows. A particular kind of block design arises from Sylvester matrices, the so-called Hadamard designs. Let H denote a $4t \times 4t$ Hadamard matrix. This means that every entry in H is a 1 or a -1 and that $HH^t = 4tI_{4t}$. We assume that such a matrix exists. (There is a longstanding open conjecture that at least one $4t \times 4t$ Hadamard matrix exists for every t. This conjecture has been verified for all $t \leq 117$. Furthermore, for infinitely many larger values of t it is known that $4t \times 4t$ Hadamard matrices do exist).

We suppose that H has been normalized so that its first row and first column consist entirely of 1's. We construct a new $4t-1 \times 4t-1$ matrix $\overline{H}$, all of whose entries are either 0 or 1 as follows. First we delete the first row and first column (consisting of all 1's) from H, and then we convert all the -1's in the remaining matrix to 0's. The resulting matrix is $\overline{H}$. This matrix is the incidence matrix of a block design with $v = 4t$, $k = 2t-1$ and $\lambda = t-1$. This design is called a Hadamard 2-design.

For each row w of $\overline{H}$ we define a hash function h_w that maps a $(4t - 1)$ vector to its dot product with the row w. These $4t-1$ different hash functions satisfy the avalanche criterion as well as the other desirable conditions listed above.

If t is odd, then these $4t-1$ linear hash functions are linearly independent. This fails if t is even. (However, in that case a large subset of the $4t-1$ hash functions is linearly independent).

Suppose that $n \not\equiv 3 \pmod 4$. Then there do not exist any Hadamard designs of size n. In this case we may choose the least integer $n' > n$ with $n' \equiv 3 \pmod 4$ and then we extend our input strings to length n' by padding on the right with (at most 3) zeroes. This gives us n' hash functions, which are linearly dependent.

24.11 Concluding Remarks

We have not given any detailed discussion of a fundamental problem for all cryptosystems, whether they be based on symmetric encryption or on the public key methodology. This problem is that of transmission errors in hardware and physical communication channels. Such errors seem impossible to eliminate. (Many estimates for transmission errors have been given in various applications). For this reason, some form of checking hash functions are essential in key generation algorithms between two parties Alice and Bob to ensure that each party has the same key.

This probablistic element pervades all of cryptography and indeed all communication methods. In many cryptographic protocols the problem is handled by explicitly including a suitable error correction methodology with the encryption.

We also should point out that in a great many applications, including for example digitized speech such as in the GSM wireless protocols, the encryption is often a so-called stream cipher. In such a case, a discrepancy of a few bits between the two final secret keys used by Alice and Bob will have a negligible effect.

ASCII

ASCII, (American Standard Code for Information Interchange), is a standard code set for representing characters. It consists of 128 characters including letters, numbers, punctuation and symbols. Each character has been assigned a unique binary string.

Table 6: ASCII Table

Char	Binary	Char	Binary	Char	Binary	Char	Binary
(nul)	00000000	(sp)	00100000	@	01000001	'	01100001
(soh)	00000001	!	00100001	A	01000010	a	01100010
(stx)	00000010	"	00100010	B	01000011	b	01100011
(etx)	00000011	#	00100011	C	01000100	c	01100100
(eot)	00000100	$	00100100	D	01000101	d	01100101
(enq)	00000101	%	00100101	E	01000110	e	01100110
(ack)	00000110	&	00100111	F	01000111	f	01100111
(bel)	00000111	'	00101000	G	01001000	g	01101000
(bs)	00001000	(	00101001	H	01001001	h	01101001
(ht)	00001001	)	00101010	I	01001010	i	01101010
(nl)	00001010	*	00101011	J	01001011	j	01101011
(vt)	00001011	+	00101100	K	01001100	k	01101100
(np)	00001100	,	00101101	L	01001101	l	01101101
(cr)	00001101	-	00101110	M	01001110	m	01101110
(so)	00001110	.	00101111	N	01001111	n	01101111
(si)	00001111	/	00110000	O	01010000	o	01110000
(dle)	00010000	0	00110001	P	01010001	p	01110001
(dc1)	00010001	1	00110010	Q	01010010	q	01110010
(dc2)	00010010	2	00110011	R	01010011	r	01110011
(dc3)	00010011	3	00110100	S	01010100	s	01110100
(dc4)	00010100	4	00110101	T	01010101	t	01110101
(nak)	00010101	5	00110110	U	01010110	u	01110110
(syn)	00010110	6	00110111	V	01010111	v	01110111
(etb)	00010111	7	00111000	W	01011000	w	01111000
(can)	00011000	8	00111001	X	01011001	x	01111001
(em)	00011001	9	00111010	Y	01011010	y	01111010
(sub)	00011010	:	00111011	Z	01011011	z	01111011
(esc)	00011011	;	00111100	[	01011100	{	01111100
(fs)	00011100	<	00111101	\	01011101			01111101
(gs)	00011101	=	00111110	]	01011110	}	01111110
(rs)	00011110	>	00111111	^	01011111	~	01111111
(us)	00011111	?	01000000	_	01100000	(del)	10000000

Shannon's Entropy Table

Here we give different values for Shannon's famous entropy function: $H(p, 1-p) = p \log \frac{1}{p} + (1-p) \log \frac{1}{1-p}$

Table 7: Shannon's Entropy Table

p	1-p	H(p,1-p)	p	1-p	H(p,1-p)
0.000	1.000	0.0000	0.250	0.750	0.8113
0.010	0.990	0.0808	0.260	0.740	0.8267
0.020	0.980	0.1414	0.270	0.730	0.8415
0.030	0.970	0.1944	0.280	0.720	0.8555
0.040	0.960	0.2423	0.290	0.710	0.8687
0.050	0.950	0.2864	0.300	0.700	0.8813
0.060	0.940	0.3274	0.310	0.690	0.8932
0.070	0.930	0.3659	0.320	0.680	0.9044
0.080	0.920	0.4022	0.330	0.670	0.9149
0.090	0.910	0.4365	0.340	0.660	0.9248
0.100	0.900	0.4690	0.350	0.650	0.9341
0.110	0.890	0.4999	0.360	0.640	0.9427
0.120	0.880	0.5294	0.370	0.630	0.9507
0.130	0.870	0.5574	0.380	0.620	0.9580
0.140	0.860	0.5842	0.390	0.610	0.9648
0.150	0.850	0.6098	0.400	0.600	0.9710
0.160	0.840	0.6343	0.410	0.590	0.9765
0.170	0.830	0.6577	0.420	0.580	0.9815
0.180	0.820	0.6801	0.430	0.570	0.9858
0.190	0.810	0.7015	0.440	0.560	0.9896
0.200	0.800	0.7219	0.450	0.550	0.9928
0.210	0.790	0.7415	0.460	0.540	0.9954
0.220	0.780	0.7602	0.470	0.530	0.9974
0.230	0.770	0.7780	0.480	0.520	0.9988
0.240	0.760	0.7950	0.490	0.510	0.9997
0.250	0.750	0.8113	0.500	0.500	1.0000

Glossary

A

AES Advanced Encryption Standard. The standard block cipher algorithm designated by the NIST for symmetric key cryptography. It is also known as Rijndael code.

ASCII American Standard Code for Information Interchange. The code generally used by most computer systems to translate characters into binary numbers.

B

Block Coding Any encoding function that encodes a block of words instead of one word at a time.

BSC Binary Symmetric Channel. A communication channel transporting the binary symbols $(0, 1)$ for which the probabilities p of receiving a 1 when 0 is transmitted is equal to the probability of receiving 0 when 1 is transmitted the same error.

C

Certificate An electronic file, typically containing a public key, that is digitally signed by a Certificate Authority. Certificates are used as a means of authentication over the Internet.

Certificate Authority The trusted party that signs and distribute certificates.

CESG British Communications Electronics Security Group.

Cryptanalysis . The art and science of deciphering encrypted messages.

CRC Cyclic Redundancy Check. A widely used error detection code.

D

DES Data Encryption Standard. The standard block cipher algorithm adopted by NIST for symmetric key cryptography in the late 70's. It has been superseded by AES.

Digital Signature A protocol by which the recipient of a message can verify that the sender of the message is in possession of the private key corresponding to a given public key and that the message itself has not been tampered with. Digital signatures should be used in conjunction with certificates to avoid the possibility of impersonation.

DNA Desoxyribonucleic Acid. The molecule were the genetic information of all live organisms is encoded. DNA is split and copied within the nuclei of each cell during multiplication.

DSS Digital Signature System. A protocol used to electronically sign documents.

E

ECC Elliptic Curve Cryptography. A public key cryptographic system based on the mathematical properties of elliptic curves.

ENIGMA The generic name given to a family of mechanical-rotor based ciphering devices utilized by the German forces during World War II.

Entropy A measure of the amount of uncertainty. Entropy is directly related to the number of *a priori* possible outcomes of a given event. For example, the entropy of a random binary string of length n is $\log(2^n) = n$.

F

FIPS Federal Information Processing Standards. The collective name for a series of standards related to information processing issued by NIST.

G

GPG GNU Privacy Guard. Open source, freeware version of PGP.

GSM Global System for Mobile telecommunications. The international standard for satellite phones.

H

Hacker   Originally the terms was used for creative programmers who program in an unorthodox way but it is now used to described those who use their technical skills to gain illegal access into computer networks to steal or vandalize information. A good hacker must have cryptanalytic skills.

Hamming Distance   The number of characters in which two strings (words) differ, measured with the metric of the corresponding alphabet.

Hash Function   A function that output a shorter version or digest of an input message. For cryptographic applications the hash function must be one-way (the input cannot be easily derived from the output) and has collisions-free (low probability that two different inputs will give the same output).

K

Kerberos   A trusted server-based protocol that provides authentication and key exchange for symmetric encryption systems. It is the logical equivalent of PKI. Windows 2000 operative systems authentication is based on Kerberos.

Key Reconciliation   A protocol by which two communicating entities obtain the same secret string after exchanging information over a public network.

L

LAN   Local Area Network.

LFSR   Linear Feed-back Shift Register. A pseudo-random sequence generator based on a recurrence relation.

M

Markov Chain (or Source)   A chain (source) of symbols source such that the next symbol of the chain depends only on the current value, not on any previous symbol.

McElliece Cryptosystem   A public key cryptosystem based on linear error correcting codes.

Memoryless Source (or Channel) A source of symbols for which the probabilities of output a given symbol does not depend on any of the previously output symbols.

N

NIST National Institute of Standards and Technologies. The United States' federal agency that develops and promotes measurement, standards and technology.

NSA National Security Agency. The United States' intelligence agency responsible for the security and cryptanalysis of electronic communications. It grew from a small US Navy task group in the World War II to be the largest employer of cryptographers in the world nowadays.

O

One-time pad A perfectly secure symmetric key encryption system that uses a random, secret key of the same length of the message to transmit.

P

PGP Pretty Good Privacy. Data and message encryption computer software developed by Phil Zimmermann on the basis of standard algorithms and public key encryption.

PKI Public Key Infrastructure. A Public Key directory database, often associated with key-pair generation functions, that provides authentication over medium to large networks.

Private Key The element of an asymmetric-key system key-pair that is kept secret by each user. This part of the key-pair is used for decryption.

Public Key The element of an asymmetric-key system key-pair that is shared with other users. This part of the key-pair is used by senders to encrypt a message.

Q

Quantum cryptography A form of cryptography that employs quantum properties of photons to exchange a random key over a public channel with perfect secrecy.

R

RSA A public-key cryptosystem patented by Rivest, Shamir, and Adleman in 1976. It is based on the computational difficulty of factoring large composite numbers. It the most widely implemented of the public key algorithms, included in applications such as PGP and SSL.

S

Shannon bit The amount of information gained (or entropy removed) upon learning the answer to a question whose two possible answers were equally likely, a priori.

SNR Signal to Noise Ratio.

SSL Secure Socket Layer. A widespread protocol used mainly for authentication of Internet transactions.

Symmetric-Key Encryption

T

TCP/IP Transmission Control Protocol/Internet Protocol. Protocols used for the transmission of information within networks of computers. TCP/IP have became the *de facto* standard for all networks connected to the Internet.

Triple-DES Symmetric key encryption system based on the successive application of three DES ciphers having different keys.

Trojan Horse A malicious program that causes damage or compromises the security, but does not replicates itself.

V

Virus A program or code that replicates by attaching itself, or infecting, another program, boot sector, partition sector, or document. Viruses are mostly malicious software that can cause large amounts of damage to sensitive data.

VOIP Voice Over IP. Any of the protocols used to digitize and pack a voice channel signal to be sent over a TCP/IP link.

W

WAN   Wide Area Network.

WEP   Wired Equivalent Privacy.   IEEE protocol used for data encryption of wireless LANs.

Worm   A ,usually malicious, program that makes copies of itself, inside the same computer or by copying itself through email. Worms may do damage and compromise the security of the computer and its network.

X

X.509 Certificate   A widely used standard for digital certificates.

Bibliography

[AB76]   M. M. Ali and Aiden A. Bruen. Bib designs and the hadamard problem. *Proceedings of Statistics Days, Ball State University*, pages 68–74, April 1976.

[AB04]   T. Alderson and Aiden A. Bruen. *On Extending Linear Codes and Bruen–Silverman Codes*. To Appear, 2004.

[AKS02]   M. Agrawal, N. Kayal, and N. Saxena. Primes is in p. Internet, 2002. http://www.cse.iitk.ac.in/news/primality.html.

[Ald02]   Timothy L. Alderson. *On MDS codes and Bruen–Silverman codes, Ph.D. Thesis*. University of Western Ontario, 2002.

[AM74]   E. F. Assmus Jr. and H. F. Mattson Jr. Coding and combinatorics. *Siam Review*, 16:349–388, 1974.

[And89]   Ian Anderson. *A First Course in Combinatorial Mathematics*. Oxford, 1989.

[Ash90]   Robert B. Ash. *Information Theory*. Dover, 1990.

[Bau02]   F. L. Bauer. *Decrypted Secrets, Methods and Maxims of Cryptology*. Springer, Berlin, third edition, 2002.

[BB75]   D. Borwein and Aiden A. Bruen. A problem of Erdös. *Can. Bull. Math.*, 17:220, 1975.

[BBB+90]   Charles H. Bennett, Francois Bessette, Gilles Brassard, Louis Salvail, and John Smolin. Experimental quantum cryptography. *EUROPCRYPT 90, Aarhus, Denmark*, 253-265 1990.

[BBR88]   Charles H. Bennet, Gilles Brassard, and Jean-Marc Robert. Privacy amplification by public discussion. *Siam J. of Computing*, 17(2):210–229, 1988.

[BDF98]   Boneh, Durfee, and Frankel. An attack on rsa given a small fraction of the private key bits. *AsiaCrypt*, pages 25–34, 1998.

[BE71]   Michel Broué and Michel Enguehard. Polynomes des poids de certain codes et fonctions thêta de certains reseaux. *Ann. Sci. Ecole Normale Sup.*, 5:157–181, 1971.

[Ber67]   E. R. Berlekamp. *Algebraic Coding Theory*. McGraw-Hill, 1967.

[Beu94]   A. Beutelspacher. *Cryptology*. The Mathematic Association of America, New York, 1994.

[Beu98]   Rosenbaum Beutelspacher. *Projective Geometry*. Cambridge, 1998.

[BH88]   Aiden A. Bruen and J. W. P. Hirschfeld. Intersections in projective space ii; pencils of quadrics. *Europ. J. Combinatorics*, 9:255–270, 1988.

[BJL86]   T. Beth, D. Jungnickel, and H. Lenz. *Design Theory*. Cambridge Univerisy Press, 1986.

[BO90]   Aiden A. Bruen and U. Ott. On the p-rank of incidence matrices and a question of E. S. Lander. *Contemporary Mathematics, AMS*, pages 39–45, 1990.

[Bon99]   Dan Boneh. Twenty years of attacks on the rsa cryptosystem. *AMS Notices*, 51(3):203–213, February 1999.

[BR49]   R.H. Bruck and H.J. Ryser. The non-existence of certain finite projective planes. *Can. J. Math*, 1:88–93, 1949.

[Bri97]   *Encyclopedia Britannica*. Chicago, fifteenth edition, 1997.

[Bro00]   J. Brown. *The Quest for the Quantum Computer*. Touchstone Books, New York, 2000.

[Bru70]   Aiden A. Bruen. Baer subplanes and blocking sets. *Bulletin American Math. Society*, 76:342–344, 1970.

[Bru71a]   Aiden A. Bruen. Blocking sets in finite projective planes. *Siam. J. Applied Math.*, 21:380–392, 1971.

[Bru71b]   Aiden A. Bruen. Partial spreads and replaceable nets. *Can. J. Math*, 20:381–391, 1971.

[Bru73]   Aiden A. Bruen. The number of lines determined by n^2 points. *J. Combinatorial Theory*, 15:225–241, 1973.

[Bru84]   Aiden A. Bruen. Arcs and multiple blocking sets. *Symposia Mathematica*, 28:15–29, 1984.

[Bru92]   Aiden A. Bruen. Polynomial multiplicities over finite fields and intersection sets. *J. Comb. Theory A*, 60:19–33, 1992.

[BS83]   Aiden A. Bruen and Robert Silverman. On the non-existence of certain mds codes and projective planes. *Math. Z.*, 183:171–175, 1983.

[BS88]   Aiden A. Bruen and Robert Silverman. On extendable planes, mds codes and hyperovals in $pg(2, q)$, $q = 2^t$. *Geom. Ded.*, 28:31–43, 1988.

[BS90]   E. Biham and A. Shamir. Differential cryptanalysis of DES-like cryptosystems. *Advances in Cryptology*, pages 2–21, 1990.

[BTB88]   Alden A. Bruen, J. A. Thas, and A. Blokhuis. On MDS codes, arcs in $pg(n, q)$ with q even and a solution of three fundamental problems of b. seqre. *Inventiones Math*, 92:441–459, 1988.

[Buc00]   J. Buchmann. *Introduction to Cryptography*. Springer, 2000.

[Bus52]   K. A. Bush. Orthogonal arrays of index unity. *Ann. Math. Stat.*, 23:426–434, 1952.

[BW99]   Aiden A. Bruen and D. Wehlau. Long binary linear codes. *Designs Codes and Cryptography*, 17:37–60, 1999.

[Cam94]   Peter J. Cameron. *Combinatorics*. Cambridge, 1994.

[Cel]   The first cell-phone worm emerges. Internet. http://www.newscientist.com/news/print.jsp?id=ns99995111.

[Cic65]   M. Cicchese. Sulle cubiche di un piano di Galois. *Rend. Mat. e Appl*, 24:291–330, 1965.

[Cic71]   M. Cicchese. Sulle cubiche di un piano lineare $s_{2,1}$ con q mod 3. *Rend. Mat*, 4:249–283, 1971.

[Cle80]   C. Herbert Clemens. *A Scrapbook of Complex Curve Theory*. Plenum, 1980.

[Coo87]   N. G. Cooper. Reflections on the life and legacy of Stanislaw Ulam. *Los Alamos Science Special Issue, Los Alamos National Laboratory*, 1987.

[Cop94]   D. Coppersmith. The data encryption standard (DES) and its strenght against attacks. *IBM J. of Research and Development*, 38(3), May 1994.

[Cov02]   Cover. Claude Elwood Shannon (1916-2001). *Notices of the American Mathematical Society*, January 2002.

[CvL80]   P. J. Cameron and J. H. van Lint. *Graphs, Codes and Designs*, volume 43. Cambridge, 1980.

[dB88]   Bert den Boer. Cryptanalysis of F.E.A.L. *EUROCRYPT*, pages 293–299, 1988.

[Den69]   R. H. F. Denniston. Nonexistence of a certain projective plane. *J. Australian Math. Society*, 10:214–218, 1969.

[dP69]   Jane W. di Paola. On minimum blocking coalitions in small projective plane games. *Siam J. Applied Math.*, 17:378–392, 1969.

[Eco03]   Uncrackable beams of light. *The Economist[Technology Quarterly]*, Sept 2003.

[Fel50]   W. Feller. *Introduction to Probability Theory*. Wiley, 1950.

[Fey99]   Richard P. Feynmann. *Feynmann Lectures on Computation*. Perseus, 1999.

[FS03]   N. Ferguson and B. Schneier. *Practical Cryptography*. John Wiley & Sons Inc., 2003.

[Gar04]   Paul Garrett. *The Mathematics of Coding Theory*. Prentice Hall, 2004.

[Gil95]   J. Gillogly. Ciphertext-only cryptanalysis of enigma. *Cryptologia*, 14(4), October 1995.

[Gol49]   M. J. E. Golay. Notes on digital coding. *Proc. IEEE*, 37:657, 1949.

[Gol82]   S. Golomb. *Shift Register Sequences*. Aegean Park Press, second edition, 1982.

[Gol02]   S. Golomb. Claude Elwood Shannon (1916-2001). *Notices of the American Mathematical Society*, January 2002.

[GP91]   Charles M. Goldie and Richard G. E. Pinch. *Communication Theory*. Cambridge, 1991.

[Gra00]   J. C. Graff. *Cryptography and E-commerce: Cryptography Basics for Non-technical Managers Working with E-business Products and Services (Wiley Tech Brief S.)*. John Wiley & Sons Inc., 2000.

[Gui77]   Silvin Guiasu. *Information Theory with Applications*. McGraw-Hill, 1977.

[Guy82]   Richard Guy. Sets of integers whose subsets have distinct sums. *Annals of Discrete Mathematics*, 12:141–154, 1982.

[Guy88]   Richard Guy. The strong law of small numbers. *American Math Monthly*, 95:697–792, 1988.

[Hay01]   Simon Haykin. *Commnications Systems*. Wiley, 2001.

[Hil86]   Raymond Hill. *A First Course in Coding Theory*. Oxford, 1986.

[HLL+92] D. G. Hoffman, D. A. Leonard, C. C. Lindner, K. T. Phelps, C. A. Rodger, and J. R. Wall. *Coding Theory, the Essentials*. Marcel Dekker, 1992.

[Hon85]   Ross Honsberger. *Mathematical Gems III*. The Mathematical Association of America, New York, 1985.

[Hor90]   J. Horgan. Claude E. Shannon, unicyclist, juggler and father of information theory. *Scientific American*, January 1990.

[HR91]   H. Hamalainen and S. Rankinen. Upper bounds for football pool problems and mixed covering codes. *J. of Combinatorial Theory A*, 56:84–95, 1991.

[HW03]   Darel Hardy and Carol Walker. *Applied Algebra:Codes, Ciphers, and Discrete Algorithms*. Prentice Hall, 2003.

[JJ00]   Gareth A. Jones and J. Mary Jones. *Information and Coding Theory*. Springer, 2000.

[Kah67]   D. Kahn. *The Codebreakers*. Macmillan, New York, 1967.

[Koc96]   Paul C. Kocker. Timing attacks on implementations of diffie-hellman, rsa, dss, and other systems. *Proc. CRYPTO*, pages 104–113, 1996.

[Lan83]   E. S. Lander. *Symmetric Design: an Algebraic Approach*. Cambridge, 1983.

[Lea96]   T. P. Leary. *Cryptology in the 16th and 17th Centuries*. Cryptologia, 1996.

[Len87]   H. W. Lenstra Jr. Factoring integers with elliptic curves. *Annals of Mathematics*, 126:649–673, 1987.

[LLL82]   A. K. Lenstra, H. W. Lenstra, and L. Lovasz. Factoring polynomials with rational coefficients. *Math. Ann.*, 261:515–534, 1982.

[Lom98]   S. Lomonaco. *Quick Glance at Quantum Cryptography*. American Mathematical Society Lecture Series, 1998.

[Lom99]   Samuel J. Lomonaco. A quick glance at quantum cryptography. *Cryptologia*, 23(1):1–41, January 1999.

[Lov79]   Laci Lovasz. On the Shannon capacity of a graph. *IEEE Trans. Information Theory*, 25:1–7, 1979.

[LTS89]   C. Lam, L. Thiel, and S. Swiercz. The nonexistence of finite projective planes of order 10. *Canadian J. Math.*, 41:1117–1123, 1989.

[Mas69]   James L. Massey. Shift-register synthesis and bch decoding. *IEEE Transactions on Information Theory*, 15:122–127, 1969.

[Mas02]   James L. Massey. Shannon and the development of cryptography. *AMS Notices*, 49(1), 2002.

[Mau93]   M. Maurer. Secret key agreement by public discussion from common information. *IEEE Trans. Inf. Theory*, 39:733–742, 1993.

[McC90]   K. McCurley. The discrete logarithm problem in cryptology and computational number theory. *AMS Proceedings in Applied Mathematics*, 42:49–74, 1990.

[McE78]   R. J. McEliece. *The Theory of Information and Coding*. Addison-Wesley, 1978.

[Mol00]   Richard A. Mollin. *Introduction to Cryptography*. Chapman & Hall/CRC Press, 2000.

[Mol02]   Richard A. Mollin. *RSA and Public-Key Cryptography*. Chapman & Hall/CRC Press, 2002.

[MS78]   F. J. MacWilliams and N. J. A. Sloane. *The Theory of Error-Correcting Codes*. North Holland, 1978.

[MST73]   F. J. MacWilliams, N. J. A. Sloane, and J. G. Thompson. On the existence of a projective plane of order 10. *J. Comb. Theory A*, 14:66–79, 1973.

[MSW02]   K. D. Mitnick, W. L. Simon, and S. Wozniak. *The Art of Deception: Controlling the Human Element of Security*. Wiley Publishing, Indianapolis, 2002.

[NC00]   Michael A. Nielsen and Isaac L. Chung. *Quantum Computation and Quantum Information*. Cambridge University Press, 2000.

[Nic99]   R. K. Nichols, editor. *ICSA Guide to Cryptography*. McGraw Hill, 1999.

[Pel70]   J. Pelikán. Properties of balanced incomplete block designs. *Combinatorial Theory and its Applications*, pages 869–889, 1970.

[Pie79]   John R. Pierce. *An Introduction to Information Theory*. Dover, 1979.

[PM02]   M. Piper and S. Murphy. *Cryptography: A Very Short Intorduction*. Oxford University Press, 2002.

[Poe93]   Edgar Allan Poe. *Tales of Mystery and Imagination*. Wordsworth, 1993.

[Pol74]   J. M. Pollard. Theorems on factorization and primality testing. *Proc. Cambridge Phil. Soc.*, 76:521–528, 174.

[PW72]   Wesley Peterson and E. J. Weldon. *Error-correction Codes*. M.I.T. Press, second edition, 1972.

[Ram12]   S. Ramanujan. Note on a set of simultaneous questions. *J. Indian Math. Society*, 4:94–96, 1912.

[Ruc87]   H. Ruck. A note on elliptic curves over finite fields. *Math. Comp*, 49:301–304, 1987.

[Rue86]   R. Rueppel. *Analysis and Design of Stream Ciphers*. Springer-Verlag, New York, 1986.

[Sch96]   B. Schneier. *Applied Cryptography: Protocols, Algorithms, and Source Code in C*. John Wiley & Sons, 1996.

[Sch03]   B. Schneier. *Beyond Fear*. Copernicus Books, 2003.

[Sha48]   C. E. Shannon. A mathematical theory of communication. *Bell Systems Tech J.*, 27:379–423, 623–656, 1948.

[Sha49]   C. E. Shannon. Communication theory of secrecy systems. *Bell Systems Tech J.*, 28:656–715, 1949.

[Sha77]   I. R. Shafarevich. *Basic Algebraic Geometry*. Springer, 1977.

[Smi02]   R. E. Smith. *Authentication: From Passwords to Public Keys*. Addison Wesley, 2002.

[SR85]   J. G. Semple and L. Roth. *Introduction to Algebraic Geometry*. Oxford: Clarendon Press, 1985.

[Sti84]   D. Stinson. A short proof of the nonexistence of a pair of orthogonal latin squares of order 6. *J. Combinatorial Theory*, 36:373–376, 1984.

[SW49]   C. E. Shannon and W. Weaver. *The Mathematical Theory of Communication*. University of Illinois Press, 1949.

[Syv97]   P. Syverson. A taxonomy of replay attacks. *Proceedings of the Computer Security Foundations Workshop (CSFW97)*, pages 187–191, June 1997.

[Tho83]   T. Thompson. From error-correcting codes through sphere packings to simple groups. *Carus Mathematical Monographys*, 21, 1983.

[Tur76]   P. Turan. *Selected Papers of A. Renyi.* Akademiai Kiado, 1976.

[TW01]   Wade Trappe and Lawrence Washington. *Introduction to Cryptography with Coding Theory.* Prentice Hall, 2001.

[vL98]   J. H. van Lint. *Introduction to Coding Theory.* Springer, third edition, 1998.

[Wal01]   M. M. Waldrop. Reluctant father of the digital age, Claude Shannon. *Technology Review*, page 64, July/August 2001.

[Wat69]   W. C. Waterhouse. Abelian varieties over finite fields. *Ann. Sci. École Norm. Sup.*, 4(2):521–560, 1969.

[Wel88]   Dominic Welsh. *Codes and Cryptography.* Oxford, 1988.

[Whi15]   E. T. Whittaker. On the functions which are represented by the expansion of the interpolation theory. *Proc. Roy. Soc. Edin.*, 35:181–194, 1915.

[Wic95]   Stephen B. Wicker. *Error Control Systems for Digital Communication and Storage.* Prentice Hall, 1995.

[Wie90]   M. J. Wiener. Cryptanalysis of short rsa secret exponents. *IEEE Transactions on Information Theory*, 36(3):188–190, May 1990.

[Wol61]   J. Wolfowitz. *Coding Theorems of Information Theory.* Prentice-Hall, New Jersey, 1961.

[WW46]   E. T. Whittaker and G. N. Watson. *A Course of Modern Analysis.* Cambridge, 1946.

Index

WILEY-INTERSCIENCE
SERIES IN DISCRETE MATHEMATICS AND OPTIMIZATION

ADVISORY EDITORS

RONALD L. GRAHAM
University of California at San Diego, U.S.A.

JAN KAREL LENSTRA
Department of Mathematics and Computer Science,
Eindhoven University of Technology, Eindhoven, The Netherlands

JOEL H. SPENCER
Courant Institute, New York, New York, U.S.A.

AARTS AND KORST • Simulated Annealing and Boltzmann Machines: A Stochastic Approach to Combinatorial Optimization and Neural Computing

AARTS AND LENSTRA • Local Search in Combinatorial Optimization

ALON, SPENCER, AND ERDŐS • The Probabilistic Method, Second Edition

ANDERSON AND NASH • Linear Programming in Infinite-Dimensional Spaces: Theory and Application

ARLINGHAUS, ARLINGHAUS, AND HARARY • Graph Theory and Geography: An Interactive View E-Book

AZENCOTT • Simulated Annealing: Parallelization Techniques

BARTHÉLEMY AND GUÉNOCHE • Trees and Proximity Representations

BAZARRA, JARVIS, AND SHERALI • Linear Programming and Network Flows

BRUEN AND FORCINITO • Cryptography, Information Theory, and Error-Correction: A Handbook for the 21st Century

CHANDRU AND HOOKER • Optimization Methods for Logical Inference

CHONG AND ŻAK • An Introduction to Optimization, Second Edition

COFFMAN AND LUEKER • Probabilistic Analysis of Packing and Partitioning Algorithms

COOK, CUNNINGHAM, PULLEYBLANK, AND SCHRIJVER • Combinatorial Optimization

DASKIN • Network and Discrete Location: Modes, Algorithms and Applications

DINITZ AND STINSON • Contemporary Design Theory: A Collection of Surveys

DU AND KO • Theory of Computational Complexity

ERICKSON • Introduction to Combinatorics

GLOVER, KLINGHAM, AND PHILLIPS • Network Models in Optimization and Their Practical Problems

GOLSHTEIN AND TRETYAKOV • Modified Lagrangians and Monotone Maps in Optimization

GONDRAN AND MINOUX • Graphs and Algorithms *(Translated by S. Vajdā)*

GRAHAM, ROTHSCHILD, AND SPENCER • Ramsey Theory, Second Edition

GROSS AND TUCKER • Topological Graph Theory

HALL • Combinatorial Theory, Second Edition

HOOKER • Logic-Based Methods for Optimization: Combining Optimization and Constraint Satisfaction

IMRICH AND KLAVŽAR • Product Graphs: Structure and Recognition

JANSON, LUCZAK, AND RUCINSKI • Random Graphs

JENSEN AND TOFT • Graph Coloring Problems

KAPLAN • Maxima and Minima with Applications: Practical Optimization and Duality

LAWLER, LENSTRA, RINNOOY KAN, AND SHMOYS, Editors • The Traveling Salesman Problem: A Guided Tour of Combinatorial Optimization

LAYWINE AND MULLEN • Discrete Mathematics Using Latin Squares

LEVITIN • Perturbation Theory in Mathematical Programming Applications

MAHMOUD • Evolution of Random Search Trees

MAHMOUD • Sorting: A Distribution Theory

MARTELLI • Introduction to Discrete Dynamical Systems and Chaos

MARTELLO AND TOTH • Knapsack Problems: Algorithms and Computer Implementations